New analytical approaches for verifying the origin of food

Related titles:

Food chain integrity: A holistic approach to food traceability, safety, quality and authenticity
(ISBN 978-0-85709-068-3)

Improving traceability in food processing and distribution
(ISBN 978-1-85573-959-8)

Food authenticity and traceability
(ISBN 978-1-85573-526-2)

Details of these books and a complete list of titles from Woodhead Publishing can be obtained by:

- visiting our web site at www.woodheadpublishing.com
- contacting Customer Services (e-mail: sales@woodheadpublishing.com; fax: +44 (0) 1223 832819; tel.: +44 (0) 1223 499140 ext. 130; address: Woodhead Publishing Limited, 80 High Street, Sawston, Cambridge CB22 3HJ, UK)
- in North America, contacting our US office (e-mail: usmarketing@woodheadpublishing.com; tel.: (215) 928 9112; address: Woodhead Publishing, 1518 Walnut Street, Suite 1100, Philadelphia, PA 19102-3406, USA)

If you would like e-versions of our content, please visit our online platform: www.woodheadpublishingonline.com. Please recommend it to your librarian so that everyone in your institution can benefit from the wealth of content on the site.

We are always happy to receive suggestions for new books from potential editors. To enquire about contributing to our Food Science, Technology and Nutrition series, please send your name, contact address and details of the topic/s you are interested in to nell.holden@woodheadpublishing.com. We look forward to hearing from you.

The team responsible for publishing this book:

Commissioning Editor: Sarah Hughes
Publications Coordinator: Anneka Hess
Project Editor: Anneka Hess
Editorial and Production Manager: Mary Campbell
Production Editor: Adam Hooper
Project Manager: Newgen Knowledge Works Pvt Ltd
Copyeditor: Newgen Knowledge Works Pvt Ltd
Proofreader: Newgen Knowledge Works Pvt Ltd
Cover Designer: Terry Callanan

Woodhead Publishing Series in Food Science, Technology and Nutrition:
Number 245

New analytical approaches for verifying the origin of food

Edited by
Paul Brereton

Oxford Cambridge Philadelphia New Delhi

Published by Woodhead Publishing Limited,
80 High Street, Sawston, Cambridge CB22 3HJ, UK
www.woodheadpublishing.com
www.woodheadpublishingonline.com

Woodhead Publishing, 1518 Walnut Street, Suite 1100, Philadelphia,
PA 19102-3406, USA

Woodhead Publishing India Private Limited, 303 Vardaan House, 7/28 Ansari Road,
Daryaganj, New Delhi – 110002, India
www.woodheadpublishingindia.com

First published 2013, Woodhead Publishing Limited

British Library Cataloguing in Publication Data
A catalogue record for this book is available from the British Library.

Library of Congress Control Number: 2013938662

ISBN 978-0-85709-274-8 (print)
ISBN 978-0-85709-759-0 (online)
ISSN 2042-8049 Woodhead Publishing Series in Food Science, Technology and Nutrition (print)
ISSN 2042-8057 Woodhead Publishing Series in Food Science, Technology and Nutrition (online)

The publisher's policy is to use permanent paper from mills that operate a sustainable forestry policy, and which has been manufactured from pulp which is processed using acid-free and elemental chlorine-free practices. Furthermore, the publisher ensures that the text paper and cover board used have met acceptable environmental accreditation standards.

Typeset by Newgen Knowledge Works Pvt Ltd, India
Printed by Lightning Source

Contents

Contributor contact details

(* = main contact)

Editor and Chapter 1
Paul Brereton
Food and Environment Research Agency
Sand Hutton
York
YO41 1LZ
UK

E-mail: paul.brereton@fera.gsi.gov.uk

Chapter 2
Mark Woolfe
(Food Standards Agency – retired, Head of the Food Authenticity Programme)
Thames Ditton
KT7 0UJ
UK

E-mail: mjwoolfe@gmail.com

Chapter 3
Andreas Roßmann
Isolab GmbH Laboratory for Stable Isotopes
Woelkestrasse 9/I
85301 Schweitenkirchen
Germany

E-mail: isolab_gmbh@t-online.de

Chapter 4
Grishja van der Veer
RIKILT
Wageningen University and Research Centre
Akkermaalsbos 2
Building 123
P.O. Box 230
6700 AE Wageningen
The Netherlands

E-mail: grishja.vanderveer@wur.nl

Chapter 5
Hermann Broll
Federal Institute for Risk Assessment, BfR
Max-Dohrn-Str. 8–10
10589 Berlin
Germany

E-mail: hermann.broll@bfr.bund.de

Chapter 6

Gerard Downey
Teagasc Food Research Centre Ashtown
Department of Food Chemistry and Technology
Ashtown
Dublin 15
Ireland

E-mail: gerard.downey@teagasc.ie

Chapter 7

B. G. M. Vandeginste
VICIM, p/a. Eleonoraweg 7
6523 MX Nijmegen
The Netherlands

E-mail:b.vandeginste.vicim@science.ru.nl

Chapter 8

B. Médina*, M. H. Salagoïty, F. Guyon and J. Gaye
Service Commun des Laboratoires
Laboratoire de Bordeaux
3 av du Dr Albert Schweitzer
336008 Pessac cedex
France

E-mail: bernard.medina@scl.finances.gouv.fr; labo33@scl.finances.gouv.fr

P. Hubert
CENBG/IN2P3 Université de Bordeaux
Centre d'Etudes Nucléaires de Bordeaux Gradignan
Chemin du Solarium
Le Haut Vigneau
BP 120, F-33175 Gradignan cedex
France

E-mail: hubertp@cenbg.in2p3.fr

François Guillaume
ISM: Institut des Sciences Moléculaires
Université de Bordeaux
351 cours de la Libération
France 33405 Talence

E-mail: f.guillaume@ism.u-bordeaux1.fr

Chapter 9

J. Martinsohn
Maritime Affairs Unit
Institute for the Protection and Security of the Citizen
European Commission Joint Research Centre
Via Enrico Fermi nr.2749
21027 Ispra (VA)
Italy

E-mail: jann.martinsohn@jrc.ec.europa.eu

Chapter 10

Kiri McComb
Department of Chemistry
University of Otago
P.O. Box 56
Dunedin 9054
New Zealand

E-mail: kmcomb@chemistry.otago.ac.nz

R. Frew*
FAO/IAEA Division of Nuclear Techniques in Food and Agriculture
P.O. Box 100
A-1400 Vienna
Austria

E-mail: r.frew@iaea.org

Woodhead Publishing Series in Food Science, Technology and Nutrition

1 **Chilled foods: A comprehensive guide** *Edited by C. Dennis and M. Stringer*
2 **Yoghurt: Science and technology** *A. Y. Tamime and R. K. Robinson*
3 **Food processing technology: Principles and practice** *P. J. Fellows*
4 **Bender's dictionary of nutrition and food technology Sixth edition** *D. A. Bender*
5 **Determination of veterinary residues in food** *Edited by N. T. Crosby*
6 **Food contaminants: Sources and surveillance** *Edited by C. Creaser and R. Purchase*
7 **Nitrates and nitrites in food and water** *Edited by M. J. Hill*
8 **Pesticide chemistry and bioscience: The food-environment challenge** *Edited by G. T. Brooks and T. Roberts*
9 **Pesticides: Developments, impacts and controls** *Edited by G. A. Best and A. D. Ruthven*
10 **Dietary fibre: Chemical and biological aspects** *Edited by D. A. T. Southgate, K. W. Waldron, I. T. Johnson and G. R. Fenwick*
11 **Vitamins and minerals in health and nutrition** *M. Tolonen*
12 **Technology of biscuits, crackers and cookies Second edition** *D. Manley*
13 **Instrumentation and sensors for the food industry** *Edited by E. Kress-Rogers*
14 **Food and cancer prevention: Chemical and biological aspects** *Edited by K. W. Waldron, I. T. Johnson and G. R. Fenwick*
15 **Food colloids: Proteins, lipids and polysaccharides** *Edited by E. Dickinson and B. Bergenstahl*
16 **Food emulsions and foams** *Edited by E. Dickinson*
17 **Maillard reactions in chemistry, food and health** *Edited by T. P. Labuza, V. Monnier, J. Baynes and J. O'Brien*
18 **The Maillard reaction in foods and medicine** *Edited by J. O'Brien, H. E. Nursten, M. J. Crabbe and J. M. Ames*
19 **Encapsulation and controlled release** *Edited by D. R. Karsa and R. A. Stephenson*
20 **Flavours and fragrances** *Edited by A. D. Swift*
21 **Feta and related cheeses** *Edited by A. Y. Tamime and R. K. Robinson*
22 **Biochemistry of milk products** *Edited by A. T. Andrews and J. R. Varley*
23 **Physical properties of foods and food processing systems** *M. J. Lewis*
24 **Food irradiation: A reference guide** *V. M. Wilkinson and G. Gould*
25 **Kent's technology of cereals: An introduction for students of food science and agriculture Fourth edition** *N. L. Kent and A. D. Evers*
26 **Biosensors for food analysis** *Edited by A. O. Scott*

27 **Separation processes in the food and biotechnology industries: Principles and applications** *Edited by A. S. Grandison and M. J. Lewis*
28 **Handbook of indices of food quality and authenticity** *R. S. Singhal, P. K. Kulkarni and D. V. Rege*
29 **Principles and practices for the safe processing of foods** *D. A. Shapton and N. F. Shapton*
30 **Biscuit, cookie and cracker manufacturing manuals Volume 1: Ingredients** *D. Manley*
31 **Biscuit, cookie and cracker manufacturing manuals Volume 2: Biscuit doughs** *D. Manley*
32 **Biscuit, cookie and cracker manufacturing manuals Volume 3: Biscuit dough piece forming** *D. Manley*
33 **Biscuit, cookie and cracker manufacturing manuals Volume 4: Baking and cooling of biscuits** *D. Manley*
34 **Biscuit, cookie and cracker manufacturing manuals Volume 5: Secondary processing in biscuit manufacturing** *D. Manley*
35 **Biscuit, cookie and cracker manufacturing manuals Volume 6: Biscuit packaging and storage** *D. Manley*
36 **Practical dehydration Second edition** *M. Greensmith*
37 **Lawrie's meat science Sixth edition** *R. A. Lawrie*
38 **Yoghurt: Science and technology Second edition** *A. Y. Tamime and R. K. Robinson*
39 **New ingredients in food processing: Biochemistry and agriculture** *G. Linden and D. Lorient*
40 **Benders' dictionary of nutrition and food technology Seventh edition** *D. A. Bender and A. E. Bender*
41 **Technology of biscuits, crackers and cookies Third edition** *D. Manley*
42 **Food processing technology: Principles and practice Second edition** *P. J. Fellows*
43 **Managing frozen foods** *Edited by C. J. Kennedy*
44 **Handbook of hydrocolloids** *Edited by G. O. Phillips and P. A. Williams*
45 **Food labelling** *Edited by J. R. Blanchfield*
46 **Cereal biotechnology** *Edited by P. C. Morris and J. H. Bryce*
47 **Food intolerance and the food industry** *Edited by T. Dean*
48 **The stability and shelf-life of food** *Edited by D. Kilcast and P. Subramaniam*
49 **Functional foods: Concept to product** *Edited by G. R. Gibson and C. M. Williams*
50 **Chilled foods: A comprehensive guide Second edition** *Edited by M. Stringer and C. Dennis*
51 **HACCP in the meat industry** *Edited by M. Brown*
52 **Biscuit, cracker and cookie recipes for the food industry** *D. Manley*
53 **Cereals processing technology** *Edited by G. Owens*
54 **Baking problems solved** *S. P. Cauvain and L. S. Young*
55 **Thermal technologies in food processing** *Edited by P. Richardson*
56 **Frying: Improving quality** *Edited by J. B. Rossell*
57 **Food chemical safety Volume 1: Contaminants** *Edited by D. Watson*
58 **Making the most of HACCP: Learning from others' experience** *Edited by T. Mayes and S. Mortimore*
59 **Food process modelling** *Edited by L. M. M. Tijskens, M. L. A. T. M. Hertog and B. M. Nicolaï*
60 **EU food law: A practical guide** *Edited by K. Goodburn*
61 **Extrusion cooking: Technologies and applications** *Edited by R. Guy*
62 **Auditing in the food industry: From safety and quality to environmental and other audits** *Edited by M. Dillon and C. Griffith*
63 **Handbook of herbs and spices Volume 1** *Edited by K. V. Peter*

64 **Food product development: Maximising success** *M. Earle, R. Earle and A. Anderson*
65 **Instrumentation and sensors for the food industry Second edition** *Edited by E. Kress-Rogers and C. J. B. Brimelow*
66 **Food chemical safety Volume 2: Additives** *Edited by D. Watson*
67 **Fruit and vegetable biotechnology** *Edited by V. Valpuesta*
68 **Foodborne pathogens: Hazards, risk analysis and control** *Edited by C. de W. Blackburn and P. J. McClure*
69 **Meat refrigeration** *S. J. James and C. James*
70 **Lockhart and Wiseman's crop husbandry Eighth edition** *H. J. S. Finch, A. M. Samuel and G. P. F. Lane*
71 **Safety and quality issues in fish processing** *Edited by H. A. Bremner*
72 **Minimal processing technologies in the food industries** *Edited by T. Ohlsson and N. Bengtsson*
73 **Fruit and vegetable processing: Improving quality** *Edited by W. Jongen*
74 **The nutrition handbook for food processors** *Edited by C. J. K. Henry and C. Chapman*
75 **Colour in food: Improving quality** *Edited by D. MacDougall*
76 **Meat processing: Improving quality** *Edited by J. P. Kerry, J. F. Kerry and D. A. Ledward*
77 **Microbiological risk assessment in food processing** *Edited by M. Brown and M. Stringer*
78 **Performance functional foods** *Edited by D. Watson*
79 **Functional dairy products Volume 1** *Edited by T. Mattila-Sandholm and M. Saarela*
80 **Taints and off-flavours in foods** *Edited by B. Baigrie*
81 **Yeasts in food** *Edited by T. Boekhout and V. Robert*
82 **Phytochemical functional foods** *Edited by I. T. Johnson and G. Williamson*
83 **Novel food packaging techniques** *Edited by R. Ahvenainen*
84 **Detecting pathogens in food** *Edited by T. A. McMeekin*
85 **Natural antimicrobials for the minimal processing of foods** *Edited by S. Roller*
86 **Texture in food Volume 1: Semi-solid foods** *Edited by B. M. McKenna*
87 **Dairy processing: Improving quality** *Edited by G. Smit*
88 **Hygiene in food processing: Principles and practice** *Edited by H. L. M. Lelieveld, M. A. Mostert, B. White and J. Holah*
89 **Rapid and on-line instrumentation for food quality assurance** *Edited by I. Tothill*
90 **Sausage manufacture: Principles and practice** *E. Essien*
91 **Environmentally-friendly food processing** *Edited by B. Mattsson and U. Sonesson*
92 **Bread making: Improving quality** *Edited by S. P. Cauvain*
93 **Food preservation techniques** *Edited by P. Zeuthen and L. Bøgh-Sørensen*
94 **Food authenticity and traceability** *Edited by M. Lees*
95 **Analytical methods for food additives** *R. Wood, L. Foster, A. Damant and P. Key*
96 **Handbook of herbs and spices Volume 2** *Edited by K. V. Peter*
97 **Texture in food Volume 2: Solid foods** *Edited by D. Kilcast*
98 **Proteins in food processing** *Edited by R. Yada*
99 **Detecting foreign bodies in food** *Edited by M. Edwards*
100 **Understanding and measuring the shelf-life of food** *Edited by R. Steele*
101 **Poultry meat processing and quality** *Edited by G. Mead*
102 **Functional foods, ageing and degenerative disease** *Edited by C. Remacle and B. Reusens*
103 **Mycotoxins in food: Detection and control** *Edited by N. Magan and M. Olsen*
104 **Improving the thermal processing of foods** *Edited by P. Richardson*

105 **Pesticide, veterinary and other residues in food** *Edited by D. Watson*
106 **Starch in food: Structure, functions and applications** *Edited by A.-C. Eliasson*
107 **Functional foods, cardiovascular disease and diabetes** *Edited by A. Arnoldi*
108 **Brewing: Science and practice** *D. E. Briggs, P. A. Brookes, R. Stevens and C. A. Boulton*
109 **Using cereal science and technology for the benefit of consumers: Proceedings of the 12th International ICC Cereal and Bread Congress, 24–26th May, 2004, Harrogate, UK** *Edited by S. P. Cauvain, L. S. Young and S. Salmon*
110 **Improving the safety of fresh meat** *Edited by J. Sofos*
111 **Understanding pathogen behaviour: Virulence, stress response and resistance** *Edited by M. Griffiths*
112 **The microwave processing of foods** *Edited by H. Schubert and M. Regier*
113 **Food safety control in the poultry industry** *Edited by G. Mead*
114 **Improving the safety of fresh fruit and vegetables** *Edited by W. Jongen*
115 **Food, diet and obesity** *Edited by D. Mela*
116 **Handbook of hygiene control in the food industry** *Edited by H. L. M. Lelieveld, M. A. Mostert and J. Holah*
117 **Detecting allergens in food** *Edited by S. Koppelman and S. Hefle*
118 **Improving the fat content of foods** *Edited by C. Williams and J. Buttriss*
119 **Improving traceability in food processing and distribution** *Edited by I. Smith and A. Furness*
120 **Flavour in food** *Edited by A. Voilley and P. Etievant*
121 **The Chorleywood bread process** *S. P. Cauvain and L. S. Young*
122 **Food spoilage microorganisms** *Edited by C. de W. Blackburn*
123 **Emerging foodborne pathogens** *Edited by Y. Motarjemi and M. Adams*
124 **Benders' dictionary of nutrition and food technology Eighth edition** *D. A. Bender*
125 **Optimising sweet taste in foods** *Edited by W. J. Spillane*
126 **Brewing: New technologies** *Edited by C. Bamforth*
127 **Handbook of herbs and spices Volume 3** *Edited by K. V. Peter*
128 **Lawrie's meat science Seventh edition** *R. A. Lawrie in collaboration with D. A. Ledward*
129 **Modifying lipids for use in food** *Edited by F. Gunstone*
130 **Meat products handbook: Practical science and technology** *G. Feiner*
131 **Food consumption and disease risk: Consumer–pathogen interactions** *Edited by M. Potter*
132 **Acrylamide and other hazardous compounds in heat-treated foods** *Edited by K. Skog and J. Alexander*
133 **Managing allergens in food** *Edited by C. Mills, H. Wichers and K. Hoffman-Sommergruber*
134 **Microbiological analysis of red meat, poultry and eggs** *Edited by G. Mead*
135 **Maximising the value of marine by-products** *Edited by F. Shahidi*
136 **Chemical migration and food contact materials** *Edited by K. Barnes, R. Sinclair and D. Watson*
137 **Understanding consumers of food products** *Edited by L. Frewer and H. van Trijp*
138 **Reducing salt in foods: Practical strategies** *Edited by D. Kilcast and F. Angus*
139 **Modelling microorganisms in food** *Edited by S. Brul, S. Van Gerwen and M. Zwietering*
140 **Tamime and Robinson's Yoghurt: Science and technology Third edition** *A. Y. Tamime and R. K. Robinson*
141 **Handbook of waste management and co-product recovery in food processing Volume 1** *Edited by K. W. Waldron*
142 **Improving the flavour of cheese** *Edited by B. Weimer*
143 **Novel food ingredients for weight control** *Edited by C. J. K. Henry*

144 **Consumer-led food product development** *Edited by H. MacFie*
145 **Functional dairy products Volume 2** *Edited by M. Saarela*
146 **Modifying flavour in food** *Edited by A. J. Taylor and J. Hort*
147 **Cheese problems solved** *Edited by P. L. H. McSweeney*
148 **Handbook of organic food safety and quality** *Edited by J. Cooper, C. Leifert and U. Niggli*
149 **Understanding and controlling the microstructure of complex foods** *Edited by D. J. McClements*
150 **Novel enzyme technology for food applications** *Edited by R. Rastall*
151 **Food preservation by pulsed electric fields: From research to application** *Edited by H. L. M. Lelieveld and S. W. H. de Haan*
152 **Technology of functional cereal products** *Edited by B. R. Hamaker*
153 **Case studies in food product development** *Edited by M. Earle and R. Earle*
154 **Delivery and controlled release of bioactives in foods and nutraceuticals** *Edited by N. Garti*
155 **Fruit and vegetable flavour: Recent advances and future prospects** *Edited by B. Brückner and S. G. Wyllie*
156 **Food fortification and supplementation: Technological, safety and regulatory aspects** *Edited by P. Berry Ottaway*
157 **Improving the health-promoting properties of fruit and vegetable products** *Edited by F. A. Tomás-Barberán and M. I. Gil*
158 **Improving seafood products for the consumer** *Edited by T. Børresen*
159 **In-pack processed foods: Improving quality** *Edited by P. Richardson*
160 **Handbook of water and energy management in food processing** *Edited by J. Klemeš, R.. Smith and J.-K. Kim*
161 **Environmentally compatible food packaging** *Edited by E. Chiellini*
162 **Improving farmed fish quality and safety** *Edited by Ø. Lie*
163 **Carbohydrate-active enzymes** *Edited by K.-H. Park*
164 **Chilled foods: A comprehensive guide Third edition** *Edited by M. Brown*
165 **Food for the ageing population** *Edited by M. M. Raats, C. P. G. M. de Groot and W. A Van Staveren*
166 **Improving the sensory and nutritional quality of fresh meat** *Edited by J. P. Kerry and D. A. Ledward*
167 **Shellfish safety and quality** *Edited by S. E. Shumway and G. E. Rodrick*
168 **Functional and speciality beverage technology** *Edited by P. Paquin*
169 **Functional foods: Principles and technology** *M. Guo*
170 **Endocrine-disrupting chemicals in food** *Edited by I. Shaw*
171 **Meals in science and practice: Interdisciplinary research and business applications** *Edited by H. L. Meiselman*
172 **Food constituents and oral health: Current status and future prospects** *Edited by M. Wilson*
173 **Handbook of hydrocolloids Second edition** *Edited by G. O. Phillips and P. A. Williams*
174 **Food processing technology: Principles and practice Third edition** *P. J. Fellows*
175 **Science and technology of enrobed and filled chocolate, confectionery and bakery products** *Edited by G. Talbot*
176 **Foodborne pathogens: Hazards, risk analysis and control Second edition** *Edited by C. de W. Blackburn and P. J. McClure*
177 **Designing functional foods: Measuring and controlling food structure breakdown and absorption** *Edited by D. J. McClements and E. A. Decker*
178 **New technologies in aquaculture: Improving production efficiency, quality and environmental management** *Edited by G. Burnell and G. Allan*
179 **More baking problems solved** *S. P. Cauvain and L. S. Young*

180 **Soft drink and fruit juice problems solved** *P. Ashurst and R. Hargitt*
181 **Biofilms in the food and beverage industries** *Edited by P. M. Fratamico, B. A. Annous and N. W. Gunther*
182 **Dairy-derived ingredients: Food and neutraceutical uses** *Edited by M. Corredig*
183 **Handbook of waste management and co-product recovery in food processing Volume 2** *Edited by K. W. Waldron*
184 **Innovations in food labelling** *Edited by J. Albert*
185 **Delivering performance in food supply chains** *Edited by C. Mena and G. Stevens*
186 **Chemical deterioration and physical instability of food and beverages** *Edited by L. H. Skibsted, J. Risbo and M. L. Andersen*
187 **Managing wine quality Volume 1: Viticulture and wine quality** *Edited by A. G. Reynolds*
188 **Improving the safety and quality of milk Volume 1: Milk production and processing** *Edited by M. Griffiths*
189 **Improving the safety and quality of milk Volume 2: Improving quality in milk products** *Edited by M. Griffiths*
190 **Cereal grains: Assessing and managing quality** *Edited by C. Wrigley and I. Batey*
191 **Sensory analysis for food and beverage quality control: A practical guide** *Edited by D. Kilcast*
192 **Managing wine quality Volume 2: Oenology and wine quality** *Edited by A. G. Reynolds*
193 **Winemaking problems solved** *Edited by C. E. Butzke*
194 **Environmental assessment and management in the food industry** *Edited by U. Sonesson, J. Berlin and F. Ziegler*
195 **Consumer-driven innovation in food and personal care products** *Edited by S. R. Jaeger and H. MacFie*
196 **Tracing pathogens in the food chain** *Edited by S. Brul, P. M. Fratamico and T. A. McMeekin*
197 **Case studies in novel food processing technologies: Innovations in processing, packaging, and predictive modelling** *Edited by C. J. Doona, K. Kustin and F. E. Feeherry*
198 **Freeze-drying of pharmaceutical and food products** *T.-C. Hua, B.-L. Liu and H. Zhang*
199 **Oxidation in foods and beverages and antioxidant applications Volume 1: Understanding mechanisms of oxidation and antioxidant activity** *Edited by E. A. Decker, R. J. Elias and D. J. McClements*
200 **Oxidation in foods and beverages and antioxidant applications Volume 2: Management in different industry sectors** *Edited by E. A. Decker, R. J. Elias and D. J. McClements*
201 **Protective cultures, antimicrobial metabolites and bacteriophages for food and beverage biopreservation** *Edited by C. Lacroix*
202 **Separation, extraction and concentration processes in the food, beverage and nutraceutical industries** *Edited by S. S. H. Rizvi*
203 **Determining mycotoxins and mycotoxigenic fungi in food and feed** *Edited by S. De Saeger*
204 **Developing children's food products** *Edited by D. Kilcast and F. Angus*
205 **Functional foods: Concept to product Second edition** *Edited by M. Saarela*
206 **Postharvest biology and technology of tropical and subtropical fruits Volume 1: Fundamental issues** *Edited by E. M. Yahia*
207 **Postharvest biology and technology of tropical and subtropical fruits Volume 2: Açai to citrus** *Edited by E. M. Yahia*
208 **Postharvest biology and technology of tropical and subtropical fruits Volume 3: Cocona to mango** *Edited by E. M. Yahia*

209 **Postharvest biology and technology of tropical and subtropical fruits Volume 4: Mangosteen to white sapote** *Edited by E. M. Yahia*
210 **Food and beverage stability and shelf life** *Edited by D. Kilcast and P. Subramaniam*
211 **Processed Meats: Improving safety, nutrition and quality** *Edited by J. P. Kerry and J. F. Kerry*
212 **Food chain integrity: A holistic approach to food traceability, safety, quality and authenticity** *Edited by J. Hoorfar, K. Jordan, F. Butler and R. Prugger*
213 **Improving the safety and quality of eggs and egg products Volume 1** *Edited by Y. Nys, M. Bain and F. Van Immerseel*
214 **Improving the safety and quality of eggs and egg products Volume 2** *Edited by F. Van Immerseel, Y. Nys and M. Bain*
215 **Animal feed contamination: Effects on livestock and food safety** *Edited by J. Fink-Gremmels*
216 **Hygienic design of food factories** *Edited by J. Holah and H. L. M. Lelieveld*
217 **Manley's technology of biscuits, crackers and cookies Fourth edition** *Edited by D. Manley*
218 **Nanotechnology in the food, beverage and nutraceutical industries** *Edited by Q. Huang*
219 **Rice quality: A guide to rice properties and analysis** *K. R. Bhattacharya*
220 **Advances in meat, poultry and seafood packaging** *Edited by J. P. Kerry*
221 **Reducing saturated fats in foods** *Edited by G. Talbot*
222 **Handbook of food proteins** *Edited by G. O. Phillips and P. A. Williams*
223 **Lifetime nutritional influences on cognition, behaviour and psychiatric illness** *Edited by D. Benton*
224 **Food machinery for the production of cereal foods, snack foods and confectionery** *L.-M. Cheng*
225 **Alcoholic beverages: Sensory evaluation and consumer research** *Edited by J. Piggott*
226 **Extrusion problems solved: Food, pet food and feed** *M. N. Riaz and G. J. Rokey*
227 **Handbook of herbs and spices Second edition Volume 1** *Edited by K. V. Peter*
228 **Handbook of herbs and spices Second edition Volume 2** *Edited by K. V. Peter*
229 **Breadmaking: Improving quality Second edition** *Edited by S. P. Cauvain*
230 **Emerging food packaging technologies: Principles and practice** *Edited by K. L. Yam and D. S. Lee*
231 **Infectious disease in aquaculture: Prevention and control** *Edited by B. Austin*
232 **Diet, immunity and inflammation** *Edited by P. C. Calder and P. Yaqoob*
233 **Natural food additives, ingredients and flavourings** *Edited by D. Baines and R. Seal*
234 **Microbial decontamination in the food industry: Novel methods and applications** *Edited by A. Demirci and M.O. Ngadi*
235 **Chemical contaminants and residues in foods** *Edited by D. Schrenk*
236 **Robotics and automation in the food industry: Current and future technologies** *Edited by D. G. Caldwell*
237 **Fibre-rich and wholegrain foods: Improving quality** *Edited by J. A. Delcour and K. Poutanen*
238 **Computer vision technology in the food and beverage industries** *Edited by D.-W. Sun*
239 **Encapsulation technologies and delivery systems for food ingredients and nutraceuticals** *Edited by N. Garti and D. J. McClements*
240 **Case studies in food safety and authenticity** *Edited by J. Hoorfar*
241 **Heat treatment for insect control: Developments and applications** *D. Hammond*
242 **Advances in aquaculture hatchery technology** *Edited by G. Allan and G. Burnell*

243 **Open innovation in the food and beverage industry** *Edited by M. Garcia Martinez*
244 **Trends in packaging of food, beverages and other fast-moving consumer goods (FMCG)** *Edited by N. Farmer*
245 **New analytical approaches for verifying the origin of food** *Edited by P. Brereton*
246 **Microbial production of food ingredients, enzymes and nutraceuticals** *Edited by B. McNeil, D. Archer, I. Giavasis and L. Harvey*
247 **Persistent organic pollutants and toxic metals in foods** *Edited by M. Rose and A. Fernandes*
248 **Cereal grains for the food and beverage industries** *E. Arendt and E. Zannini*
249 **Viruses in food and water: Risks, surveillance and control** *Edited by N. Cook*
250 **Improving the safety and quality of nuts** *Edited by L. J. Harris*
251 **Metabolomics in food and nutrition** *Edited by B.C. Weimer and C. Slupsky*
252 **Food enrichment with omega-3 fatty acids** *Edited by C. Jacobsen, N. Skall Nielsen, A. Frisenfeldt Horn and A.-D. Moltke Sørensen*
253 **Instrumental assessment of food sensory quality: A practical guide** *Edited by D. Kilcast*
254 **Food microstructures: Microscopy, measurement and modelling** *Edited by V. J. Morris and K. Groves*
255 **Handbook of food powders: Processes and properties** *Edited by B. R. Bhandari, N. Bansal, M. Zhang and P. Schuck*
256 **Functional ingredients from algae for foods and nutraceuticals** *Edited by H. Domínguez*
257 **Satiation, satiety and the control of food intake: Theory and practice** *Edited by J. E. Blundell and F. Bellisle*
258 **Hygiene in food processing: Principles and practice Second edition** *Edited by H. L. M. Lelieveld, J. Holah and D. Napper*
259 **Advances in microbial food safety Volume 1** *Edited by J. Sofos*
260 **Global safely of fresh produce: A handbook of best practice, innovative Commercial solutions and case studies** *Edited by J. Hoorfar*
261 **Human milk biochemistry and infant formula manufacturing technology** *Edited by M. Guo*

Part I

Introduction

1

Verifying the origin of food: an introduction

P. Brereton, The Food and Environment Research Agency, UK

DOI: 10.1533/9780857097590.1.3

Abstract: This chapter provides an introduction into why analytical methods and systems are needed to verify the origin of our food, in particular the drivers for the work, in terms of regulatory and more recently consumer and industry requirements. Some of the difficulties in satisfying regulatory aspects are described as well as a brief look at how some of the most recent scientific developments not included in the main body of this book will impact on the way consumers can be assured about the origin of the food they eat.

Key words: food origin, food fraud, food traceability, food provenance, food verification.

1.1 Introduction: the importance of food origin

Consumers have always shown an interest in where their food comes from and their familiarity of their local supply chain has historically given them confidence in the quality and wholesomeness of the food they consumed. For thousands of years consumers have had intimate knowledge of the origin of their food, whether home grown or purchased from their local producer/market. More recently consumer preference for continual (year round) access to a wide variety of fresh food has transformed supply chains from local to global ones, thereby decreasing consumer knowledge about who produced their food and how. As a result the control bodies and food industry have introduced regulation, traceability systems and marketing strategies aimed at reassuring the consumer about the quality, safety and provenance, that is, integrity, of their food.

1.2 Historical context

Consumer preference for 'high value' foods is nothing new and where there is a consumer with money to spend on such products, there have always been fraudsters wishing to exploit consumer demand for premium goods. Ancient Athens had a public inspector of wines, presumably due to concerns about quality of the wine. Pliny the Elder had concerns about the authenticity of Roman wine (Robinson, 2006) which was frequently adulterated with lead. This adulteration is likely to have resulted from a common practice where lead containing additives were used to sweeten and preserve sour wines leading to endemic and deadly *colica pictonum* (Eisinger, 1982) in the patrician population. Whereas this ancient example was probably not due to malicious adulteration but rather ignorance of lead toxicity, it did sow the seeds of consumer preference for 'natural' food, free from additives and preservatives, that is a common consumer request today. It was not until the middle ages that awareness of lead poising was reported and lead additives proscribed through legislation, with German authorities making such adulteration punishable by death (Lessler, 1988). In fact food fraud rarely involves overtly malicious adulteration, that is, intention to harm. More commonly it involves an ignorance of toxicity issues together with a blatant disregard for general consumer safety in deference to money making activities. Such ignorance and disregard are still present in twenty-first century frauds. It is unclear how many of the actors in the Chinese baby milk scandal of 2008 (Ingelfinger, 2008) were aware of the toxicity of melamine and cyanuric acid, though the US pet food incident involving the same adulterants should have sensitised the whole community to the dangers of using melamine to enhance the apparent protein content of food products. What is clear is that where there is considerable money to be made then concomitant resources and levels of sophistication will be applied, if necessary, by the fraudsters to achieve their goals.

Analytical methods have a clear role in helping identify food fraud. Unfortunately the resources that have historically been made available to antifraud agencies have always been limited in deference to those that are available to protect public health. While this is as one would expect, the picture changes when food fraud has a direct impact on Government revenue. Alcohol, for example, was first taxed in Britain in 1643 and is now globally the most highly regulated and monitored 'food' commodity due to the lucrative revenues that taxation can bring to Government coffers. Premium products continue to be major targets for fraudsters with adulteration of olive oil, milk, honey and saffron being some of the most reported (Moore *et al.*, 2012).

1.3 The impact of the Common Agricultural Policy

The advent of the Common Agricultural Policy (CAP) within the European Union (EU) in the twentieth century was a means of subsidising European agriculture and was a furtive environment for agri-food fraud. The substantial differences between European and world markets together with a complex system of tariffs and duties created a utopia for fraudsters where, in some cases, fraudulently mislabelled products could be recycled to exploit the differential between import duty and export tariff. For example butter produced within the EU attracted a significant export refund due to the lower butter prices outside of the EU. In one case the exported butter was then relabelled by unscrupulous traders who then imported back into the EU. The significant differences between export refund and the import tariff meant that, at times, significant additional profits of over £1 per kilo of butter could be made (Balling and Rossmann, 2004; Kelly *et al.*, 2005).

The extremes of the CAP in the late twentieth century had a major impact on research into analytical methods that could identify such frauds. Concern grew not only about the profligacy of CAP in terms 'wine lakes' and 'sugar mountains' but also from the European tax payer in terms of authenticity of the contents of this surplus production. The European Commission embarked on an investment in antifraud measures: significant investment in research, development of sophisticated food standards and integrating analytical methods into legislation that put Europe at the frontier of research and surveillance into food fraud/authenticity. Each major food commodity had a suite of prescribed methods (vertical commodity standards) in an attempt to assure its authenticity, with additional methods being adopted as new fraudulent practices were uncovered. European Union legislation for olive oil (EEC No. 2568/91) and wine (EEC No. 2676/90), for example, prescribed extensive specifications that required an array of analytical methods to ensure compliance. Of the many innovative methods developed to identify olive oil fraud, one of the most significant was developed by Lanzón *et al.* (1994) to identify adulteration of extra virgin olive oils (EVOOs) with refined olive oil (ROO). At the time fraudsters could make considerable money by extending EVOO, the premium product, with a cheaper refined substitute. Researchers identified a marker (stigmasta-3,5-diene) of the refining process (Fig. 1.1) and developed a method that was very successful at detecting this type of adulteration.

Another method that has left a significant legacy and owes its development and initial exploitation to the EU CAP agenda is that of SNIF© NMR. This method was developed primarily as a means of identifying the origin of sugar and sugar derived alcohol by identifying site specific information about the

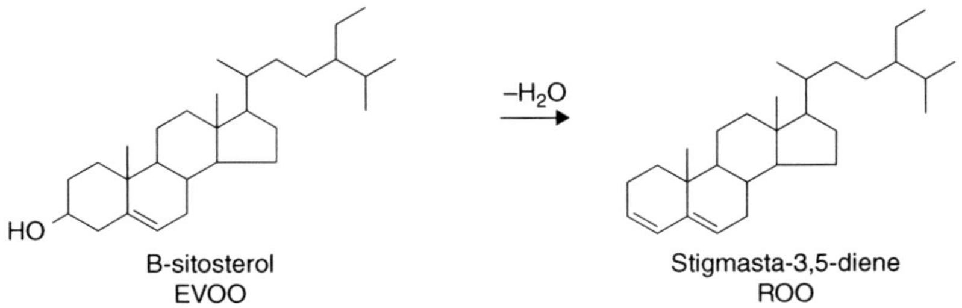

HO

B-sitosterol
EVOO

$-H_2O$

Stigmasta-3,5-diene
ROO

Fig. 1.1 During the refining process low grade virgin olive oil is refined using deodorisation and bleaching processes that dehydrate B-sitosterol forming stigmasta-3,5-diene which Lanzón *et al.* (1994) identified as a marker of refined olive oil. Diagram courtesy of Colin Crews.

deuterium–hydrogen ratio (Martin *et al.*, 1988). One of the advantages of the methodology was that it could identify beet sugar derived alcohol; as a result it was strongly embraced by the European Commission as a means of assuring the authenticity of chaptalised wine and specifically that put into storage under the Common Agriculture Policy. The method was adopted into Regulation (EC 2676/90) and Member States were supplied with instrumentation and training in the technique. The method is still used today to control wine and has many other applications, for example, fruit juices and natural flavours.

The EU has continued to be a major sponsor of research into food authenticity primarily throughout its Framework Programmes, the largest being the €20M TRACE project (TRACE, 2005–2009). With CAP reform there has been a greater focus in the twenty-first century on protecting an ever more discerning European consumer from being sold fraudulently mislabelled goods. Through consumer demand for perceived added value foods there has been a drive for strengthening of the legislation and standards around the labelling of food, particularly with regard to provenance. Examples include the General food law Regulation ((EC) No. 178/2002) which specifically included the need for food traceability and the Food information for consumers Regulation ((EU) No. 1169/2011) which significantly strengthens the food labelling requirements.

Alongside these developments the EU developed Protected Denomination of Origin in an attempt to formally recognise foods of a particular geographical origin or traditional production process (Regulation (EC) No. 510/2006). The retail sector has begun to exploit consumer preference for regional foods through extensive marketing of provenance and other perceived value added attributes, for example, fair trade, ethical production, production process, organic, free range etc.

1.4 Food assurance

While consumer interests will always be paramount there is also a realisation that the honest producer is also a major victim either directly, through damage of brand integrity, or loss of market share, to unfair competition from inferior quality products. Food assurance schemes that transparently demonstrate the integrity of European food will feature heavily in the EU strategy of adding resilience to the European agri-food economy in the face of food security pressures for mass produced food from South America, China and Eastern Europe.

1.5 Analytical procedures for verifying authenticity

Determining the authenticity can require a range of verification approaches depending on the level of sophistication of the fraud itself. For the purposes of this chapter I will refer to them as crude and sophisticated (in terms of the solutions needed to uncover them).

1.5.1 Crude frauds

These are usually instances where the authentic product is completely replaced with a substitute product. The fraud is not too difficult to identify, the fraudsters realise that the fraud will eventually be discovered but the lucrative nature of the fraud makes it worthwhile in their eyes. This type of fraud usually involves high cost products such as alcohol and branded goods. Although sometimes easy to detect, the time during which these crude imitations are on the market can have fatal consequences for consumers such as the instance of fraudulent Czech alcohol in 2012 that killed over 25 people due to the lethal levels of methanol in the substitute product (Economist 17 September 2012).

1.5.2 Sophisticated frauds

These are usually expected, by the fraudsters, to be longer lasting frauds owing to the difficulties in their detection or even legal interpretation. They often involve product extension – the addition of a similar cheaper matrix/ingredient to extend the more expensive one. Adding glycerol to wine, watering of wine and milk, extending EVOO with ROO, extending fruit juice with pulp wash all involve adding natural components back into the product. As a result new methodology is often required to differentiate the premium product from its inferior replacement. Major frauds often have a considerable lifespan due to the fraudsters having intimate knowledge of control methodology. Whereas some implicated in the melamine tragedy may have pleaded ignorance of the toxicity of melamine and related compounds, what is clear is that those who masterminded the fraud had full comprehension of a control methodology that relied on indirect methods that obtained the protein values from nitrogen content via a conversion factor. This loophole was ruthlessly exploited through the addition of a high nitrogen containing compound, melamine, to enhance the apparent protein content of products, with tragic results.

Sometimes scientific breakthroughs can alter the authenticity landscape back in favour of the honest producer by producing robust detection technologies that make the fraud relatively easy to detect, a good example is the breakthrough in DNA technology that can now easily provide species determination for most animal tissue.

1.5.3 Uncertainty

One of the difficulties in differentiating similar matrices is the degree of uncertainty in the result (Fig. 1.2). Usually many samples of authentic samples are

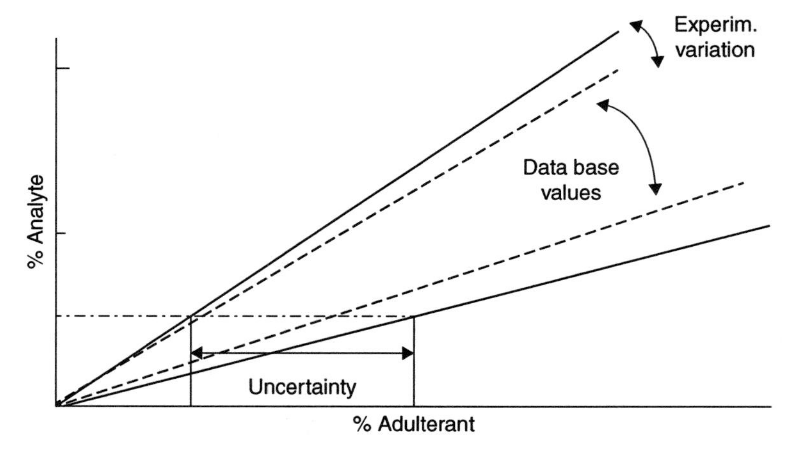

Fig. 1.2 The effect of natural variation of a marker analyte in a food commodity and analytical variation (precision) on the determination of % adulterant.

required to build a reliable database with which to challenge the suspect sample. Chemometrics are often needed to provide interpretation and demonstrate the uncertainty of the result. Such uncertainty does not sit well within the legal process and can reduce the effectiveness of such methodology, for example, geographical origin using stable isotopes.

1.6 Future trends

1.6.1 Databases and warehouses

Because food fraud incidents invariably have a negative impact on brand integrity producers and national governments can be reticent to highlight the issue and databases describing the authentic product specification are rarely shared. Instead industry experts will often proactively help control authorities by providing intelligence and their product expertise. This means that developing comprehensive databases can be problematic, yet establishing specifications for food products is an essential component in developing robust analytical methods, as is the procurement of authentic reference materials. Both are in short supply and provision of these would be a major step in combating food fraud and establishing authentic markers of provenance.

1.6.2 Exploitation of omic technologies

One of the new developments at the chemistry–biology interface that could not be included at the time of writing this book is how omic technologies can be applied to food origin problems. It is clear that the step changes that are taking place in our ability to more quickly interpret the genomic information pertaining to our food will have profound implications for industry and consumer alike. Next Generation Sequencing (NGS) (Mardis, 2008) and associated bioinformatics have the potential to rapidly provide genomic information that could be related to the origin of a particular food sample. Recently published work has shown the potential to use microfloral 'fingerprints' as a means of assessing the geographical origin of food commodities. Using polymerase chain reaction (PCR) Denaturing Gradient Gel Electrophoresis (DGGE) fish bacteria profiles were obtained and successfully used to identify the origin (district) of the different aquaculture farms within An Giang province in Vietnam (Le Nguyen, 2008). Similar approaches have been used to determine the origin of nuts, fruits and cheese by profiling their fungal and bacterial communities respectively (El Sheikha, 2009, 2010; Arcuri, 2013). Although further research is required to understand the half-life and reproducibility of the microfloral fingerprints, this innovative approach has considerable potential, especially when combined with the added speed and information-rich processes that NGS could provide.

Similarly, the use of ^{1}H-NMR and state of the art High Resolution Time of Flight Mass Spectrometry (HR-ToF-MS) will have a major impact on our ability to identify non-conforming products and, importantly, more easily identify markers of adulteration. Metabolite profiling, metabolomics, proteomics and genomics all have huge potential to elucidate the subtle differences that often exist between genuine and fraudulent products (Charlton *et al.*, 2002; Alonso-Salces *et al.*, 2011). Such methods can adopt a 'non-targeted approach' and identify unexpected compounds in food products. It is hoped that methods such as these will provide the analysts with a much improved chance of identifying the next 'melamine'.

1.7 Conclusion

Significant investment is being made in developing intelligent packaging, holograms and other security features that, when combined with sophisticated traceability systems, it is hoped will strengthen consumer confidence in the food we eat. However as most of these systems focus on the packaging and the consumer's primary interest will always be in the contents, analytical methods will continue to be needed to assure us about the provenance and authenticity of our food.

1.8 References

ALONSO-SALCES R M, MARGARET V, HOLLAND M V and GUILLOU C, '1H-NMR fingerprinting to evaluate the stability of olive oil', *Food Control*, (2011), **22**, 2041–2046.

ARCURI E F, EL SHEIKA A, RYCHLIK T, PIRO-MÉTAYER I and MONTET D, 'Determination of cheese origin by using 16S rDNA fingerprinting of bacteria communities by PCR-DGGE: preliminary application to traditional Minas cheese', *Food Control*, (2013), **30**, 1–6.

BALLING H P and ROSSMANN A, 'Countering fraud via isotope analysis – Case report', *Kriminalistik*, (2004), **58**(1), 44–47.

CALLON C, DELBÈS C, DUTHOIT F and MONTEL M C, 'Application of SSCP–PCR fingerprinting to profile the yeast community in raw milk Salers cheeses', *Systematic and Applied Microbiology*, (2006), **29**(2) 172–180.

CHARLTON A J, FARRINGTON W H and BRERETON P, 'Application of H NMR and multivariate statistics for screening complexed mixtures: quality control and authenticity of instant coffee', *Journal of Agricultural and Food Chemistry*, (2002), **50**(11), 3098–3103.

Commission Regulation (EEC) No. 2676/90 of 17 September 1990 determining Community methods for the analysis of wines. *Official Journal of Legislation* (1990), **272**, 1–192.

Commission Regulation (EEC) No. 2676/90 of 17 September 1990 determining Community methods for the analysis of wines: Annex 16: detecting enrichment of grape musts, concentrated grape musts and wines by application of nuclear magnetic resonance of deuterium (SNIF NMR). *Official Journal of European Communities*, (1990), **L272**, 64–73.

Commission Regulation (EEC) No. 2568/91 of 11 July 1991 on the characteristics of olive oil and olive-residue oil and on the relevant methods of analysis, *Official Journal of Legislation* (1991), **248**, 1–112.

Council Regulation (EC) No. 510/2006 of 20 March 2006 on the protection of geographical indications and designations of origin for agricultural products and foodstuffs. *Official Journal of Legislation*, (2006), **93**, 12–25.

Council Regulation (EC) No. 178/2002 of the European Parliament and of the Council of 28 January 2002 laying down the general principles and requirements of food law, establishing the European Food Safety Authority and laying down procedures in matters of food safety. *Official Journal*, (2002), **L31**, 1–24.

Council Regulation (EU) No. 1169/2011 of the European Parliament and of the Council of 25 October 2011 on the provision of food information to consumers, amending Regulations (EC) No. 1924/2006 and (EC) No. 1925/2006 of the European Parliament and of the Council, and repealing Commission Directive 87/250/EEC, Council Directive 90/496/EEC, Commission Directive 1999/10/EC, Directive 2000/13/EC of the European Parliament and of the Council, Commission Directives 2002/67/EC and 2008/5/EC and Commission Regulation (EC) No. 608/2004. *Official Journal*, (2011), **L304**, 18–63.

Economist 17 September 2012, Lethal Hooch, available from http://www.economist.com/blogs/easternapproaches/2012/09/czech-politics.

EISINGER J, 'Lead and wine. Eberhard Gockel and the colica Pictonum', *Medical History*, (1982), **26**(3), 279–302.

EL SHEIKHA A F, 'Determination of the geographical origin of fruits by using 26S rDNA fingerprinting of yeast communities by PCR-DGGE: an application to Shea tree fruits', *Journal of Life Sciences*, (2010), **4**, 9–15.

EL SHEIKHA A, CONDUR A, MÉTAYER I, LE NGUYEN D D, LOISEAU G and MONTET D, 'Determination of fruit origin by using 26S rDNA fingerprinting of yeast communities by PCR-DGGE: preliminary application to Physalis fruits from Egypt', *Yeast*, (2009), **26**, 567–573.

INGELFINGER J R, 'Melamine and the global implications of food contamination', *The New England Journal of Medicine*, (2008), **359**, 2745–2748.

KELLY S, HEATON K and HOOGEWERFF J, 'Tracing the geographical origin of food: The application of multi-element and multi-isotope analysis', *Trends in Food Science and Technology*, (2005), **16**(12), 555–567.

LANZÓN A, ALBI T, CERT A and GRACIÁN J, 'The hydrocarbon fraction of virgin olive oil and changes resulting from refining', *Journal of the American Oil Chemists Society*, (1994), **71**(3), 285–291.

LE NGUYEN D D, HA NGOC H, DIJOUX D, LOISEAU G and MONTET D, 'Determination of fish origin by using 16S rDNA fingerprinting of bacterial communities by PCR-DGGE: An application on Pangasius fish from Viet Nam', *Food Control*, (2008), **19**, 454–460.

LESSLER M A, 'Lead and lead poisoning from antiquity to modern times', *Ohio Journal of Science*, (1988), **88**(3), 78–84.

MARDIS E R, 'The impact of next generation sequencing technology on genetics', *Trends in Genetics*, (2008), **24**(3), 133–141.

MARTIN G, GUILLOU C, MARTIN M L, CABANIS M, TEP Y and AERNY J, 'Natural factors of isotope fractionation and the characterisation of wines', *Journal of Agricultural and Food Chemistry*, (1988), **36**, 316–322.

MOORE J, SPINK J and LIPP M, 'Development and application of a database of food ingredient fraud and economically motivated adulteration from 1980 to 2010', *Journal of Food Science*, (2012), **77**(4), 118–126.

ROBINSON J (ed), *The Oxford Companion to Wine*, Third Edition, pp. 4 and 26–27. Oxford University Press, USA, (2006).

TRACE: delivering integrated traceability systems that will enhance consumer confidence in the authenticity of food, Framework Programme 6, Integrated project contract no. 006942, 2005–2009 (www.trace.eu.org).

2

Food origin labelling legislation and standards

M. Woolfe, Thames Ditton (Food Standards Agency-retired), UK

DOI: 10.1533/9780857097590.1.12

Abstract: Among the main drivers of developing methods to determine origin are checking and enforcement of mandatory labelling rules on country of origin, methods of production and named species. These rules and standards have been increasing as consumers require more information on the origin of their food. The chapter outlines the international rules defining origin and protecting named origin, as well as international standards in which country of origin is a requirement. In the European Union, a considerable number of Community rules require origin labelling especially for meat and seafood, but also for honey, wines and olive oil. Other countries including the USA and Australia have also brought in mandatory country of origin labelling on particular foods.

Key words: WTO, USA and Australia origin rules, Codex Standards, UN/ECE Standards, EU rules on protected origin, country of origin and production origin labelling.

2.1 Introduction

There are several drivers for developing methods to verify the origin of foods. One is to ensure the integrity of the food chain and assist in the traceability of foods. Another is the monitoring and enforcement of mandatory requirements of origin labelling or verification of claims, as part of food standards enforcement programmes and the prevention of food fraud. This chapter will examine the current international legislation and other relevant standards that deal with origin of foods. Origin in this context covers not only geographical origin, but production method and species or varietal origin as well.

2.1.1 Country of origin labelling

The mandatory requirement to give information on the country of origin of a food is relatively recent, and in general only applies to certain specific foods. The reasons for making this information obligatory are quite diverse. In some cases it has arisen because of a serious food safety issue of a food from certain countries. In others, it has been industry pressure to give protection of local food production from cheaper imports. However, more recently it has been consumer driven; with increased awareness of sustainability and purchasing local food, animal welfare and good hygienic or manufacturing practice, consumers are demanding to know where their food comes from. As this demand grows, there is pressure on regulators to widen the requirement for origin labelling, which may be relatively easy for single or mainly single ingredient foods, but much more difficult for composite foods.

2.1.2 Geographical indication

Foods with names, which are linked directly or indirectly to a designated geographic origin, on the other hand, have been around for many years. Most notable are wines for example, champagne, which in many wine producing countries follow France's example of *appellation d'origine contrôlée* (AOC). Similar systems control wines in Spain, Germany, Portugal, Italy, USA, Canada, Austria, South Africa and Switzerland. In order for a wine to have AOC status, it must be produced to an agreed rigorous standard, and be made from ingredients (including specified grape varieties) derived from a specific geographical location. Consumers know that a wine has met AOC or other similar national status because, apart from its name, it carries a label or seal issued by the enforcing authorities on the bottle. Many spirit drinks also have names that identify their place of origin, for example, Scotch whisky and Tequila, and although these are protected names in national legislation, like wines some also have protected status at international level.

The food products with a long tradition of designated geographic origin are cheeses, and in France these have also been given AOC status, which has now been replaced by a European based system of protected denomination of origin (PDO) and protected geographic indication (PGI) (see Section 2.5)

2.2 International legislation and standards

The United Nations has set up two organisations to facilitate trade between its members. This is achieved principally by agreeing minimum standards. In addition, the World Trade Organisation has been established to arbitrate in trade disputes between members. It also has agreements to protect consumers and intellectual property rights of traders.

2.2.1 Codex Alimentarius general standard for the labelling of prepackaged foods

Codex Alimentarius is a joint FAO/WHO Commission, whose purpose is to facilitate international trade by setting standards, protecting consumer health and ensuring fair trade practices.[1] It is not mandatory for countries to adopt its standards and guidance. However, under the WTO-SPS (World Trade Organization Sanitary and Phytosanitary) Agreement,[2] countries cannot impose health or safety requirements on imports that are stricter than Codex standards or guidance, and similarly under the WTO-TBT (Technical Barriers to Trade)[2] Agreement, higher requirements cannot be imposed on imports than exist in international standards (Codex being one of the main sources). This has given Codex standards a much higher profile than they enjoyed previously.

The General Standard for the Labelling of Prepacked Foods (CODEX STAN 1-1985) is similar to the EU Labelling Directive 2000/13/EC[3] in detailing the basic information required on the labels of prepacked foods. Although the address of manufacturer or packer should be given, it is not necessary to give the country of origin unless it would be misleading to omit it.

2.2.2 Codex Alimentarius standards

There are over 250 published Codex standards[4] dealing with meat products, fish products, fresh and processed fruits and vegetables, milk products, cereals and pulses, chocolate, quick-frozen foods, packaged drinking water and some regional foods. In general, standards only require the country of origin to be declared on non-retail packages, but the exceptions are 16 Codex standards for named cheeses such as Mozzarella, Cheddar, Danbo, etc., which all require the country where the cheese was manufactured or underwent its last major transformation, to be declared on the retail pack. Table 2.1 gives a list of all these cheese standards. The standards permit the milk ingredient to be derived from cows and/or buffalo milk, except cream cheese, which just states the ingredient as 'milk'.

2.2.3 United Nations Economic Commission for Europe (UNECE) agricultural standards

The UNECE is one of the five UN regional commissions set up to promote pan-European economic integration. The commercial quality standards developed by the UNECE Working Party on Agricultural Quality Standards (WP.7) help facilitate international trade, encourage high-quality production, improve profitability and protect consumer interests. Although the drafting of standards is hosted by UNECE, any member country of the UN can attend the working party on an equal basis, hence the standards are regarded as international standards. UNECE standards are used by governments, producers, traders, importers and exporters, and other international

Table 2.1 List of Codex Cheese Standards requiring country of origin labelling on retail packs (nearly all Standards were revised in 2010)

Name of Cheese	Codex Standard	Name of Cheese	Codex Standard
Mozzarella	262–2006	Tilsiter	270–1968
Cheddar	263–1966	Saint Paulin	271–1968
Danbo	264–1966	Provolone	272–1968
Edam	265–1966	Cottage or Creamed Cottage Cheese	273–1968
Gouda	266–1966	Coulommiers	274–1969
Havarti	267–1966	Cream Cheese	275–1973
Samsø	268–1966	Camembert	276–1973
Emmental	269–1967	Brie	277–1973

organisations. There are 88 standards[5] covering a wide range of agricultural products, including fresh fruit and vegetables, dried fruit and nuts, seed potatoes, meat and poultry, and eggs.

Nearly all the standards define requirements which apply at the export/control stage and must meet the legal requirements of the importing country. In general, information on origin, species, etc., is obligatory for bulk packaging or documentation, but not necessarily for consumer information. Since the foods are agricultural produce rather than composite foods, the country of origin for meat/poultry is defined as where the animal is born, reared, slaughtered and processed. Only the Standard for Eggs in Shell (2010)[6] require that information on the country of origin be given to the consumer.

2.2.4 World Trade Organisation (WTO) definition of origin

As mentioned in Section 2.2.1, the WTO has agreements to protect consumers in the area of food safety, and plant and animal health and safety, while at the same time trying to avoid trade protectionism. There are no requirements to declare the origin of products, however, there are non-preferential rules, which are used to define and determine the country of origin for the purposes of quotas, anti-dumping, anti-circumvention, statistics or origin labelling. The 'country of origin' is defined where a product is wholly obtained or produced completely within one country, the product shall be deemed as having origin in that country. For a product which has been produced in more than one country, the 'country of origin' is defined as the country where the last substantial transformation took place. A substantial transformation is regarded as happening when:

- there is a change of tariff classification or;
- the product is subject to a value added rule or
- the product is subject to a special processing rule, and the minimum transformation is described.

The conditions above mean that according to the non-preferential rules, a product always has only one country of origin. However, the different definitions of substantial transformation mean that the country of origin may differ from country to country, and the same product may have different origins depending on which country's rules are applied. Usually it is the rules of the country into which a product is being imported that apply.

There are also WTO preferential rules of origin, where a country is part of a free trade area or given preferential trading status (e.g., tariff concessions). In this case the rules of origin determine what products can benefit from the tariff concession or preference.

2.2.5 WTO protected Geographical Indications (GIs)

When negotiations on the WTO Agreement on Trade-Related Aspects of Intellectual Property Rights ('TRIPS') were concluded in 1994,[7] the governments of all WTO member countries (now 159 countries as of March 2013) agreed to set certain basic standards for the protection of GIs. There are, in effect, two basic obligations on WTO member governments relating to GIs in the TRIPS agreement:

1. Article 22 of the TRIPS Agreement says that all governments must provide legal opportunities in their own laws for the owner of a GI registered in that country, to prevent the use of marks that mislead the public as to the geographical origin of the food. This includes prevention of use of a geographical name, which although literally true 'falsely represents' that the product comes from somewhere else.
2. Article 23 of the TRIPS Agreement says that all governments must provide the owners of GI the right, under their laws, to prevent the use of a geographical indication identifying wines not originating in the place indicated by the geographical indication. This applies *even where the public is not being misled*, where there is no unfair competition and where the true origin of the good is indicated or the geographical indication is accompanied by expressions such as 'kind', 'type', 'style', 'imitation' or the like. Similar protection must be given to geographical indications identifying spirits.

Article 22 of TRIPS also says that governments may refuse to register a trademark or may invalidate an existing trademark (if their legislation permits or at the request of another government) if it misleads the public as to the true origin of a good. Article 23 says governments may refuse to register or may invalidate a trademark that conflicts with a wine or spirits GI whether the trademark misleads or not.

Article 24 of TRIPS provides a number of exceptions to the protection of geographical indications that are particularly relevant for wines and spirits (Article 23). For example, Members are not obliged to bring a geographical indication under protection where it has become a generic term for describing the product in question. Measures to implement these provisions should not prejudice prior trademark rights that have been acquired in good faith; and, under certain circumstances – including long-established use – continued use of a geographical indication for wines or spirits may be allowed on a scale and nature as before.

Negotiations[8] are in progress to draw up a multilateral register of wines and spirits that have protection under TRIPS, and the transition periods to implement full protection. The discussions are also around whether to expand the register to cover other protected geographical indications.

2.3 European Union (Community) legislation and standards

The European Union (EU) has grown from the original six members in 1957 to 27 Member States. It has established a Single Market by laying down common safety and food standards regulations, marketing regulations for agricultural commodities and fisheries, and general and specific food rules.

2.3.1 Food Labelling Directive

Council Directive 2000/13/EC[3] regulates the description and labelling of foods for the ultimate consumer at retail and in catering. The Directive details, the mandatory information required for prepacked foods, and the minimum information required for food sold in catering. For prepacked foods, the label must include the name and address of the manufacturer or packer, or EU seller, so that consumers have an address they can contact for advice, more information or complaints. In terms of labelling the place of origin of the food, this is only required if omission would mislead consumers to a material degree with regard to its true origin of provenance. Therefore, origin labelling is still open to interpretation, and is easier for single ingredients foods rather than composite foods, which have many ingredients from different sources, and which are often changed according to market conditions. The WTO definition of origin applies, which may lead to some confusion in the declaration of country of origin. Under the WTO definition, British bacon, for example, is bacon manufactured in Britain, but the pork may not necessarily be British. The Food Standards Agency has published guidance on country of origin labelling,[9] recommending that labelling be more transparent in the above case, and it should give the origin of the pork as well as the bacon. This advice has been reinforced by the results of a study, which indicate that consumers do not recognise the WTO definition of origin, but want to know the origin of the ingredients.[10]

Annex 1 of the Labelling Directive also contains a list of categories of ingredients, that is, generic names for example, ‘fish’ or ‘oil’, where the ingredients can be a mixture of different species of fish and different types of oil.

The category ingredient name 'meat' is not including in the Annex. This means that where a mixture of different species of meat is used in a product, each species has to be declared separately in the list of ingredients.

Genetically Modified Organisms (GMOs) and irradiated ingredient
Directive 2000/13/EC also requires a separate declaration on the label where the food has been irradiated or contains irradiated ingredients, and if any of the ingredients are produced from GMOs. A declaration of the use of GMO ingredients is also required for food sold 'loose', and must be displayed next to the description of the food. GMOs have to be authorised, and non-authorised GMOs cannot be used in food or animal feed. The GM Food and Feed Regulation (EC) No 1829/2003[11] lays down more detailed rules to cover the labelling of all GM food and animal feed, regardless of the presence of any GM material in the final product. Labelling is not required if the GMO is present in amounts less than 0.9% of a food ingredient or of a single ingredient food, and if this presence is adventitious or technically unavoidable. The rules extend labelling to flour, oils and glucose syrups derived from GMO ingredients. However, products such as enzymes or vitamins produced by GMO microorganisms do not need to be labelled or does meat, milk or eggs derived from animals fed with feed containing GMOs.

2.3.2 Food Information Regulation

In January 2008, the European Commission published a proposal for an EU Food Information Regulation[12] that would replace the Food Labelling Directive 2000/13/EC. After four years of discussion, the European Council and Parliament agreed a revised Regulation, which was published in November 2011.[13] Businesses will have three years to implement the labelling changes. Apart from consolidating all the amendments and changes that have previously occurred to the Labelling Directive, and removing any differences in national provisions on labelling in different EU Member States, there will be some significant changes in origin labelling. More foods and ingredients will have to give country of origin labelling where at present there is only guidance to do so. Hence origin labelling will be mandatory:

1. If failure to do so might mislead the consumer as to the true origin of the food, in particular if the information accompanying the food or the label would otherwise imply that the food has a different origin.
2. Where the country of origin of a food is given, and where it is not the same as that of its primary ingredient (i.e., an ingredient that makes up more than 50% of the food), the country of origin of the primary ingredient in question shall also be given, and indicated as being different to that of the food. This will address the issue of British bacon being manufactured from non-British pork.

3. Country of origin for pork, mutton/lamb, goat meat and poultry meat. Beef origin labelling (see Section 2.3.4) already has its own rules, and whether the same information (place of birth, rearing and slaughter) has to be given for the other meat species is under consideration.

The details of above labelling requirements will be drawn up in Commission Regulations, which need to be completed within two years that is, by 13 December 2013. There will also be consideration of more detailed rules of origin labelling for milk and milk ingredients, meat ingredients, ingredients making up more than 50% of the food, unprocessed and single ingredient foods within three years, that is, by 13 December 2014.

2.3.3 Health or identification marks for animal products

All premises dealing with the basic preparation of animal-based products, that is, abattoirs, cutting plants, premises preparing minced meat and meat preparations, dairies and cheese manufacturers, fish processing plants and egg packers have to be approved and meet certain standards.[14] The approved processing plant is identified by a health or identification mark, which must accompany the animal based product down the food chain. Therefore, consumers purchasing prepacked basic animal based products, for example, red meat and poultry cuts, minced meat, burgers, sausages, milk, cheese, fish fillets and smoked fish, will be able to identify the country where the last major transformation process (e.g., mincing, smoking or cooking) occurred. The health mark is normally an oval identifying the member state of the EU, and licence number of the approved premise as shown in Fig. 2.1.

2.3.4 Beef labelling

The concerns about health risks in the transmission of bovine spongiform encephalopathy (BSE) from cattle to humans and other animals prompted the compulsory origin information that enabled beef to be traced back to its source. Council Regulation (EC) 1760/2000[15] replaced earlier regulations and established a system for beef labelling, and Commission Directive (EC) 1825/2000[16] gives the detailed rules on the labelling of beef. The most recent amendment (Commission Regulation (EC) 275/2007[17]) gives more flexibility

Fig. 2.1 Example of a health mark.

Fig. 2.2 Example of a beef label.

on the labelling of trimmings and mince from batches prepared from a mixed origin of beef cuts. The rules apply to all sales of raw beef, whether chilled or frozen, beef mince including uncooked beef burgers (without any added ingredients), and require information on where the animal was born, reared, slaughtered and cut up.

This information is required to be printed on prepacked beef labels, but can be displayed on a notice for beef sold loose in butchers, for example. Figure 2.2 gives an example of a beef label and the compulsory origin information that has to be provided. Each beef product is given a reference number or code which serves as a batch code, and permits the product to be linked back to the source animal, group of animals or batches of beef used in the trimmings for example for minced beef.

2.3.5 Labelling of honey

Honey is one of the food products where there are specific rules[18] requiring that consumers receive information about the country where the honey was harvested. As country is not defined necessarily as the EU Member State, then individual countries, for example, Scotland and Wales in the UK, could be used. Where honey is blended from more than one country, then there is the option of simplifying the declaration by using one of the following statements:

- A 'blend of EC honeys'; or 'blend of non-EC honeys'; or 'blend of EC and non-EC honeys'.

More specific geographical or regional origins can be given, provided the honey is all from that source. In addition, the floral origin of the honey (except for filtered and bakers honey) can be given if the honey is derived wholly or mainly from the indicated source, but this is optional. The term 'mainly' is used because bees forage nectar and pollen over a wide area, and the honey will nearly always be a mixture of plant nectars even if one floral type predominates.

2.3.6 Labelling of fish and fishery products

Fish is another food where specific rules have been introduced so that consumers receive information about the species and geographic origin of the fish. Council Regulation (EC) 104/2000,[19] which controls the marketing of fishery and aquaculture products, makes the following requirements for describing fish and shellfish:

- the commercial designation of the fish/shellfish (i.e., an agreed commercial name for that species of fish/shellfish);
- the production method (i.e., whether it is farmed or wild, and if wild, whether caught at sea or in inland waters);
- the catch area (i.e., an area of the ocean in the case of fish caught at sea, or country of production in the case of farmed fish or fish caught in inland waters).

These requirements only cover fish and shellfish products in chapter 3 of EU Customs Code of Combined Nomenclature (CN codes). Whether a product has to follow these rules sometimes appears inconsistent because it depends on its presentation, and whether it falls within the chapter 3 groups of CN codes. Table 2.2 illustrates this by giving details of what is or is not included in chapter 3. More detailed rules on implementing these requirements for the above products are given in Commission Regulation (EC) 2065/2001.[20]

Commercial designations

Each Member State is required to draw up and publish a list of commercial designations for fish species accepted within their territory. The commercial designation is the common name for a species of fish described by its international, scientific (Latin) name. For example, only fish of the species *Melanogrammus aeglefinus* (L.) can be described as 'haddock' in the UK, no other fish species can use this name. The European Commission has lists of commercial designations from each Member State. The Food Standards Agency (and now Defra) is responsible for drawing up and amending the list of commercial designations for the UK. An updated list of English commercial names is published in Schedule 1 of The Fish Labelling (England) Regulations 2010.[21] The UK also has to accept commercial designations agreed in the Republic of Ireland in English and *vice versa*. In some cases, a specific commercial designation covers a family of fish species, for example, the designation 'hake' covers all species of the family *Merluccius*. Commercial designations are legal names, and hence the link between the common name

Table 2.2 Fish and shellfish subject to labelling and description rules

Fish and shellfish products included (Chapter 3 products)	Fish and shellfish products excluded (Non-chapter 3 products)
Live fish	
Fresh, chilled or frozen fish (whole, gutted, headed, fillets, steaks, minced fish) with no other ingredients except salt. Fish blocks (fillets or mince) without other ingredients except brine.	Products with added ingredients or which have been further processed, preserved, treated or cooked. Poached salmon, poached salmon fillets/slices. Composite products where the fish is an intrinsic component of the end-product: – coated, battered, breaded fish products, e.g., fish fingers, coated scampi. Ready meals, fish pies, etc., where fish or seafood is an ingredient.
Smoked fish (cold or hot smoked by natural processes) with only salt, e.g., • smoked salmon • smoked herring (e.g., Buckling), • kippers, • smoked haddock.	All smoked fish with colours, flavours etc. e.g., • smoked salmon with honey and sugar • smoked mackerel with colourings and other ingredients (e.g., peppered smoked mackerel).
Surimi (i.e., processed fish protein)	Surimi-based preparations and/or products such as crabsticks, fishsticks and similar products.
Fish with butter and/or sauce packaged separately.	Fish where butter and/or sauce is added directly on to the fish is considered a further process.
Dried fish	
Salted or brined fish e.g., salted cod, anchovies	Dried or salted fish with no other ingredient except salt and water.
Crustaceans, whether in shell or not, e.g., prawns, crabs, lobsters. Cooked in-shell crustaceans. Cooked, unpeeled crustaceans. Peeled, uncooked crustaceans.	Crustaceans which are both cooked and peeled, e.g., cooked and peeled prawns
Raw molluscs, whether in shell or not, e.g., mussels, scallops, oysters.	Cooked molluscs, e.g., cockle meat out of shell, winkle meat with or without shell.

and scientific name covers all fish products not just those in Chapter 3 of the CN codes, but food sold in catering as well. Hence any food sold as 'cod in batter' has to be made from *Gadus morhua, Gadus macrocephalus* or *Gadus ogac*. The scientific names for fish are required on documentation for traceability purposes especially in international trade, and more recently be made available to consumers (see below in '*Geographical origin*').

Production method

The method of production has to be given with the commercial designation and describes how the fish or seafood was 'harvested', that is, whether it was produced by aquaculture or caught 'wild' in the ocean or inland waters. The details in Regulation 2065/2001 define how the production method should be declared to the consumer:

- For fish/shellfish products of aquaculture, the terms 'farmed' or 'cultivated' must be used to indicate that they have been farmed, for example, farmed sea bass. In the UK, the term 'cultivated' is used only occasionally for certain molluscs, for example, 'cultivated oysters'.
- For products caught at sea or in freshwater, the terms 'caught' or 'caught in freshwater' must be used. However, these terms can be omitted if it is clear from the name that the fish is wild, for example, North-East Atlantic haddock.

Geographical origin

The Regulation requires that fish or shellfish produced by aquaculture has to declare the country of production where the final development took place before harvesting. In the EU, the named country has to be a Member State. Hence a food described as 'farmed Scottish trout' would also need to add the description (somewhere on the label) 'produced' or 'farmed' in the UK, because Scotland is not a Member State.

Regulation 2065/2001 also details how the catch area for fish and shellfish caught at sea should be described. The Regulation divides the oceans into 12 catch areas, based on FAO statistical classifications as shown in Table 2.3 and Fig. 2.3. Whilst this is an accurate division of the world's catch areas, many of the catch areas detailed in Table 2.3 are not so well known to the average consumer. The designated catch areas have to be declared even if more local areas are used in the description. For example, fish described as 'North Sea mackerel' would still need to mention the designated catch area 'North-East Atlantic'. The designated catch areas can be abbreviated, for example, N.E. Atlantic.

In the case where a more precise local name for farmed, or 'wild' (caught at sea or in fresh water) fish or shellfish is used, there is some flexibility as to where the mention of the country of origin or designated catch area is placed. It does not necessarily have to be next to the name, but in the case of a label can be in another position or on the back label for example. For fish and shellfish sold loose, it is possible to have a 'readily discernible' declaration on a label or ticket or on the wall of the retail outlet as a poster or notice in full view of customers, for example, 'Fish caught in the N.E. Atlantic'.

For mixtures of fish of the same species coming from a variety of production methods, the regulations require that the labelling must state each production method; for example, 'a mix of farmed Scottish cod and cod

Table 2.3 Designation of catch areas of World's oceans

Catch area	FAO identification of the area
North-West Atlantic	FAO area 21
North-East Atlantic (Excluding Baltic Sea)	FAO area 27
Baltic Sea	FAO area 27.IIId
Central-Western Atlantic	FAO area 31
Central-Eastern Atlantic	FAO area 34
South-West Atlantic	FAO area 41
South-East Atlantic	FAO area 47
Mediterranean Sea	FAO areas 37.1, 37.2 and 37.3
Black Sea	FAO area 37.4
Indian Ocean	FAO areas 51 and 57
Pacific Ocean	FAO areas 61, 67, 71, 77, 81 and 87
Antarctic	FAO areas 48, 58 and 88

Source: FAO yearbook. Fishery Statistics. Catches. Vol.86/1. 2000.

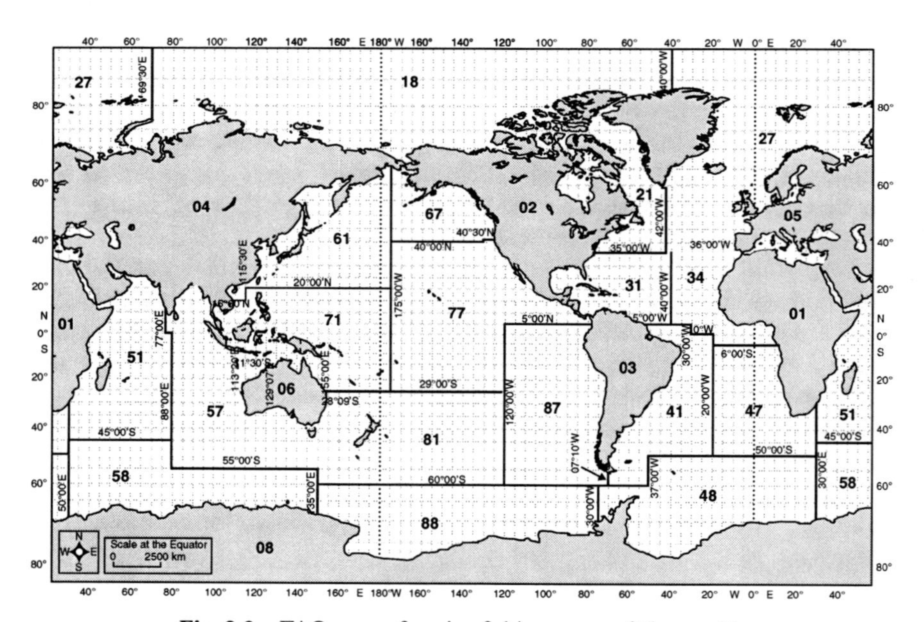

Fig. 2.3 FAO map of major fishing areas of the world.

caught in the N.E. Atlantic', in the order in which origin predominates. For mixtures of fish of the same species coming from different catch areas or fish-farming countries, the origin that is most representative of the batch in terms of quantity must be stated. Hence a batch of 'farmed salmon steaks' may originate predominantly in Scotland, but also Norway or Chile, and could be described as 'farmed salmon steaks originating from Scotland, Norway and Chile'.

An amendment to the Fish Labelling Regulation – Council Regulation 1224/2009[22] and more detailed rules – Commission Implementing Regulation (EU) 404/2011[23] ensures that all the information required by Commission Regulation 2065/2001 including the scientific name is given to the consumer at retail sale. However, there is a derogation that the scientific name can be provided by a billboard or poster, or even a list to be available on request by consumers.

2.4 European Community (EC) agricultural marketing standards

In 2007, there was a major consolidation of 21 marketing standards and Regulations into one Marketing Regulation (1234/2007).[24] The standards for poultrymeat, eggs, fresh fruit and vegetables and olive oil are of particular relevance to origin labelling, as there are rules covering country of origin, production origin or requirements to give other details of varieties or types for these foods.

2.4.1 Poultrymeat standard

The detailed rules for the poultrymeat standard are given in Commission Regulation (EC) No 543/2008.[25] The poultry species covered by the standard are chickens, turkeys, ducks, geese and guinea fowls. Country of origin labelling is required only for poultrymeat imported from non-EU countries. Where claims are made about production method, then these must follow the definitions in Regulation 543/2008, for example:

- 'Fed with … % of … ', where the poultry is fed a diet of at least 65% cereal during rearing, and if a specific cereal is mentioned it must constitute at least 35% of the cereal diet, but for maize it is 50%. In the UK, 'corn-fed' (a description derived from the USA) is a more familiar and understood term by UK consumers than the general claim wording, and is used where the poultry is fed with at least 50% maize during rearing.
- 'Free range' or 'traditional free range' must comply with certain stocking densities outlined in the regulation, and birds must have had access to open air runs during at least half their lifetime to an area with vegetation or for 'traditional free range' continuous access and minimum slaughter ages, and fed with at least 70% cereals during rearing.

2.4.2 Egg standard

The Standard applies to eggs for retail sale, and more detailed rules are given in Commission Regulation 589/2008.[26] All Class A eggs (i.e., those sold through retail and to catering) are required to be marked with a code identifying the method of production, country of origin and the production establishment:

- The first number in the code denotes the production method: '0' = Organic, '1' = Free Range, '2' = Barn, '3' = Eggs from Caged Hens. The following two letters denote the country of origin (e.g., 'UK'), this is followed by a code identifying the registered production site (numbers sometimes with letters). So for example an egg stamped with $^{1}\mathbf{UK}^{4321}$ is a free range egg from the UK establishment 4321.

Council Directive 1997/74/EC[27] defines the methods of production described or labelled on boxed eggs, and contained in the code marking.

- 'Barn' is a system which has a series of perches and feeders at different levels. The maximum stocking density is 9 birds per square metre and there must be at least 250 cm square of litter area/bird. Perches for the birds must be installed to allow 15 cm of perch per hen. There must be at least 10 cm of feeder/bird and at least one drinker/10 birds. There must be one nest for every 7 birds or 1 square metre of nest space for every 120 birds. Water and feeding troughs are raised so that the food is not scattered
- In 'free-range' systems, the birds are housed as described in the barn system above. In addition birds must have continuous daytime access to open runs which are mainly covered with vegetation and with a maximum stocking density of 2500 birds per hectare.

2.4.3 Standards for fresh fruits and vegetables

Recent changes in EU legislation have simplified the marketing standards covering a range of fresh fruit and vegetables. Commission Regulation (EC) 1221/2008[28] has reduced the number of specific marketing standards from 36 to 10. Produce not included in the list of specific standards has to conform to a general marketing standard in Annex 1 Part A of the Regulation. Indication of country of origin has to be given for the fresh produce covered by Regulation 1121/2008. The Regulation also requires other specific information to be given to the consumer as outlined in Table 2.4.

2.4.4 Olive oil standard

Olive oil also has provisions which require 'designation of origin'. The provisions which cover these existed in an earlier Commission Regulation (EC) No 1019/2002,[29] but only apply to extra virgin olive oil and virgin olive oil:

- Olive oils originating from a single Member State or third country shall have a reference to the Member State, the Community (i.e., the European

Table 2.4 Specific information other than country of origin for certain fresh fruit and vegetables

Fresh produce	Indication	Fresh produce	Indication
Apples	Variety or varieties	Pears	Variety
Citrus fruits	Oranges – variety and type Mandarin – variety or species Clementine – whether pips or not	Sweet Peppers	Colour Commercial type – elongated, flat etc. or name of variety
Kiwifruits	Variety (optional)	Strawberry	Variety (optional)
Lettuce	Type – Cos, Oak-leaf, Lollo rosso, etc. Little gem or mixed salad if appropriate	Table grapes	Variety or varieties
Peaches/ nectarines	Colour of flesh Variety (optional)	Tomatoes	Whether 'on the vine' Variety (optional)

Community or EC) or the third country. In the case of PDO/PGI registered olive oil, the regional origin may be given.

- If the olives were harvested in a Member State or third country other than that in which the mill where the oil was extracted is situated, the designation of origin has to contain the following wording: '(extra) virgin olive oil obtained in (the Community or the name of the Member State concerned) from olives harvested in (the Community or the name of the Member State concerned)'. In the case of blended olive oils originating from one or more Member State or third country, one of the following designations must be used:
  - 'blend of Community olive oils' or a reference to the Community
  - 'blend of non-Community olive oils' or a reference to the non-community origin
  - 'blend of Community and non-Community olive oils' or a reference to Community and non-Community origin.

2.4.5 Basmati rice

Basmati rice is a high value rice grown in the Indo-Gangetic plains of India and Pakistan (N. Punjab, Haryana and Uttar Pradesh). Its high price is due to its eating quality (aromatic, long slender grains which only elongate during cooking) and relatively low yield (only one harvest per year). The market for rice is controlled by Regulation 1234/2007,[24] which levies a duty on imported rice grown outside of the EU, and favours the milling of brown rice in the EU

Table 2.5 Approved varieties of rice that can use the description 'Basmati'

Varieties listed in Commission Regulation 972/2006	Other approved Basmati varieties
Basmati 370	Basmati 198
Dehra Dun (Type3)	Basmati 385
Basmati 217	Kasturi
Ranbir	Haryana Basmati
Taraori (HBC 19)	Mahi Sugandha
Kernel	Punjab Basmati
Basmati 386	
Super	
Pusa	

Varieties in italics are pure-line or 'traditional' varieties, whereas those in normal font are crosses or hybrid varieties.

rather than importing milled white rice. Basmati rice is permitted to enter the EU duty free provided it meets the requirements of Commission Regulation (EC) 972/2006.[30] The description 'Basmati' is a customary name (meaning the 'fragrant one'), but has been defined in a Code of Practice agreed between the UK rice industry and the UK enforcement authorities and official control laboratories.[31] It has been principally defined in terms of all the varieties of rice approved by the Indian and Pakistani authorities to be exported as 'Basmati', as listed in Table 2.5. Regulation 972/2006 only gives duty free status to nine of these varieties, which are listed in Table 2.5, and are the main commercial varieties grown in the two exporting countries. Export documents have to be accompanied by an authenticity certificate using a DNA rice variety test. Country of origin labelling on retail packs ('Product of India/Pakistan') is voluntary, however if claimed, the UK Code of Practice lays down that 97% of the grains must originate from that country. In addition, if the description 'Basmati' is used, the pack can only contain a maximum of 7% non-Basmati grains, otherwise the rice has to be described as a mixture of Basmati and non-Basmati rice.

2.4.6 Labelling of wines

The rules that cover the labelling of wine are given in Council Regulation (EC) 1493/1999[32] and Commission Regulation 753/2002.[33] All wines marketed in the EU have to state their country of origin. Auxiliary information should be given depending on the type of wine:

For table wine:

- In the case of despatch to another Member State or exporting state, the name of the Member State if the grapes are produced and made into wine in that State.

- The words 'mixture of wines from different countries of the European Community' in cases of wines resulting from a mixture of products originating in a number of Member States.
- The words 'wine obtained in … from grapes harvested in … ', supplemented by the names of the Member States concerned in the case of wines produced in a Member State from grapes harvested in another Member State.

In addition:

- Table wines with geographical indication – the name of the geographical region, the bottler or persons involved with marketing.
- Quality wines produced in specified regions – the name of the production area.
- Imported wines – the name of the country of origin and, when designated with a geographical indication, the name of the geographical area in question.

Varietal indication is optional. However, each Member State or exporting third country has to draw up a list of varieties used in their wine production, and these are listed in Annex II of Regulation 753/2002. Where a single variety is used to describe the wine, 85% of the wine must be made from that variety. Where two or three varieties are mentioned, then 100% of the product must be from those varieties, and listed in descending order in the proportion they are used in the wine. Varieties not included in Annex II cannot be used.

2.5 Organic standards and protected descriptions of geographic origin

The description 'organic' can only be used if the food has been grown, reared or prepared according to the strict criteria laid down in EU law. It can also be used for processed products if 95% of the ingredients are produced organically. The Regulation governing organic production has been reviewed and revised, and Council Regulation (EC) No 834/2007[34] covers organic production and labelling, whereas Council Regulation (EC) 889/2008[35] gives more detail rules on the implementation and certification of organic production. Each Member State has a government body responsible for the regulation, and it approves control bodies to inspect and certify organic producers. Each control body must be accredited to EN 45011 or ISO Guide 65, and has a code number, which is displayed along with an EU logo for organic food. The code of the control body has to use the acronym of the country as set out in ISO Standard 3166.[36] Hence for the UK, the code will use the acronym GB. The European Commission publishes a list of all approved control

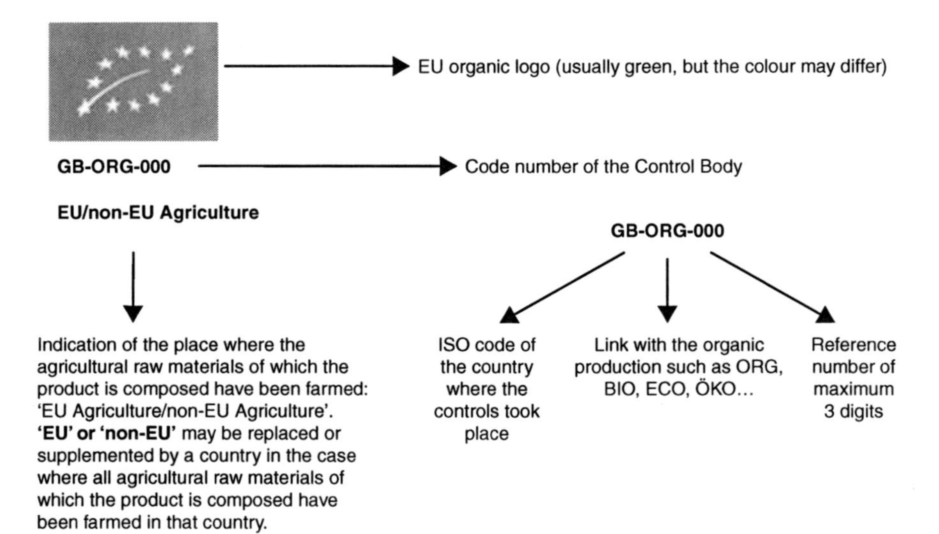

Fig. 2.4 The new EU organic logo and the compulsory information required.

bodies in Member States.[37] From 1 July 2010, the EU organic logo[38] shown in Fig. 2.4 is obligatory for all organic prepackaged food products within the EU, and a mention of the place where the agricultural products were farmed. It is also possible to use the logo on a voluntary basis for non-prepackaged organic goods produced within the EU or organic products imported from third countries.

There is control over the production of processed foods described as 'organic'. Regulation 889/2008 gives a list of permitted additives and ingredients that can be used in organic foods. Where a food contains less than 95% organic ingredients, the logo cannot be used, and the reference to the organic production method may only appear in relation to the organic ingredients. In this case, the food has to label which of the ingredients are organic, and their proportion in the list of ingredients.

In 1993, EU legislation came into force which provided for a system for the protection of food names on a geographical or a traditional recipe basis. The idea was to have a system similar to '*appellation d'origine controleé*' for wines, but applied to other foods and beverages. The scheme is open to regional and traditional foods that can guarantee their authenticity and/or origin. Once registered, the name of the food or beverage is given legal protection against imitation throughout the EU. Foods granted PDO (Protected Designation of Origin) and PGI (Protected Geographical Indication) status under Council Regulation 510/2006[39] will be linked to a stated geographical origin. On the other hand, foods given TSG (Traditional Speciality Guaranteed) status under Council Regulation 509/2006[40] will not be linked to a geographical origin, but may specify particular breeds of animal or varieties of plant foods in their traditional recipes, for example, Gloucester Old Spot pork. The procedures to apply to one of these protected schemes are detailed in Commission Regulation 1898/2006[41] for PDOs/PGIs, and Commission Regulation 1216/2007[42] for TSGs. Producers normally apply through their national organisations (in the case of the UK this is ADAS), and the European Commission publishes a full list of foods that have been registered (http://ec.europa.eu/agriculture/quality/door/list.html).

A summary of the requirements for PDO, PGI and TSG status is given below (logos printed with the permission of the European Commission):

PDO

Open to products, which are produced, processed and prepared within a particular geographical area, and with features and characteristics which must be due to the geographical area

PGI

Open to agricultural products and foodstuffs closely linked to the geographical area. At least one of the stages of production, processing or preparation takes place in the area.

TSG

Open to products which are traditional or have customary names and have a set of features which distinguish them from other similar products. These features must not be due to the geographical area where the product is produced, but highlights traditional character, either in the composition or means of production.

Table 2.6 summarises the total number of products for six major EU countries that have been registered or published as having PDO or PGI status. It highlights the difference between northern European and southern European countries, where the latter has more of a tradition of protecting foods from specific geographic origins such as meat and meat products, cheeses, olive oil and butter, and fruits, vegetables and cereals.

Table 2.6 Summary of products with registered or published PDO and PGI status for six EU countries (as of June 2011)

Country	Fresh meat	Meat products	Cheeses	Fruit, vegetables and cereals	Oils and fats (Olive oil, butter etc.)	Total PDOs and PGIs of all products
France	5 PDOs 57 PGIs	0 PDO 7 PGIs	46 PDOs 5 PGIs	16 PDOs 24 PDOs	9 PDOs 0 PGI	86 PDOs 107 PDOs (37)
Germany	2 PDOs 2 PGIs	0 PDO 13 PGIs	4 PDOs 2 PGIs	0 PDO 9 PGIs	0 PDO 1 PGI	29 PDOs 52 PGIs (18)
Italy	0 PDO 4 PGIs	22 PDOs 17 PGIs	44 PDOs 1 PGI	33 PDOs 63 PGIs	44 PDOs 1 PGI	157 PDOs 92 PGIs (24)
Poland	0 PDO 0 PGI	0 PDO 1 PGI	3 PDOs 1 PGI	1 PDO 5 PGIs	0 PDO 0 PGI	9 PDOs 16 PGIs (3)
Spain	0 PDO 15 PGIs	5 PDOs 9 PGIs	25 PDOs 1 PGI	18 PDOs 31 PGIs	26 PDOs 1 PGI	84 PDOs 72 PGIs (28)
UK	4 PDOs 4 PGIs	0 PDO 2 PGI	9 PDOs 3 PGIs	2 PDOs 1 PGI	0 PDO 0 PGI	17 PDOs 23 PGIs (7)

The numbers in brackets are the total number of PDO + PGI applications awaiting approval.

2.6 Legislation and standards in other parts of the world

Many countries follow the origin rules given in the Codex Alimentarius General Standard on the Labelling of Prepacked Foods, however, several countries have brought in their own specific rules, and USA and Australia are two examples.

2.6.1 USA legislation

There has been a long debate in the USA about introducing mandatory country of origin labelling for a range of foods. On the one hand, some sectors of the food and agricultural industry have lobbied federal lawmakers to introduce such requirements as a thinly disguised trade barrier to imported produce; on the other hand, opposing groups have argued that US consumers have a right to know where their foods comes from as they already receive such information on other consumer goods.

In 2002, an amendment to the 1946 Agricultural Marketing Act permitted the mandatory declaration of country of origin. In 2004, mandatory country of origin labelling was introduced for wild and farmed fish and shellfish. The final rule to introduce country of origin labelling for a number of other

Table 2.7 Summary of US country of origin labelling requirements

Scope	Retail sale of fresh, frozen cuts and mince of beef, lamb, pork, goat and chicken, fresh and frozen fruits and vegetables, peanuts, pecans, Macadamia nuts, ginseng, fresh and frozen wild and farmed fish and shellfish.
Exemptions	Catering sales for direct consumption, processed or composite foods e.g., roasted peanuts.
Mixed origins	Give all countries of origin – e.g., pork ribs from US, Canada and Mexico. If animal reared in one country and slaughtered in US, give where animal reared first e.g., beef steak from Canada and US.
US origin	Animal must be reared, slaughtered and cut in US. Farmed fish or shellfish must be cultivated, harvested and processed in US. Wild fish or shellfish must be caught at sea/fresh water and processed in US or in US registered boats. Fruits and vegetables must be harvested and packed in US.
Declaration	Flexibility as to how give information – can be on label or sticker, but also twist tie or sign

foods was passed in January 2009,[43] and came into force two months later. The labelling is enforced by the Agricultural Marketing Service, which carries out annual checks on retail premises. A summary of the requirements is given in Table 2.7.

2.6.2 Australian legislation and standards

Australian country of origin labelling for packaged food and consumer law
Food Standards Australia and New Zealand amalgamated several codes into a new food standards code, which became law in Australia on 8 December 2005, and have been revised in January 2013.[44] Chapter 1 of this code requires all packaged foods to carry a separate statement identifying the country where the food was produced, processed or packaged in addition to the address of the manufacturer. This rule is mandatory in Australia but voluntary in New Zealand.

In January 2011, Australia introduced the Australian Consumer Law (ACL)[45] to replace previous consumer protection legislation in fair trading acts. It is contained in a schedule to the Trade Practices Act 1974, which has been renamed the Competition and Consumer Act 2010 (CCA), and covers fair trading on all products not just food. The new ACL clarifies origin labelling and provides a defence for certain claims especially a new one for 'grown in country X'. This claim can be made provided each significant ingredient was grown in that country, and if processed (prepared), that this was also carried out in the same country. The ingredient/s in question has/ve to comprise 50% or more of the weight of the product. These rules cover all prepacked foods.

Australian country of origin labelling for unpackaged (loose) foods
The new code also required certain unpackaged foods to label the country of origin, which came into force in June 2006, and was amended in January 2012. The unpackaged foods requiring a declaration include:

- pork including cured pork products such as bacon and ham;
- beef, lamb and chicken (taking effect on 18 July 2013);
- fruits, vegetables and nuts including dried or sun-dried products and;
- fresh or frozen fish and shellfish including smoked, salted or marinated, and coated fish.

The requirement does not cover composite foods (apart from seafood), dairy products or fats and oils. Also foods sold in catering for direct consumption do not need to provide origin information. Any unpackaged product containing seafood (fish or shellfish) has to give origin of the seafood. The information can be as a sticker on the food or on a sign.

2.7 Conclusion and future trends

This chapter has demonstrated that there is now a considerable amount of legislation and standards, which require the mandatory declaration of geographical, or production origin, as part of the consumer information. The increase in global trade in food with differing standards and practices in production, coupled with a greater consumer interest in where and how food has been produced, will no doubt continue the trend for more requirements in the future.

In Europe, the discussion and detail of the country of origin requirements in the EU Food Information Regulation will lead to a significant change in meat labelling and other foods with major ingredients in excess of 50% of the food. However, the full details of what will be included in the scope and how this information is to be conveyed to the consumer is a few years away. When it is eventually implemented, it is very likely to have an influence over origin labelling in other countries outside the EU. There will certainly be a move to align the Codex General Standard on Food Labelling more closely with the Food Information Regulation, and if this is successful many other countries will introduce a mandatory requirement of country of origin labelling where specific foods or ingredients are grown, reared or harvested, and prepared or processed.

2.8 Sources of further information and advice

Geographical Indications – Information about protection of regional product names:
http://www.geographicindications.com/

WTO Rules on Origin:
  http://www.wto.org/english/tratop_e/roI_e/roi_info_e.htm
2011 TRIPS negotiations on a list of protected wines and spirits:
  http://www.wto.org/english/news_e/news11_e/trip_ss_03mar11_e.htm
EU Marketing Standards:
  http://archive.defra.gov.uk/foodfarm/food/industry/sectors/eggspoultry/faq/marketing.htm

2.9 References

1. Codex General Standard on Food Labelling. Codex Stan 1–1985, Amended 1991, 1999, 2001, 2003, 2005, and 2008. FAO/WHO Codex Alimentarius, Rome.
2. WTO Sanitary and Phytosanitary Measures Agreement and Technical Barriers to Trade Agreement: Understanding the WTO – Standards and Safety: http://www.wto.org/english/thewto_e/whatis_e/tif_e/agrm4_e.htm.
3. Directive 2000/13/EC of 20 March 2000 on the approximation of the laws of the Member States relating to the labelling, presentation and advertising of foodstuffs. *OJ L*, **109**, 2000, 29–42.
4. List of published Codex Alimentarius Standards, FAO/WHO Codex Alimentarius, Rome: http://www.codexalimentarius.net/web/standard_list.do?lang=en.
5. United Nations Economic Commission for Europe, Working Party on Agricultural Quality Standards, Geneva: http://live.unece.org/trade/agr/welcome.html.
6. UN/ECE Standard for Eggs in Shell, Geneva: http://www.unece.org/trade/agr/standard/eggs/Standards/EGG01_EggsInShell_2010E.pdf.
7. Understanding the WTO: The Agreements. Intellectual Property: protection and enforcement. TRIPS Agreement Section II, Part 3, WTO, Geneva: http://www.wto.org/english/thewto_e/whatis_e/tif_e/agrm7_e.htm; http://www.wto.org/english/docs_e/legal_e/27-trips_04b_E.htm.
8. Geographical Indications Negotiations – Chairman's Report to the Trade Negotiations Committee 22 March 2010: http://www.wto.org/english/tratop_e/trips_e/tnc_chair_report_march10_e.htm#tnip20.
9. Country of origin labelling guidance Food Standards Agency 31 October 2008: http://www.food.gov.uk/multimedia/pdfs/originlabellingguid0909.pdf.
10. Country of Origin Labelling: a synthesis of research. Food Standards Agency 14 January 2010: http://www.food.gov.uk/news-updates/news/2010/jan/coolresearch.
11. Regulation (EC) No 1829/2003 of the European Parliament and of the Council of 22 September 2003 on Genetically Modified Food and Feed. *OJ L* **268**, 2003, 1–23.
12. Proposal for a Regulation of the European Parliament and of the Council on the provision of food information to consumers: http://ec.europa.eu/food/food/labellingnutrition/foodlabelling/publications/proposal_regulation_ep_council.pdf.
13. Regulation (EU) No 1169/2011 of the European Parliament and of the Council of 25 October 2011 on the provision of food information to consumers, amending Regulations (EC) No 1924/2006 and (EC) No 1925/2006 of the European Parliament and of the Council, and repealing Commission Directive 87/250/EEC, Council Directive 90/496/EEC, Commission Directive 1999/10/EC, Directive 2000/13/EC of the European Parliament and of the Council, Commission Directives 2002/67/EC and 2008/5/EC and Commission Regulation (EC) No 608/2004. *OJ L* **304**, 2011, 18–63.
14. Regulation (EC) No 853/2004 of the European Parliament and of the Council of 29 April 2004 laying down specific hygiene rules for food of animal origin. *OJ L* **139**, 2004, 55–205.

15. Regulation (EC) No 1760/2000 of the European Parliament and of the Council of 17 July 2000 establishing a system for the identification and registration of bovine animals and regarding the labelling of beef and beef products and repealing Council Regulation (EC) No 820/97. *OJ L* **204**, 2000, 1–10.
16. Commission Regulation (EC) No 1825/2000 of 25 August 2000 laying down detailed rules for the application of Regulation (EC) No 1760/2000 of the European Parliament and of the Council as regards the labelling of beef and beef products. *OJ L* **216**, 2000, 8–12.
17. Commission Regulation (EC) No 275/2007 of 15 March 2007 amending Regulation (EC) No 1825/2000 laying down detailed rules for the application of Regulation (EC) No 1760/2000 of the European Parliament and of the Council as regards the labelling of beef and beef products. *OJ L* **76**, 2007, 12–15.
18. Council Directive of 2001/110/EC of 20 December 2001 relating to honey. *OJ L* **10**, 2002, 47–52.
19. Council Regulation (EC) No 104/2000 of 17 December 1999 on the common organisation of the markets in fishery and aquaculture products. *OJ L* **17**, 2000, 22–51. (Article 4 on 'consumer information').
20. Commission Regulation (EC) No 2065/2001 of 22 October 2001 laying down detailed rules for the application of Council Regulation (EC) No 104/2000 as regards informing consumers about fishery and aquaculture products. *OJ L* **278**, 2001, 6–8.
21. The Fish Labelling (England) Regulations 2010. SI 210 No 420. HMSO: http://www.legislation.gov.uk/uksi/2010/420/pdfs/uksi_20100420_en.pdf.
22. Council Regulation (EC) No 1224/2009 of 20 November 2009 establishing a Community control system for ensuring compliance with the rules of the common fisheries policy. *OJ L* **343**, 2009, 1–50.
23. Commission Implementing Regulation (EU) No 404/2011 of 8 April 2011 laying down detailed rules for the implementation of Council Regulation (EC) No 1224/2009 establishing a Community control system for ensuring compliance with the rules of the Common Fisheries Policy. *OJ L* **112**, 2011, 1–153.
24. Council Regulation (EC) No 1234/2007 of 22 October 2007 establishing a common organisation of agricultural markets and on specific provisions for certain agricultural products (Single CMO Regulation). *OJ L*, **299**, 2007, 1–242.
25. Commission Regulation (EC) No 543/2008 of 16 June 2008 laying down detailed rules for the application of Council Regulation (EC) No 1234/2007 as regards the marketing standards for poultrymeat. *OJ L*, **157**, 2008, 46–87.
26. Commission Regulation (EC) No 589/2008 of 23 June 2008 laying down detailed rules for implementing Council Regulation (EC) No 1234/2007 as regards marketing standards for eggs. *OJ L* **163**, 2008, 6.
27. Council Directive 1999/74/EC of 19 July 1999 laying down minimum standards for the protection of laying hens. *OJ L* **203**, 1999, 53–57.
28. Commission Regulation (EC) No 1221/2008 of 5 December 2008 amending Regulation (EC) No 1580/2007 laying down implementing rules of Council Regulations (EC) No 2200/96, (EC) No 2201/96 and (EC) No 1182/2007 in the fruit and vegetable sector as regards marketing standards. *OJ L* **336**, 2008, 1–80.
29. Commission Regulation (EC) No 1019/2002 of 13 June 2002 on marketing standards for olive oil. *OJ L* **155**, 2002, 27–31.
30. Commission Regulation (EC) No 972/2006 of 29 June 2006 laying down special rules for imports of Basmati rice and a transitional control system for determining their origin. *OJ L* **176**, 2006, 53–62.
31. Code of Practice of Practice on Basmati Rice, July 2005. The Rice Association, British Rice Millers Association, and British Retail Consortium. http://www.riceassociation.org.uk/content/1/4/documents.html.
32. Council Regulation (EC) No 1493/1999 of 17 May 1999 on the common organisation of the market in wine. *OJ L* **179**, 1999, 1–83.

33. Commission Regulation (EC) No 753/2002 of 29 April 2002 laying down certain rules for applying Council Regulation (EC) No 1493/1999 as regards the description, designation, presentation and protection of certain wine sector products. *OJ L* **118**, 2002, 1–54.
34. Council Regulation (EC) No 834/2007 of 28 June 2007 on organic production and labelling of organic products and repealing Regulation (EEC) No 2092/91. *OJ L*, **189**, 2007, 1–23.
35. Commission Regulation (EC) No 889/2008 of 5 September 2008 laying down detailed rules for the implementation of Council Regulation (EC) No 834/2007 on organic production and labelling of organic products with regard to organic production, labelling and control. *OJ L* **250**, 2008, 1–84.
36. ISO Standard 3166 – Country Names and Codes. http://www.iso.org/iso/country_codes.htm.
37. List of Control Bodies and Control Authorities in Charge of Controls in The Organic Sector Provided for in Article 35(B) of Council Regulation (EC) No 834/2007. http://ec.europa.eu/agriculture/organic/files/consumer-confidence/inspection-certification/EU_control_bodies_authorities_en.pdf.
38. The EU Organic Farming Logo. www.organic-farming.europa.eu.
39. Council Regulation (EC) No 510/2006 of 20 March 2006 on the protection of geographical indications and designations of origin for agricultural products and foodstuffs. *OJ L* **93**, 2006, 12–25.
40. Council Regulation (EC) No 509/2006 of 20 March 2006 on agricultural products and foodstuffs as traditional specialities guaranteed. *OJ L*, **93**, 2006, 1–11.
41. Commission Regulation (EC) No 1898/2006 of 14 December 2006 laying down detailed rules of implementation of Council Regulation (EC) No 510/2006 on the protection of geographical indications and designations of origin for agricultural products and foodstuffs. *OJ L*, **369**, 2006, 1–19.
42. Commission Regulation (EC) No 1216/2007 of 18 October 2007 laying down detailed rules for the implementation of Council Regulation (EC) No 509/2006 on agricultural products and foodstuffs as traditional specialities guaranteed. *OJ L* **275**, 2007, 3–15.
43. Mandatory Country of Origin Labeling of Beef, Pork, Lamb, Chicken, Goat Meat, Wild and Farm-Raised Fish and Shellfish, Perishable Agricultural Commodities, Peanuts, Pecans, Ginseng, and Macadamia Nuts. US Department of Agriculture, Agricultural Marketing Service. *Federal Register* **74**(10), 2009, 2658–2707.
44. Country of Origin Labelling, Food Standards Australia and New Zealand factsheet http://www.foodstandards.gov.au/consumerinformation/labellingoffood/countryof-originlabel5598.cfm.
45. Australian Consumer Law 1 January 2011, http://www.accc.gov.au/content/index.phtml/itemId/964551.

Part II

Analytical methods

3

New approaches for verifying the geographical origin of foods

A. Roßmann, Isolab GmbH Laboratory for Stable Isotopes, Germany

DOI: 10.1533/9780857097590.2.41

Abstract: Stable isotope methods have been used for food authenticity investigation for 20 years, but their analysis for geographical origin checks developed more recently. Stable isotope ratio analysis (SIRA) and multi-element pattern are increasingly being applied in control of geographical origin of food, and both methods are often and successfully used in combination. Basic requirements for the use of the stable isotope methodology in particular and practical examples are presented in this paper. Recently developing profiling methods such as non-targeted ^{1}H-NMR analysis will be applied for questions of geographical provenance of food products in combination with stable isotopes and multi-element pattern also.

Key words: stable isotopes, geographical origin, multi-element analysis, reference materials, data evaluation.

3.1 Introduction: the commercial importance of the geographical origin of foods

Consumers are increasingly interested in the provenance of the foods they consume, especially following several recent problems such as the bovine spongiform enzephalopathie (BSE) crisis; the incidence of avian influenza; the scandal of melamine in Chinese milk products; the detection of residues of pharmaceutical products in food products; and very recently the contamination of vegetables with microorganisms. Other reasons for this interest in provenance range from (a) patriotism; (b) specific culinary or organoleptic qualities associated with regional products; (c) decreased confidence in the quality and safety of products produced outside their local region, country, the EU, or those of unknown provenance; or (d) concern about animal welfare and production methods perceived as 'environmentally friendly' and more often adopted by smaller regional producers. The upshot is that products from highly esteemed geographical origins can be sold at markedly higher prices than similar products of other or unknown provenance. For certain products, such as wines, indicating geographical origin has been an established practice for many years, for others this kind of marketing is more recent. In some European countries, marking certain products with a geographical name or brand has a long tradition (e.g., cheeses in France and Italy), and today in Europe a number of products with a specific geographical origin and production method are officially protected (protected denomination of origin = PDO; protected geographic indication = PGI). The PDO rules demand not only that food must be processed according to traditional methods in a particular region, but even the raw materials must be supplied from a defined area. As these products are sold at higher prices, there is reasonable suspicion that unscrupulous producers might buy cheaper raw materials from other regions and illegally sell the product with the PDO label. The control bodies have a duty to investigate premium products and ascertain whether they have been produced from raw materials from within or without the allowed area. Investigation of this sort is required both to avoid consumers being cheated, and to protect honest producers. Such producers are frequently small companies in less favoured areas who need protecting from big competitors who could squeeze them from the market. This objective is mainly accomplished by controlling the certification papers and the flow of raw materials and products, but until now the use of objective analytical criteria has been infrequent. Analytical methods rely largely on determinations of chemical composition, which may be quite similar even where the same materials come from different geographical areas. Attempts have been made by determining some components as typical for certain areas or production methods or using multi-element pattern or trace element content as determined by inductively coupled plasma mass spectrometry (ICP-MS). Other methods that might be applied in the future include genetic methods (in a case where different strains/breeds of organisms are involved in production) or multi-parametric analyses and sophisticated statistical evaluations of obtained data (see other chapters of this book). Sometimes results of such analyses do not indicate the geographical origin, but rather the production techniques linked to certain provenances (Pillonel *et al.*, 2003), or the effects of production technology as a source of contaminants (e.g., trace element content, Gremaud *et al.*, 2004). Stable isotope analysis, especially if applied as multi-element stable isotope analysis, has been used for several years to detect the geographical origin of raw materials and premium products by linking the stable isotope pattern of products with those of the natural environments from which they are derived (Rossmann, 2001; Kelly, 2003; Kelly *et al.*, 2005). A very simple link is that of precipitation water and its stable isotope signature in a region of provenance and the same stable isotope signature in the water from authentic products (such as wine, juice, vegetables, milk and milk products, animals and meat; for example, Hegerding *et al.*, 2002). Unfortunately, the water isotope signature in food products is often changed through technological and storage processes (Thiem *et al.*, 2004). More reliable stable isotope methods therefore apply the stable isotope ratios of organic compounds of a food product (Boner and Förstel, 2004; Camin *et al.*, 2007).

3.1.1 The necessity for improved methods of verifying food origin

The methods of control by 'paperwork', consisting of inspection certificates supplied by producers, and the flow of raw materials and products, can easily be manipulated if not accompanied by objective control procedures based on chemical and physical parameters. Though well trained and experienced specialists may be able to detect fraud by simple testing of organoleptic properties, rendering this as an objective method is difficult. Chemical methods of analysis, which are very useful for control with regard to contaminants (ethylene glycol in wine, melamine in milk) or pesticide residues, are limited in their power of differentiation, because the data for the same product of different geographical provenance show overlapping ranges largely due to natural variation. The content of certain elements (such as the calcium and potassium content of orange juice for the differentiation of products from Brazil and Spain) or multi-element patterns (magnesium, boron, selenium and holmium content of basmati rice for differentiation of the original product from the Himalayan area from substitutes produced in Europe or the USA) (Kelly *et al.*, 2002) has sometimes given a very good indication of product sources, but in other cases such discrimination is insufficient (Pillonel *et al.*, 2003); the soil or production conditions may be similar in different areas. Another problem particularly with methods using trace and ultra-trace elements such as rare earth elements, is the risk posed by production technology or even by the analysis itself. Biological or genetic analyses allow for determination, if certain strains (such as microorganisms for cheese production) or breeds (such as specific types of cattle, which are used for the production of certain PDO products) have been used, which can be, but may not always be, connected to a certain geographical origin. Typical strains or breeds may be used to produce certain types of food all over the world (e.g., the strains of microorganisms used in the production of Emmentaler cheese are used in both Finland and Australia).

3.2 Stable isotope approaches for verifying the origin of foods

Stable isotopes offer a completely new approach to the problem of differentiation of substances, as they can help to differentiate materials, which are chemically identical, by physical parameters, the stable isotope ratios of one or several elements.

3.2.1 Introduction

Originally, stable isotopes were used in biogeochemistry to unravel the origin and history of organic and inorganic components found in rocks (Faure, 1986). One of the first questions to which stable isotopes were applied was whether organic components, which are today parts of living organisms and which have been found in very old sediments (more than three billion years old), were produced by abiotic processes or by living organisms. It has been possible to confirm the origin of those materials from living systems, and to trace life on Earth nearly four billion years into the past using carbon isotope ratios of such materials, because all living systems possess depleted ^{13}C content of their products. In ecological research, stable isotopes have been used for more than 30 years to elucidate food webs or to assign organisms to certain environments (particularly carbon and nitrogen isotopes). It has been possible to detect animal migration routes based, for example, on their hydrogen isotope ratio of proteins (Fry, 2006; Lajtha and Michener, 2007; Hobson and Wassenar, 2008). In environmental and hydrological investigations, exploration of climatic changes or movement of groundwater have been performed, and the results have enabled the production of maps with stable isotope signatures of precipitation and groundwater (IAEA map, West *et al.*, 2010). In food control, stable isotopes have been used since around 1975 to detect adulteration of products like honey, fruit juice or maple syrup with cheaper extenders, such as sugar syrup made from maize or cane carbohydrates, industrially produced acids, or simply water (Rossmann, 2001). Those 'traditional' applications of stable isotopes in food control relied on the isotopic analysis of only one or two elements (carbon and oxygen, carbon and hydrogen) aiming to confirm if a certain component of a food material was a non-natural product but added from other plants or industrial production. Several of these methods have been officially tested and acknowledged as AOAC or CEN methods (the NMR and IRMS methods for wine have been accepted by EU and OIV) (Table 3.1).

To address the question of geographical origin, the methods developed and used for testing food authenticity are useful as a starting point, because they already confirm that stable isotope data can be measured reliably in routine work, and compared successfully in different laboratories. Nevertheless, the methods, instrumentation, reference materials and data evaluation must be adapted to enable the checking of geographical provenance. Authenticity checks can often work successfully with one or two stable isotope parameters (see Table 3.1), while geographical origin determinations usually require as many stable isotope parameters as possible, which means the stable isotope data for hydrogen, carbon, nitrogen, oxygen and sulphur (the light elements, which form biological materials, so called 'bio elements'), and sometimes heavier elements, such as Sr, Pb or Nd and Os, and isotopic ratios (Hölzl *et al.*, 2004; Kelly *et al.*, 2005). The heavier elements (or 'geo elements', because of their importance in geochemical research) are non-essential or at

Table 3.1 Officially acknowledged methods for food quality control based on stable isotope ratio analysis by isotope ratio mass spectrometry (IRMS) or ^{2}H nuclear magnetic resonance spectroscopy (^{2}H-NMR)

Foodstuff	Detection of	Isotopic ratio	Country, institution and year[a]
Fruit juice	S	$^{13}C/^{12}C$ (sugars)	EU – CEN 1995
Fruit juice	S	$^{13}C/^{12}C$ (sugars)	USA – AOAC 1981
Fruit juice	S	$^{13}C/^{12}C$ (sugars and pulp)	EU – CEN 1998
Fruit juice (concentrate)	S	$^{18}O/^{16}O$ (water)	USA – AOAC 1992
Fruit juice	S	$^{2}H/^{1}H$ (ethanol)[b]	EU,USA;CEN/ AOAC 1996
Fruit juice	W	$^{18}O/^{16}O$ (water) $^{2}H/^{1}H$ (water)	EU – CEN 1995
Honey	S	$^{13}C/^{12}C$ (honey)	USA – AOAC 1978
Honey	S	$^{13}C/^{12}C$ (honey and protein)	USA – AOAC 1991
Wine	S	$^{2}H/^{1}H$ (ethanol)[b]	EU 1991
Wine	W	$^{18}O/^{16}O$ (water)	EU 1996
Wine	S	$^{13}C/^{12}C$ (ethanol)	EU 2003
Maple syrup	S	$^{2}H/^{1}H$ and $^{13}C/^{12}C$ (ethanol)[b]	USA-AOAC 2001
Vinegar	S	$^{2}H/^{1}H$ and $^{13}C/^{12}C$ (ethanol)[b]	

[a] = year of official acknowledgement; [b] = ^{2}H-NMR. S = sugar-addition; W = water-addition; CEN = European Commission for Normalization; AOAC = Association of Official Analytical Chemists (USA).

least not quantitatively important components of living organisms, nor are they actively integrated into the structure of life, and are thereby unaffected by measurable isotopic fractionation. However, they can be very important indicators of the ground (rocks, soils), where living matter comes from, and sometimes of anthropogenic contamination as well. These elements will be discussed separately below.

3.2.2 Modern methods of measurement and instrumentation

Multi-element stable isotope analysis of the 'bio elements' HCNOS in food control can only be routinely applied if several conditions have been fulfilled. In the early days of stable isotope analysis, determination of one stable isotope ratio for an organic material was a time consuming and laborious process, as the substance had to be converted 'offline' into the pure measuring gas required (CO_2, N_2, SO_2, H_2, CO) by manual preparation, using high vacuum glass apparatus and freezing of gases with liquid nitrogen or conversion of degradation products with poisonous or difficult to handle chemicals. The isotopic determinations were carried out on the gases prepared using the dual inlet stable isotope measurement (repeated comparison of pure sample and reference gas, Rossmann, 2001; Kelly, 2002; Meier-Augenstein, 2010). Such procedures, developed for basic scientific research in geochemistry or biological sciences, are not suitable for routine control methods or the processing of high sample numbers. Since the late 1980s, instrument suppliers have developed automated systems, based on elemental analysers capable of delivering the required gases separately, by fully automated procedures, online and without the need for liquid nitrogen or problematic reagents, and for analysing isotopic ratios of three different elements (CNS) from one sample simultaneously, and isotopic data from two different elements (H, O) from another sample of the same material. It is even possible to determine the stable isotope ratios of four elements (HCNS) from one sample (Sieper *et al.*, 2006; Fry, 2007), but in practice this may not be the best way, because of different requirements in sample preparation ('comparative equilibration' for hydrogen isotope analysis, Wassenar and Hobson, 2003; Kelly *et al.*, 2009) and weight. The repeatability of such online multi-element stable isotope analyses in 'continuous flow mode' (sample gases are carried into the source of the mass spectrometer by a continuous helium carrier gas flow) is comparable with, if not better than, earlier single element measurements using the manual, offline preparations and dual inlet stable isotope analysis. In hydrogen and oxygen isotope analysis, the newly developed instrumentation, which applies automated high temperature conversion of organic compounds into H_2 and CO as analytes (often described as 'pyrolysis'), have enabled scientists to produce large amounts of isotopic data for those elements for organic materials. It is possible to determine hydrogen and oxygen isotopic ratios simultaneously from one sample, but in practice, requirements for sample preparation, weight and specific modifications of the high temperature conversion system may make single element analysis preferable. The separation of the combustion or high temperature conversion gases initially achieved by chromatographic methods (a GC column), is now possible using trapping and subsequent release of gases from specific traps (Sieper *et al.*, 2006). This has improved the chances of reaching baseline separation, especially of carbon dioxide and sulphur dioxide, or for nitrogen gas and carbon monoxide (Sieper *et al.*, 2010). Nevertheless, the oxygen isotope analysis of organic substances containing high amounts of nitrogen and sulphur is still a challenge (Qi *et al.*, 2011). To conclude, many more laboratories are able to conduct those multi-element analyses in routine work today, reducing the time and price (Fig. 3.1). An estimation of the cost shows a significant reduction in price with automated multi-element stable isotope analysis about 20–25% of what had to be paid for the same analyses done with single element offline preparation and measurement. It can be expected, therefore, that the progress in instrumentation development, resulting in more rapid and more precise simultaneous multi-element stable isotope determinations, will continue.

Sample preparation	
Casein preparation from cheese (cutting, drying, grinding, lipid extraction)	1 h per sample
methionin from casein (hydrolysis, derivatisation, preparative HPLC)	3 h per sample
Off-line preparation of measuring gases	
for hydrogen isotope analysis	2 h per sample
for carbon isotope analysis	0.25 h per sample
for nitrogen isotope analysis	2 h per sample
for oxygen isotope analysis	2 h per sample
for sulphur isotope analysis	3 h per sample
Off-line measurement including calibration:	2.5 h per sample for 5 elements
Total working time casein isotopic analysis (H,C,N,S)	*13.25 h per sample*
Total work time on-line measurement CNS weighing	1.5 h measurement + 1.25 h preparation +
Total work time on-line measurement D/H	0.5 h measurement + 0.25 h preparation + weighing
Sum for C,N,S + D/H of casein (separate)	3.5 h per sample
C,N,S + D/H of casein simultaneous	*2.75 h total working time*

Fig. 3.1 Working time for sample preparation and analysis with earlier offline and single element analyses and online simultaneous multi-element stable isotope determination.

3.2.3 Reference materials and methods

Unfortunately, the situation is not so positive with regard to the reference materials which are required for this type of stable isotope analyses: there are international reference materials available, including from International Atomic Energy Agency (IAEA) or from National Institute of Standards and Technology (NIST), even organic materials such as sugars, oils or amino acids, but they are not suited to the simultaneous stable isotope measurement of food or food components as plant dry matter or proteins. The main problem is the different elemental composition of the standard materials and the real samples, especially with regard to their sulphur content, which makes it very difficult to use them for direct calibration purposes. As an example, sulphur standards as S-1 are inorganic compounds (AgS) containing high amounts of sulphur (10–15%), while our samples (proteins or plant matter) contain between 0.1% and 1% of sulphur. Until now organic standards have not offered a real solution to these problems, as amino acids such as methionine or other organic compounds with relatively high sulphur content (15–20%) do not fit well for analysis of the above mentioned real samples. As a consequence, TRACE (www.trace.eu.org) partners in the working group covering stable isotopes of light elements (WP 1) have chosen a selection of internal reference materials such as collagen, casein, wheat flour and others, for the inter-calibration of multi-element stable isotope analysis

Table 3.2 Inter-laboratory Comparison Materials (ICMs) distributed to the TRACE WP1 laboratories

Collagen
Extra virgin olive oil
Honey (protein)
Wheat flour (defatted)
Lamb (freeze-dried and defatted)
Frozen fresh minced lamb (defatted)
Sieved and ground soil
NIST Bovine Liver CRM 1577b
NIST Bovine Muscle CRM 8414
Barium sulphate solid ICM
Barium sulphate solid CSL
Freeze-dried poultry meat (defatted)

between partner laboratories (Table 3.2). These substances were calibrated against international reference materials at the beginning of the project. With these inter-calibration and additional control substances, the results of different laboratories have been found comparable, an important requirement if data from different laboratories is to be used for a common database. At present it seems problematic to have such materials acknowledged as official reference substances by IAEA or Bureau Communautaire de Références (BCR), as they are not chemically well defined and there has been concern that the materials are not very homogeneous and may not be stable if stored over a longer time period. From our experience, this cannot be confirmed, at least in taking into account a period of six years for several reference materials from TRACE as mentioned above, and of more than 10 years for the reference material 'casein' (which had been chosen as a common reference material for another EU project about milk products in 1998). In addition, other stable isotope laboratories have applied and still use similar substances, which were initially selected for multi-element determinations, such as bovine muscle (NIST 8414) or bovine liver (NIST 1577b) powder (Fry, 2007), for use as internal reference material, but unfortunately these substances have been available only in small quantities, and now are often exhausted. This standardization and inter-calibration is an especially important issue for hydrogen isotope analysis, which is performed by applying the 'comparative equilibration technique' (Wassenar and Hobson, 2003; Kelly *et al.*, 2009), as a means to account for exchangeable hydrogen atoms in most organic compounds and food components such as proteins. Air humidity in different regions of the world possesses different stable isotope ratios of hydrogen and oxygen; exchanges with the hydrogen atoms bound to oxygen, nitrogen or sulphur in organic substances are unavoidable. In consequence, different raw hydrogen isotope ratios are found for the same materials in different laboratories (at different locations) and even in the same laboratory at different times (e.g., summer and winter). These variations must be corrected using a working reference material of commonly agreed hydrogen isotopic ratio. As

Table 3.3 Inter-laboratory reproducibility for the freeze-dried lamb ICM (lamb ICM 050818)

	Lab 1	Lab 2	Lab 3	Lab 4	Lab 5	Lab 6	Mean	SR
$\delta^{13}C$ (‰)$_{V\text{-}PDB}$	−26.62	−26.90	−26.65	−27.01	−26.83	−26.62	**−26.66**	**0.16**
$\delta^{15}N$ (‰)$_{AIR}$	7.56	7.51	7.95	7.67	7.50	7.52	**7.53**	**0.2**
$\delta^{2}H$ (‰)$_{V\text{-}SMOW}$	−102.6	−96.4	−96.2	−95.9	−98.4	−94.3	**−97.7**	**4.4**
$\delta^{34}S$ (‰)$_{V\text{-}CDT}$	5.16					5.30	**5.28**	**0.2**

SR = standard deviation of reproducibility.

an example, the results for lamb meat protein reference material from different laboratories, as measured during the TRACE project after such correction, are given in Table 3.3. Similar problems have been identified with measurement of oxygen isotopic ratios in organic substances, such as carbohydrates and lipids (olive oil), but additional work is required for this element to provide reliable inter-calibration, and further improvement of the instrumentation will need to be introduced (Qi *et al*., 2011). With regard to sulphur isotope analysis, problems occurred regarding the oxygen-18 contribution to mass 66 in continuous flow sulphur isotope analyses (assuming that masses 64 and 66 are measured for determination of the $\delta^{34}S$ value of organic substances). This can be overcome using organic reference materials with at least two different $\delta^{34}S$ values, and instrumentation which makes the oxygen isotopic ratio of SO_2 from combustion the same for all samples, either that of the oxygen gas used for combustion, or using an 'oxygen exchange furnace' after the combustion process (Fry *et al*., 2002; Sieper *et al*., 2006). Using inorganic sulphur reference materials (sulphates) for isotopic analysis of organically bound sulphur in the continuous flow mode cannot be recommended. This has been the traditional method, where all sulphur from organic materials is first converted 'offline' into barium sulphate, which is then thermally degraded to form SO_2 for the sulphur isotope analysis.

Assuming the inter-calibration problems can be at least partially resolved, the methods developed for sample preparation and multi-element stable isotope analysis through TRACE can be subjected to inter-laboratory tests, and would finally be suitable candidates for official methods of analysis of, for example, meat or cheese proteins, oils, cereals, or honey and honey proteins. In principle, all those preparations (Fig. 3.2) are based on the removal of water and subsequently of lipids, because they contain hydrogen and carbon with isotopic ratios very different from those of proteins or carbohydrates, and because they 'dilute' the nitrogen and sulphur content of the sample, increasing the necessary sample weights for isotopic analysis of those elements. Such an inter-comparison of sample preparation and multi-element stable isotope analysis has been conducted with regard to the multi-element stable isotope analysis (HCNS) of cheese protein, but results have not yet been published, and no such methods have been acknowledged officially. Another remark may be justified regarding sample weights: due

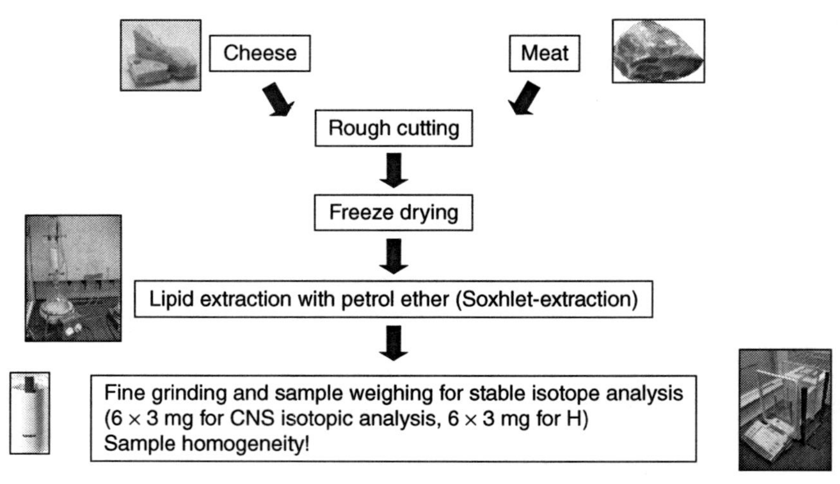

Fig. 3.2 Sample preparation scheme for multi-element stable isotope analysis of food products.

to instrument sensitivity, the sample weights required even for simultaneous stable isotope analysis of several elements could be reduced to some 100 μg depending on the elemental contents and instrumentation. However, for real samples, homogeneity often becomes the real limitation of weight reduction. As an example, it is possible to perform a stable isotope analysis of hydrogen isotopes from samples of about 100 μg, but it is difficult to produce sufficiently homogeneous samples from meat protein (by use of a ball mill), which enable stable isotope data with the same reproducibility as specified by the suppliers of the isotope ratio mass spectrometry (IRMS) instrumentation.

3.2.4 Data evaluation and databases

Geographical origin checks cannot be carried out by simply determining stable isotope ratios of one or two elements; in most cases a reliable indication requires the combination of data from four (HCNS) or five (HCNOS) elements, especially if differentiation of more than two possible regions of provenance is necessary. The easily applied and explained two- or three-dimensional plot of stable isotope results (e.g., Rossmann, 2001; Kelly *et al.*, 2002) cannot, therefore, be used to visualize and describe the data. Statistical methods such as principal components analysis (PCA) and linear discriminant analysis (LDA) must be involved (see Camin *et al.*, 2007), which are less easily understood by people unfamiliar with these specific data interpretation processes. As these are statistical methods, their reliable application and interpretation is only possible if a sufficient number of representative results for authentic samples is available. Taking into account stable isotope ratios of only four elements and ten different geographical regions, the 'statistically desirable' number would be about 400 samples. If this is extended into various smaller areas, which PDO regions often are, this would rapidly lead to databases of several thousand samples for each product, in order to obtain statistically sound conclusions. Obviously, this is something which neither the food industry nor the public control authorities would be able or willing to pay for. Another question with regard to databases of authentic samples, of which the European wine data bank is a good example, is the question of seasonal effects on the stable isotope data of a certain region, and over a longer time period, the possibility of systematic shifts in stable isotope data, due, for example, to 'global climatic change' and other natural or anthropogenic shifts in the environment (as with the increase of carbon dioxide content and the decrease of the ^{13}C content in the air as a result of burning of fossil fuels and coal). Consequently, any database of authentic samples must be updated annually. The only question is, how many samples need be analysed for such an update? And do all or only selected parameters (stable isotope data of certain elements) need updating (e.g., stable isotope data of hydrogen and oxygen, which are influenced by the stable isotope composition of precipitation and their seasonal changes)? Another, more formal, but nevertheless important, question is: who should have access to such a database, and under what conditions? All these questions have led to an attempt to describe the stable isotope data or the stable isotope pattern of material from certain regions by models which are able to predict the expected range of data based on an input of more simple, generally accessible information such as mean annual temperature, distance from the sea, elevation above sea level, geographical and geological situation of a provenance region, human activity or certain agricultural practices (as an example, wine water oxygen isotope data Hermann and Voerkelius, 2008). Such models should at least reduce the necessity for authentic data collection and number thereof. They have been developed for stable isotope data prediction, for example of water isotopes in precipitation developed by the project TRACE (Van der Veer *et al.*, 2008), mineral water data (Voerkelius *et al.*, 2009), and attempts have been made to describe the carbon isotope pattern of European cereals (Van der Veer, 2009; Goitom Asfaha *et al.*, 2011) and the sulphur isotope pattern of hair from central USA (Valenzuela *et al.*, 2011). These 'stable isotope maps', which are based on models which attempt to describe stable isotope ratios for certain elements, and finally the multi-element stable isotope pattern expected for certain regions, are named 'isoscapes' (West *et al.*, 2010). From our practical experience, not only with food samples, but in environmental studies as well, these 'isoscapes' can be very useful for gaining initial information about what is expected for a certain area. For a final decision, however, on whether a defined sample is of a labelled provenance or not, it is still necessary to have authentic samples from a certain area for comparison, not model-driven data alone. Nevertheless, trials to develop such models have driven the research to find and understand factors responsible for shifts in the stable isotope composition of food products, and they can help to estimate in advance whether a differentiation of products from a certain region from those of other regions may be possible, and what stable isotope parameters might be more suitable for a differentiation. As a very general indication, hydrogen and oxygen isotopic data for organic matter in food will indicate a link to the hydrogen and oxygen isotope data for water from the source region; nitrogen and carbon isotopes are related to the climate and the agricultural practices; and sulphur isotopes are affected by geology, volcanism, distance from the sea and certain anthropogenic effects. Taking into account what has been achieved with the data from TRACE areas of research and the models derived, the next step would be to test if those models can be applied to other regions of the Earth, and what must be modified to make them fit for other food production areas. This has already been partially tried by associated partners from China and Argentina, but to date nothing has been published on these results. In practical work, due to a limited number of available databank samples and difficulties in obtaining reliable separations for very small areas, it can be useful to obtain an assignment of unknown samples from a larger region (e.g., Central Europe, Southern Europe, North-Western Europe) as a first step (Goitom Asfaha *et al.*, 2011).

3.3 Stable isotope analysis of heavy ('geo') elements

The application, analysis and importance of stable isotope ratios of light ('bio') elements HCNOS has been described and discussed. In addition, during the last few years the determination of stable isotope ratios of heavy elements, such as strontium, lead and others, has been applied, or at least suggested for application (Rossmann *et al.*, 2000; Hölzl *et al.*, 2004; Kelly *et al.*, 2005; Manton, 2010). The most important differences to the light elements are that isotopic ratios of the heavy elements must be analysed with different instrumentation (thermal ionization mass spectrometry = TI-MS, inductively coupled plasma mass spectrometry = ICP-MS), as the analytes to be measured are not gases, but salts. Further, the isotopic ratios of those 'geo elements' vary as a result of geology, mainly in age and rock mineral composition, and subsequently of soils and living matter growing there. As examples, strontium (^{87}Sr), neodymium (^{143}Nd) and lead (^{207}Pb) isotope data used for geographical origin assignments are products of radioactive decay of their mother nuclides with a very long half-life time (billions of years), which shift the respective isotopic ratio slowly during the Earth's history. Consequently, there is already reasonable knowledge about the possible stable isotope ratios for different geographical areas from geochemical research work, and there is, unlike with 'bio elements', no measurable fractionation of these elements during biochemical processes. Nevertheless, there are certain specific problems linked to the application of those elements to food origin assignment as well, one being the relatively low and sometimes highly variable content of such elements in organic matter, another one being the risk of introducing contamination. Further, organic samples are often very different to geological samples (rocks, minerals). Quantitative conversion into dissolved rather than bound form, and the separation of certain elements as cations by chromatographic methods is difficult. The measurement of the analytes coming from degradation of organic materials often raises additional problems because sometimes interfering masses from other, even trace components, have to be taken into account and corrected for. This is especially the case if the stable isotope determination is performed with ICP-MS methods, as with this method even the additional separation due to differing ionization energy (which can improve the separation of required and interfering substances in case of TI-MS by stepwise heating of the sample and removal of more volatile interfering substances) cannot be used. Assuming that those problems can be resolved or at least taken into account, a remaining problem is that stable isotope analysis of those heavy elements requires more laborious and until now partially manual preparation. It is more time consuming and more expensive to analyse these compared to the stable isotope analysis of the 'bio elements'. The advantage of those 'geo elements' would be that products which may originate from areas with very similar environmental conditions (climate, agricultural practices), but different underlying geology, would be easy to differentiate, from their strontium isotope ratio, for example. For lamb or beef analysed during TRACE, the light element stable isotope data are very similar, for example for Franconia, a region in Bavaria, and the Massif Central in France (Camin *et al.*, 2007), but are well differentiated using strontium isotope data (Hölzl *et al.*, pers. comm.). For fruit juices, a good example of the usefulness of strontium isotope ratios is the detection of added Brazilian rediluted orange concentrates, or South African oranges, to single strength orange juices labelled as originating from Spain (Rummel *et al.*, 2010). Such differentiation is not as easy to achieve using conventional analytical methods, or with only the isotopic data of the 'bio elements'. For other regions strontium isotope ratios are often very similar and close to the $^{87}Sr/^{86}Sr$ ratio of sea water (0.7092), therefore the additional information from strontium isotopes is not always significant. Another valuable indication from strontium isotope ratios can be obtained for PDO cheeses from certain North Italian regions; each of them being typical for a very small area in alpine valleys of Trentino, Veneto, Piemonte and Friuli Veneto Giulia; each of these valleys has a very specific geology not found in other valleys close by (Bontempo *et al.*, 2011).

3.4 Trace element approaches for verifying the origin of foods

In general, the description 'trace elements' means not only the application of rare earth elements or other elements occurring only at very low abundance. The elemental content of certain elements (such as sodium, calcium, potassium) of heavier metals (such as barium, strontium, iron, nickel, copper) and rare elements (such as holmium, iridium, gold) have been found indicative of certain geographical areas. In general, the high content of sodium or chlorine (always related to the data of the same product from different areas), may indicate marine provenance, high content of potassium may indicate an origin from granitic or gneiss underlying rocks, and high calcium content a provenance from sedimentary or basaltic rocks. In the case of very rare elements (nickel, vanadium, chromium), a typical signature might be a result of technological processes (Manton, 2010). In TRACE we have until now described combined results of stable isotope analyses and mineral content determinations of olive oils from Mediterranean countries (Camin *et al.*, 2010), which applied the content of 14 elements, mainly alkaline (K, Rb, Cs) and earth alkali (Mg, Ca, Sr) metals, and isotopic ratios of hydrogen, carbon and oxygen. While the stable isotopes enabled a classification success of 78%, this improved to 95%, when the elemental content of the above-mentioned 14 elements was included. Mineral content without stable isotopes enabled assignment to three geological classes with 76% classification success, but not to individual geographical areas with sufficient reliability. In a study of alpine cheeses from Italian regions, a reclassification success of 94% was achieved using stable isotope data (C, N, O) and mineral content (Ba, Ca, K, Mg, Rb) results (Bontempo *et al.*, 2011). A study on basmati rice has indicated a reliable differentiation of products from Europe, the Himalayan region, and the US, using carbon and oxygen isotopes and element content, especially of Mg, B, and W (Kelly *et al.*, 2002). Similar results could be reported from cereals, as investigated in TRACE (Goitom Asfaha *et al.*, 2011), or from fruit juices (Nikdel, 1995; Amarasiriwardena and Barnes 1995; Hofsommer, 1995). The conclusion from practical experience would thus be that element and trace element pattern could indicate an origin from certain geological areas, but not reliably assign samples to one defined region of provenance. Nevertheless, the geographical origin assignment based on stable isotopes can be remarkably improved in combination with certain element content data.

3.5 Alternative new methods for verifying the origin of foods

As already mentioned, other new methods of analysis, as genetic methods or so called 'fingerprinting methods' are reported elsewhere in this volume. In most cases fingerprinting methods rely on spectroscopic analyses and sophisticated evaluation of the huge amount of individual data produced. The data for unknown samples have to be compared with the same results for authentic samples held in a database. One of the most advanced examples for this application is the 'SGF-Profiling' (Spin Generated Fingerprint Profiling) of fruit juices based on ^{1}H-NMR spectroscopy (Spraul *et al.*, 2009). This method is now routinely applied to screen fruit juices. The results indicate the matching of the unknown sample with the dataset of authentic samples of particular provenance, either a full agreement or a deviation from the data from authentic samples. Even certain analytical parameters, such as the content of sugars or acids, or of certain other components, can be derived from the NMR signals. The presence of substances which should not usually occur in the relevant products, can be detected. This method enables rapid testing of high sample numbers, and even detection of certain adulterants or impurities, but cannot always guarantee indication of true provenance or the exact reasons for non-matching results. An indication of a deviation for a certain sample requires detailed control using very specialized methods of analysis, such as chromatographic (HPLC analysis) investigations or stable isotope determination, to confirm that a sample has been adulterated by addition of components from other sources, that the geographical origin has been improperly attributed, or that a seasonal or technological reason can be found for the observed discrepancies.

3.6 Conclusion and future trends

For practical purposes, analytical methods can be differentiated into rapid, but non-specific screening methods, and very specific methods of analysis, such as stable isotope analyses, which are more time consuming and require greater resources. The screening (fingerprinting) methods are usually applied for preselection of these samples, which will then be submitted to specific analytical tests. No single method can be labelled as the 'future tool', but it can be expected that an appropriate combination of methods are likely to offer the most promising way forward. As new types of adulterations and new additives and/or impurities (such as the melamine problem) may probably occur, combining analytical methods will be important. The same might be true for a combination of genetic methods and stable isotope analyses, which has already been discussed for determining the provenance of wood (tropical wood sourced only from licensed and/or sustainable production) or for questions of forensic investigation. Stable isotope analyses themselves will undergo further development with improved analytical instrumentation, delivering higher sensitivity and stability of the isotope ratio mass spectrometers; even better linearity of the source; a higher dynamic range of the analysers; improved performance of the elemental analysers with regard to sample stability on the analyser; more rapid sample conversion with lower consumption of reagents and carrier gas; better separation of the gases to be analysed from each other and from by-products of the reaction; and longer lifetimes of high temperature furnaces due to new materials and systems layout. With regard to standards and methods, international organizations (IAEA, BCR) should proceed with their work to provide reference materials for practical work in the field of organic substances and food components. This would help to accelerate the introduction of official analytical methods on the basis of multi-element stable isotope analysis, as for milk products or meat. At the same time, better standardization and inter-calibration opportunities can help to extend existing databases (such as from TRACE) towards other regions of the world. Expanded databases and research efforts may help to improve understanding of the factors which influence the stable isotope pattern of products, and drive the improvement of models for worldwide data prediction. An important question is whether the physiological conditions of an organism could change its multi-element stable isotope pattern in metabolic products significantly, even where feeding is constant. Such questions would require extended international collaboration and projects, or at least a framework for an exchange of know-how and data to avoid redundancy.

An additional element, which may be used for future stable isotope work, besides the elements already mentioned as 'geo elements' (lead, osmium, neodymium), could be the light 'bio-element', boron. Some applications of boron isotopes have already been published (rice or coffee), nevertheless use of boron isotopes in routine isotope analysis has not been widespread. The reason is that boron is usually present in natural products only in very small amounts (a few ppm), the sample preparation requires laborious offline sample conversion and separation of borate from other components, there is a reasonable risk of boron contamination and the determination of boron isotope ratios is only possible by TI-MS or ICP-MS, as there is no easily accessible, gaseous boron compound, which can be analysed in gas isotope ratio mass spectrometers. Unlike all other light elements, in boron the heavier isotope ^{11}B is more abundant than the light ^{10}B, with a ratio of $^{11}B/^{10}B$ of about 4 in natural substances. The indications from boron isotopes could in several cases be similar to those from sulphur isotopes, as the two most important reservoirs for boron, with very typical and different stable isotope ratios, are marine boron and evaporites, and terrestrial boron deposits. The first applications of this element may be in the field of drinking water research and contamination control (boron from wastewater as it is an important component of washing agents), or for vegetables (rice, Oda *et al.*, 2002) and expensive food products such as coffee (Serra *et al.*, 2005) and tea. For animal products no application has been reported, one reason for this might be the even lower content of boron in animal products as compared to plant materials. It is well known that boron is an essential element for plants, but until now its importance for animals has not been so evident.

Assuming that improved methods of stable isotope analysis supply greater amounts of analytical data, then the data management and evaluation systems to work with them and to perform food origin and authenticity control on their basis, must also be developed. This not only means producing stable isotope maps ('isoscapes') from data, but providing programmes for data comparison for commercial samples and databank results on the basis of statistical methods (Wachter *et al.*, 2009) and visualizing the results in a way that non-specialized readers may understand.

3.7 Sources of further information and advice

A few review papers have been published within the last ten years describing stable isotope application or a combination of stable isotope and other methods of analysis for food authenticity and geographical origin control (Rossmann, 2001; Kelly, 2003; Kelly *et al.*, 2005). Nevertheless, they have not yet included the results obtained during TRACE and published during the last two or three years, and the application especially of hydrogen and sulphur isotope determinations. Several papers especially from the US have presented the concept of 'isoscapes', recently even a book *Isoscapes* has been published (West *et al.*, 2010) where their use especially for forensic investigations was discussed. Several interesting books have been written on multi-element stable isotope analysis and its application, not only for food origin control and forensic investigation, but for questions of environmental and biological research as well, including the following titles:

- Faure, G., *Principles of Isotope Geology*;
- Fry, B., *Stable Isotope Ecology*;
- Hobson, K.A. Wassenar, L.I., *Tracking Animal Migration with Stable Isotopes*;
- Lajtha, K. Michener, R., *Stable Isotopes in Ecology and Environmental Science*;
- Meier-Augenstein, W., *Stable Isotope Forensics*; and
- West, J.B., Bowen, G.J., Dawson, T.E. Tu, K.P., *Isoscapes*.

A very useful book for basic information, even if already some years old, is *Principles of Isotope Geology* of Faure (1986), which contains much stable isotope information, mainly with regard to geological research, but is of reasonable interest for the field of environmental, biological, forensic and food research as well.

3.8 References

AMARASIRIWARDENA CJ and BARNES RM. Detection of orange juice adulteration by means of Inductively Coupled Plasma Mass Spectrometry. In: S Nagy and RL Wade (Ed), *Methods to Detect Adulteration of Fruit Juice Beverages*, Volume **1** (1995), 66–83. Auburndale (FL): AGSCIENCE, INC.

BONER M and FÖRSTEL H, Stable isotope variation as a tool to trace the authenticity of beef. *Anal Bioanal Chem* **378** (2004), 301–310.

BONTEMPO L, LARCHER R, CAMIN F, HÖLZL S, ROSSMANN A, HORN P and NICOLINI G. Elemental and isotopic characterisation of typical Italian alpine cheeses. *Int Dairy J* **21** (2011), 441–446.

CAMIN F, BONTEMPO L, HEINRICH K, HORACEK M, KELLY SD, SCHLICHT C, THOMAS F, MONAHAN FJ, HOOGEWERFF J and ROSSMANN A. Multi-element (H,C,N,S) stable isotope characteristics of lamb meat from different European regions. *Anal Bioanal Chem* **389** (2007), 309–320.

CAMIN F, LARCHER R, NICOLIN G, CONTEMPO L, BERTOLDI D, PERINI M, SCHLICHT C, SCHELLENBERG A, THOMAS F, HEINRICH K, VOERKELIUS S, HORACEK M, UECKERMANN H, FROESCHL H, WIMMER B, HEISS G, BAXTER, M, ROSSMANN A and HOOGEWERFF J. Isotopic and elemental data for tracing the origin of European olive oils. *J Agric Food Chem* **58** (2010), 570–577.

FAURE G, *Principles of Isotope Geology*, Second Edition (1986). New York, Chichester, Brisbane, Toronto, Singapore: John Wiley & Sons.

FRY B, SILVA SR, KENDALL C and ANDERSON RK. Oxygen isotope corrections for on-line δ^{34}S analysis. *Rapid Comm Mass Spectrom* **16** (2002), 854–858.
FRY B, *Stable Isotope Ecology* (2006). New York: Springer.
FRY B and COUPLED N. C and S stable isotope measurements using a dual-column gas chromatography system. *Rapid Comm Mass Spectrom* **21** (2007), 750–756.
GOITOM ASFAHA D, QUÉTEL CR, THOMAS F, HORACEK M, WIMMER B, HEISS G, DEKANT C, DETERS-ITZELSBERGER P, HOELZL S, RUMMEL S, BRACH-PAPA C, VAN BOCXSTAELE M, JAMIN E, BAXTER M, HEINRICH K, KELLY S, BERTOLDI D, BONTEMPO L, CAMIN F, LARCHER R, PERINI M, ROSSMANN A, SCHELLENBERG A, SCHLICHT C, FROESCHL H, HOOGEWERFF J and UECKERMANN H. Combining isotopic signatures of n(87Sr)/n(86Sr) and light stable elements (C, N, O, S) with multi-elemental profiling for the authentication of provenance of European cereal samples. *J Cereal Sci* **53** (2011), 170–177.
GREMAUD G, QUAILE S, PIANTINI U, PFAMMATTER E and CORVI C. Characterization of Swiss vineyards using isotopic data in combination with trace elements and classical parameters. *Eur Food Res Technol* **219** (2004), 97–104.
HEGERDING L, SEIDLER D, DANNEEL H-J, GESSLER A and NOWAK, B. Sauerstoffisotopen-V erhältnisanalyse zur Herkunftsbestimmung von Rindfleisch. *Fleischwirtschaft* **82**(4) (2002), 95–100.
HERMANN A and VOERKELIUS S. Research note: Meteorological impact on oxygen isotope ratios of German wines. *Am J Enol Vitic* **59**(2) (2008), 194–199.
HOBSON KA and WASSENAR LI, *Tracking Animal Migration with Stable Isotopes* (2008). London: Elsevier.
HOFSOMMER H-J. Analytical methodologies of fruit juice adulteration detection as practiced in Europe. In: S Nagy and RL Wade (Ed), *Methods to Detect Adulteration of Fruit Juice Beverages*, Volume **1** (1995), 336–355. Auburndale (FL): AGSCIENCE, INC.
HÖLZL S, HORN P, ROSSMANN A and RUMMEL S. Isotope-abundance ratios of light (bio) and heavy (geo) elements in biogenic tissues: Methods and applications. *Anal Bioanal Chem* **378**(2) (2004), 270–272.
KELLY JF, BRIDGE ES, FUDICKAR AM and WASSENAAR LI. A test of comparative equilibration for determining non-exchangeable stable hydrogen isotope values in complex organic materials. *Rapid Comm Mass Spectrom* **23** (2009), 2316–2320.
KELLY SD, BAXTER M, CHAPMAN S, RHODES C, DENNIS J and BRERETON P. The application of isotopic and elemental analysis to determine the geographical origin of premium long grain rice. *Eur Food Res Technol* **214** (2002), 72–78.
KELLY SD. Using stable isotope ratio mass spectrometry (IRMS) in food authentication and traceability. In: M Lees (Ed), *Food Authenticity and Traceability* (2003), 156–183. Cambridge: Woodhead Publishing Limited.
KELLY SD, HEATON K and HOOGEWERFF J. Tracing the geographical origin of food: The application of multi-element and multi-isotope analysis. *Trends Food Sci Tech* **16** (2005), 555–567.
LAJTHA K and MICHENER R. *Stable Isotopes in Ecology and Environmental Science*, Second Edition (2007). Oxford: Wiley-Blackwell.
MANTON WI. Determination of the provenance of cocoa by soil protolith ages and assessment of anthropogenic lead contamination by Pb/Nd and lead isotope ratios. *J Agric Food Chem* **58** (2010), 713–721.
MEIER-AUGENSTEIN W. *Stable Isotope Forensics* (2010). Chichester: Wiley-Blackwell.
NIKDEL S, Artificial neural networks and trace minerals: application to Florida Orange Juice and Orange pulpwash. In: S Nagy and RL Wade (Ed), *Methods to Detect Adulteration of Fruit Juice Beverages*, Volume **1** (1995), 52–65. Auburndale (FL): AGSCIENCE, INC.
ODA HW, KAWASAKI A and HIRATA T. Determining the rice provenance using binary isotope signatures along with cadmium content. Symposium paper No.2018, Proceedings 17th WCSS 14–21 August 2002, Thailand.
PILLONEL L, BADERTSCHER R, FROIDEVAUX P, HABERHAUER G, HOELZL S, HORN P, JAKOB A, PFAMMATTER E, PIANTINI U, ROSSMANN A, TABACCHI R and BOSSET JO. Stable isotope

ratios, major, trace and radioactive elements in Emmental cheeses of different origins. *Lebensm-Wiss und -Technol* **36** (2003), 615–623.

QI H, COPLEN TB and WASSENAAR LI. Improved online $\delta^{18}O$ measurements of nitrogen- and sulphur bearing organic materials and a proposed analytical protocol. *Rapid Comm Mass Spectrom* **25** (2011), 2049–2058.

ROSSMANN A. HABERHAUER G, HÖLZL S, HORN P, PICHLMAYER F and VOERKELIUS S. The potential of multi element stable isotope analysis for regional origin assignment of butter. *Eur Food Res Technol* **211** (2000), 32–40.

ROSSMANN A. Determination of stable isotope ratios in food analysis. *Food Rev Int* **17**(3) (2001), 347–381.

RUMMEL S, HOELZL S, HORN P, ROSSMANN A and SCHLICHT S. The combination of stable isotope abundance ratios of H, C, N and S with 87Sr/86Sr for geographical origin assignment of orange juices. *Food Chem* **118** (2010), 890–900.

SERRA F, GUILLOU C, RENIERO F, BALLARIN L, CANTAGALLO M I, WIESER M, IYER S S, HEBERGER K and VAN HAECKE F. Determination of the geographical origin of green coffee by principal component analysis of carbon, nitrogen and boron stable isotope ratios. *Rapid Comm Mass Spectrom* **19** (2005), 2111–2115.

SIEPER H-P, KUPKA H-J, WILLIAMS T, ROSSMANN A, RUMMEL S, TANZ N and SCHMIDT H-L. A measuring system for the fast simultaneous isotope ratio and elemental analysis of carbon, hydrogen, nitrogen and sulphide in food commodities and other biological material. *Rapid Comm Mass Spectrom* **20** (2006), 2521–2527.

SIEPER H-P, KUPKA H-J, LANGE L, ROSSMANN A, TANZ N and SCHMIDT H-L. Essential methodological improvements in the oxygen isotope ratio analysis of N-containing organic compounds. *Rapid Comm Mass Spectrom* **24** (2010), 2849–2858.

SPRAUL M, SCHÜTZ B, RINKE P, KOSWIG S, HUMPFER E, SCHÄFER H, MÖRTTER M, FANG F, MARX UC and MINOJA A. NMR-based multi parametric quality control of fruit juices: SGF profiling. *Nutrients* **1** (2009), 148–155.

THIEM I, LÜPKE M and SEIFERT H. Factors influencing the 18O/16O-ratio in meat juices. *Isotopes Environ Health Stud* **40**(3) (2004), 191–197.

VALENZUELA LO, CHESSON LA, O'GRADY SP, CERLING TE and EHLERINGER JR. Spatial distributions of carbon, nitrogen and sulfur isotope ratios in human hair across the central United States. *Rapid Comm Mass Spectrom* **25** (2011), 861–868.

VAN DER VEER G, VOERKELIUS S, LORENZ G, HEISS G and HOOGEWERFF JA. Spatial interpolation of the deuterium and oxygen-18 composition of global precipitation using temperature as ancillary variable. *J Geochemical Exploration* **101**(2) (2009), 175–184.

VOERKELIUS S, LORENZ G, RUMMEL S, QUÉTEL C R., HEISS G, BAXTER M,, BRACH-PAPA C, DETERS-ITZELSBERGER P, HOELZL S, HOOGEWERFF J, PONZEVERA E, VAN BOCXSTAELE M and UECKERMANN H. Strontium isotopic signatures of natural mineral waters, the reference to a simple geological map and its potential for authentication of food. *Food Chem* **118** (2010), 933–940.

VAN DER VEER G *et al.* TRACE final report (2009).

WACHTER H, CHRISTOPH N and SEIFERT S. Verifying authenticity of wine by Mahalanobis distance and hypothesis testing of stable isotope pattern – a case study using the EU wine databank. *Mitteilungen Klosterneuburg* **59** (2009), 237–249.

WASSENAR LI and HOBSON KA. Comparative equilibration and online techniques for determination of non-exchangeable hydrogen of keratin for use in animal migration studies. *Isotopes Environ Health Stud* **39**(3) (2003), 211–217.

WEST JB *et al.* (Eds), *Isoscapes* (2010). Dordrecht, Heidelberg, London, New York: Springer.

4

Development and application of geospatial models for verifying the geographical origin of food

G. van der Veer, RIKILT, Wageningen University and Research Centre, The Netherlands

DOI: 10.1533/9780857097590.2.60

Abstract: Geospatial modelling is increasingly used as a tool to aid verification of the geographical origin of food. Since geospatial modelling is a relatively new approach in the field of food authentication, this chapter focuses on the conceptual framework and provides suggestions related to the technical and practical aspects of model building as well as the application of these models for food authenticity testing. The geospatial modelling approach as described in this chapter provides a unique framework for geographical provenancing of food, which is equally relevant for other fields of science including forensics, archaeology and animal migration studies.

Key words: geographical provenancing, food origin authentication, spatial interpolation, geospatial modelling, area of possible origin.

4.1 Introduction

The use of geographical indications (GI) such as a Protected Designation of Origin (PDO) or Protected Geographical Indication (PGI) has become a key source of niche marketing for all sorts of food and beverages. Geographical indications provide producers not only with market recognition, they also often lead to better prizes as consumers are willing to pay for a 'piece of authenticity'. At the same time, such high-value products are more prone to fraud.

False declaration of the origin of food that has a protected GI status, or otherwise claims to be of a specific geographical region, is the most obvious case in which food fraud involves a geographical aspect. In addition, geographical origin of food often plays an implicit role in more general authenticity as well as safety issues. This includes preventing food from a certain region or country that is known to be accidentally or deliberately contaminated from entering the market.

Most methods developed for the verification of the geographical origin of food are based on a non-spatial modelling approach. Typically chemical fingerprinting or similar techniques are used in combination with chemometric modelling (see Chapter 7). The chemometric model can then be used to test whether a sample is authentic or not at a certain level of confidence based on its measured composition. This approach is sometimes referred to as the 'database approach' because the models are *exclusively* based on the underlying data set, and hence its application on a larger scale depends largely on the completeness of the variation captured by the samples in the database.

In addition to chemometric modelling, geospatial modelling is increasingly used as a tool to aid the verification of product authenticity in terms of geographical origin or source. In terms of food authenticity, such geospatial models predict the expected composition of some commodity for each location covered by the spatial extent of the model. As such, geospatial models:

- allow verification of the acclaimed origin of a sample at a certain level of confidence by testing whether the measured composition falls within the intervals predicted by the model (see Fig. 4.1a and 4.1b) and
- identify the area where the sample – given its measured composition – could possibly originate from.

Compared to the non-spatial approach, the spatial modelling approach, as defined in this chapter, has several advantages:

- It provides more accurate predictions for those production areas that have not been sampled and is more cost-efficient, especially in cases where many different production areas and/or producers are involved.
- It provides the area from where a sample could originate according to its composition (area of possible origin).
- It is generally more robust against changes in production methods as the models are based on an empirical relationship with 'static' geo-climatic factors.

The chemometric and geospatial approaches therefore have somewhat different scopes and applications, which are important to consider when making a choice between them. As a general rule of thumb it can be said that a chemometric approach will often be sufficient in the following two cases:

1. There is one expensive brand that is produced in a geographically confined area, which is to be distinguished from a limited number of cheaper look-alikes that are produced in other areas. For example, Champagne versus some similar but cheaper sparkling wines.

(a)

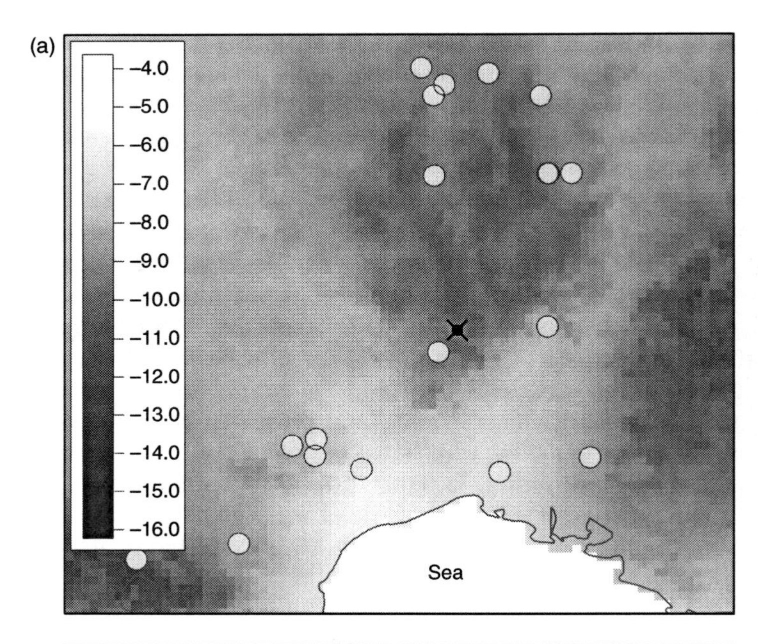

(b)

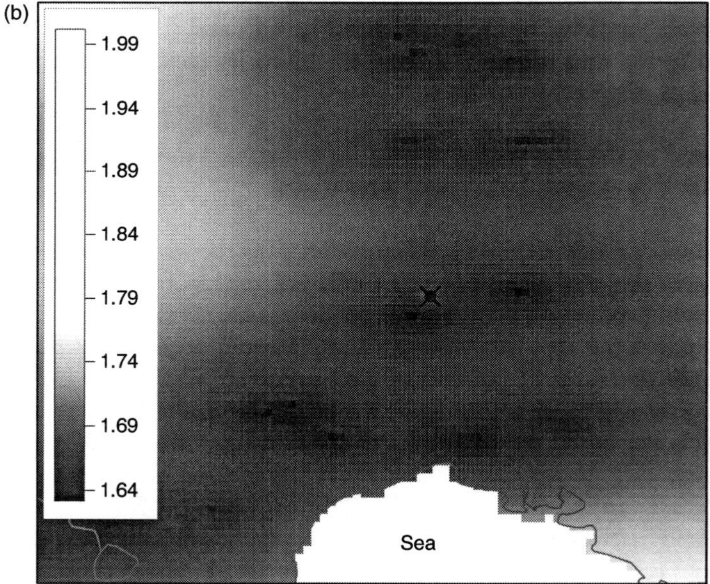

Fig. 4.1 (a) Subsection of a spatial model for oxygen-18 isotope ratios ($\delta^{18}O$) in commercial mineral water based on a kriging interpolation of $\delta^{18}O$ values in the TRACE European mineral water database (actual sample locations indicated). The predicted $\delta^{18}O$ ratio at the hypothetical acclaimed area of origin (black cross) would be $-9.1 \pm 1.7‰$. (b) The upper 95% confidence limit around the predicted values.

2. There are several brands that are produced in a number of different but confined geographical areas that need to be distinguished from each other. For example, several Italian virgin olive oils which have a different quality and price range.

So whether it concerns a 'one-versus-several-others' or a 'several-from-each-other' situation, the non-spatial approach is often sufficient in cases where the number of (production) areas that are to be distinguished is rather limited (e.g., < 10). The chemometric approach also works in principle in the case of a 'one-versus-many' question, for example Bordeaux wine versus any other red wine, although in this case the spatial modelling approach is probably more cost-efficient. In case of a 'many-versus-many' situation, or a 'where could the sample possibly originate from' question, the required sample numbers would become too large for a chemometric approach to be cost-efficient and this question therefore requires a spatial approach.

The general assumption used in both of these approaches for food provenancing is that the commodity of interest is grown or reared in, or produced from ingredients that are sourced from, one or more confined production areas. These methods are less effective for composite food products of which the ingredients are sourced from different geographical areas. This, for example, also applies to animal produce from animals that get a substantial part of their feed from non-local sources. Additionally, when the production area(s) of the commodity of interest are so large that they include substantial geo-climatic variation (e.g., province or country scale), the prediction intervals will be so wide that there will be substantial overlap with other production areas. In this case the resulting power of the model will be limited.

The term 'geospatial model' is used here as a generic term for any prediction model in geographical space. The term 'isoscapes', which was first coined by Jason West, specifically refers to geospatial prediction models of stable isotopes such as C, H, O and N in the natural environment (see e.g., Bowen, 2010). Although the concept and general approach in terms of model building is similar, for geographical authentication purposes the concept should not be confined to stable isotopes alone, and preferably be extended to include (I) any set of suitable geographical markers (e.g., isotopes of heavy elements, organic compounds, trace elements) and (II) allow combining several models of individual markers at the same time which helps to confine the area of possible origin. In this respect the geospatial modelling approach described here is an extension and generalization of the isoscape approach as well as the 'food isotope mapping' approach as described in Kelly *et al.* (2011).

Geospatial modelling is a relatively new approach in the field of food authentication and this chapter will, therefore, rather than providing a review of existing work, focus instead on the conceptual framework and requirements (Section 4.2) and the technical and practical aspects of model building (Section 4.3), as well as application of the models for authenticity testing (Section 4.4).

Much of the work and ideas elaborated in this chapter are based on results and experience gained in the European Integrated Project TRACE (2005–2010).

4.2 Conceptual framework and requirements

Methods for verifying the geographical origin of food in an objective way, that is, independent of paper trails and labelling, make use of the fact that intrinsic characteristics – so-called geographical 'markers' or 'fingerprints' – are carried along the production and distribution chain. Provided that these markers or fingerprints are sufficiently characteristic to distinguish the relevant production areas or regions, they can be used for geographical provenancing of food as well as other commodities.

The above concept applies to the non-spatial as well as the spatial modelling approach for food origin authentication. Yet the non-spatial approach often makes use of multivariate data (so-called chemical fingerprints), whereas the spatial approach makes use of (a set of) individual markers. The non-spatial approach can therefore be regarded as a kind of 'interpolation' in attribute space, whereas the spatial approach makes use of spatial interpolation in geographical space.

Like the non-spatial approach, the spatial approach requires a representative set of samples that are collected from a number of production areas. In the spatial approach, however, the measurements are then interpolated in geographic space to provide expected values for all unsampled locations. However, when the amount of locations covered by the database is small compared to the spatial extent of the model (sparse sampling), spatial interpolation should preferably be supported by so-called ancillary information.

Ancillary information in this framework refers to any digital information that is available at a much higher spatial resolution than the food sample locations, and which explains a substantial amount of the variation of the marker compound in the food samples. By using ancillary information during interpolation, the resulting model will provide more accurate estimates of the expected values in the unsampled areas.

This brings us to the first set of requirements for building a spatial model in this framework:

- There should be a clear relationship between the geographic marker compound in the food and some ancillary variable(s).
- The geographic marker compound(s)
  - should be measurable in the food commodity of interest with sufficient precision, and
  - should not be significantly affected or altered by any production process unless this occurs in a systematic way.
- The ancillary variable should be available in the form of a (high-resolution) digital map.

Any potential marker compounds for which these conditions do not apply should be used in a non-spatial framework as they might contain location- and/or production-specific information. Yet since this information is not related to any known or available ancillary variable it cannot therefore be upscaled with sufficient spatial accuracy.

Although a review of potentially suitable marker compounds falls outside the scope of this chapter, it is clear that stable isotope ratios such as deuterium (δ^2H) and oxygen-18 (δ^{18}O) as well as heavy element isotopes such as strontium (^{87}Sr/^{86}Sr) play a key role in geographical provenancing of food as well as other commodities (see e.g., Voerkelius *et al.*, 2010; Kelly *et al.*, 2011). This can mainly be ascribed to their relationship with climatic (δ^2H and δ^{18}O) or geological variables (^{87}Sr/^{86}Sr), which is relatively well understood and strongly reflected in many natural commodities. In addition, these marker compounds can be measured in most food stuffs with sufficient precision so that they meet all requirements for geospatial modelling.

In principle, the same reasoning applies to carbon-13 (δ^{13}C) and nitrogen-15 (δ^{15}N) which have been used in many food authentication studies. For carbon-13, geospatial modelling has focussed on generating global and regional isoscapes for plant δ^{13}C based on modelled C_3 and C_4 plant and crop distributions (Still *et al.*, 2003). For nitrogen-15 Bowen (2010) mentions that no single process has been identified as a dominant global control on its geographic distributions. For δ^{15}N it will therefore be necessary to probe further into which ancillary variable(s) can explain and upscale its geographical distribution in an accurate manner. Other potentially suitable geographical markers include isotopes systems of trace elements including boron, iron or lead, as well as organic compounds and their isotopic signatures.

Additional requirements concern the ancillary variable (or variables), which should explain a significant amount of the (spatial) variation of the relevant marker compound, and which should be available in the form of a high-resolution digital map that can be used in a GIS environment. As shown by the overview in Table 4.1, various global and continental scale maps are currently available for different climate variables such as mean temperature and precipitation amount, as well as for various geological variables such as rock type and age.

Instead of using one of the geo-climatic variables in the GIS maps provided in Table 4.1 as ancillary variable, one can also use existing isotope models of natural commodities such as precipitation, for example, to derive a relationship between isotopes in the natural commodity and the food commodity. Provided that there is a relationship between the (isotopic) composition of the natural commodity and the food, such isoscapes could be used in the same way as any of the geo-climatic ancillary variables. This approach has been taken by Ehleringer *et al.* (2008) to provide a model of δ^{18}O in human hair based on a geospatial model of δ^{18}O in tap water in the USA. One drawback of this approach might be, however, that by propagation of errors the uncertainties in the final model could be substantial. An overview of existing geospatial models of stable isotopes is provided by Bowen (2010).

Table 4.1 Non-exhaustive overview of GIS maps that can be used as ancillary variable(s) for developing geospatial models for geographical provenancing (30 arc seconds equals ~1 km at the equator)

Name	Coverage	Type	Resolution	Attributes
GTOPO30[a]	World	Raster	30 arc-sec.	Digital elevation model (DEM)
SRTM 90m Digital Elevation DB v4.1[b]	World	Raster	3 arc-sec.	Digital elevation model (DEM)
WorldClim[c]	World	Raster	30 arc-sec.	Mean annual and monthly climate (period 1950–2000)
CRU-TS 3.1 Climate Database[d]	World	Raster	0.5 degrees	Time series of monthly climate (period 1901–2009)
Global Aridity and PET Database[e]	World	Raster	30 arc-sec.	Global Potential Evapo-Transpiration (PET) and Global Aridity Index
USGS World Geological Maps[f]	Various continents	Vector		Geological age, rock type (limited information)
ESDB v2 Raster Archive[g]	Europe	Vector and raster	1 × 1 km	Soil related attributes (soil type, parent material, etc.)

[a]http://eros.usgs.gov/#/Find_Data/Products_and_Data_Available/GTOPO30
[b]http://www.cgiar-csi.org/data/srtm-90m-digital-elevation-database-v4–1
[c]http://www.worldclim.org/current
[d]http://badc.nerc.ac.uk/view/badc.nerc.ac.uk__ATOM__dataent_1256223773328276
[e]http://www.cgiar-csi.org/data/global-aridity-and-pet-database
[f]http://energy.usgs.gov/OilGas/AssessmentsData/WorldPetroleumAssessment/WorldGeologicMaps.aspx
[g]http://eusoils.jrc.ec.europa.eu/ESDB_Archive/ESDB/index.htm

Having reviewed and discussed the requirements for the geographical marker compounds as well as the ancillary variables, some additional concepts and further requirements will be discussed in the remaining part of this section. These involve the choice of the spatial interpolation approach and methods for confining the possible area of origin *a priori*.

An important aspect of the geospatial modelling approach is to make informed statements about the possible origin of a sample (or set of samples) for which an origin has to be verified at a certain level of confidence. In order to be able to do so, it is important to have a fair understanding of the uncertainty of the model. As such, the second set of requirements for building a spatial model in this framework is that the approach or technique used for spatial interpolation should (I) allow an estimate of the uncertainty associated with each model estimate at (x, y) to be provided. And, since we want to incorporate ancillary information, in addition to the approach or technique used for spatial interpolation should (II) be able to incorporate ancillary information. Both these points will be discussed in detail in Section 4.3.

Yet another important aspect of the geospatial modelling approach is to make informed statements about the possible origin of a sample (or set of samples) for which the origin is unknown (i.e., 'where can the sample possibly come from?'). By using geo-climatic ancillary variables during modelling it is obvious that, for areas with similar climatic or geological conditions, there is a high chance of observing similar values. This implies that unless the production area of relevance is characterized by some unique climatic or geological condition within the total spatial extent of the model, the question 'where can this sample possibly come from?' will often not provide a unique answer, but cover a larger area or a set of areas.

One way to increase the 'spatial discrimination power' of the geospatial approach is by combining several models in order to confine the extent of the combined area of possible origin. This approach has been referred to by Kelly *et al.* (2011) as the 'multiple isotope approach' and is extended here to the 'multiple marker approach'. The 'multiple marker approach' can be regarded as an analogue to fingerprinting techniques in which multiple variables contained within the 'fingerprint' are used.

The requirement for using a set of geographical markers instead of one single marker is that the markers used for the individual models should be independent or at best only weakly dependent. Although it is intuitively clear that only markers containing a substantial amount of additional information add to the spatial discrimination power of the approach, in practice it is often not realized that stable isotopes such as δ^2H and $\delta^{18}O$, and even $\delta^{13}C$, are strongly correlated in many natural commodities. A good example is the correlation of δ^2H and $\delta^{18}O$ in precipitation, which is known as the global meteoric water line (GMWL; Craig, 1961), which is often reflected in the isotopic composition of plant- and animal- derived materials. So in addition to focussing only on stable isotopes of C, H and O for geographical provenancing, it is therefore advantageous to include other types of geographical markers as well.

In addition to using a combination of geospatial models to confine the area of possible origin by including only those areas where a commodity of interest can possibly be coming from, that is, only those areas where it is known that the commodity is grown or produced (target area). This would require detailed maps of relevant production areas, or alternatively, more general maps of land use. Since this information is also required to optimize the sampling strategy, this topic will be further discussed in the next section.

4.3 Geospatial modelling

In the previous section, various requirements and boundary conditions for the successful application of the geospatial modelling approach were formulated.

Once these requirements and conditions are met, the next step is to select a set of suitable geographical marker compounds and associated ancillary information (i.e., GIS maps containing relevant attributes). Subsequently the general approach for building a (set of) geospatial model(s) in this context would consist of the following steps:

1. Setting up an appropriate sampling strategy within the predefined target area.
2. Sampling of the food commodity of interest and analysis of the relevant marker compound(s).
3. Geospatial interpolation of the marker compound(s) using the relevant ancillary information.

This section will focus on steps 1 and 3 as these are specifically linked to the spatial aspects of the modelling approach. The procedures for step 2 are similar for the spatial as well as non-spatial approach, except that in the spatial approach the samples need to be geo-referenced.

4.3.1 Sampling strategy

Although the sampling strategy significantly affects the quality of the final model its importance is often ignored. In order to set up a sampling strategy in the context of spatial sampling the following points should be considered:

- The target area, which is the area where the target population occurs within the predefined extent of the model.
- The sample size, which concerns the amount of sample taken from the target population.
- The sampling design, which determines the spatial configuration of the sample locations within the target area.

The theoretical background and mathematical descriptions of these aspects are provided in standard text books about spatial sampling (e.g., De Gruijter *et al.*, 2005). These will not be expanded upon here as our focus is on the practical aspects of setting up a sampling strategy in a food provenancing framework, whilst bearing general statistical boundary conditions in mind.

In a food provenancing framework, the definition of the target area requires spatial information on all individual areas or locations where the food commodity of interest is grown, reared or otherwise produced. As mentioned in Section 4.2, this would require a detailed map of relevant production areas, or alternatively more a general land use map.

Although global land use maps are becoming increasingly available, the definition of the classes used in these maps is usually broad and often not detailed enough for the food commodity of interest. Alternatively, administrative data about annual production statistics, such as for example provided

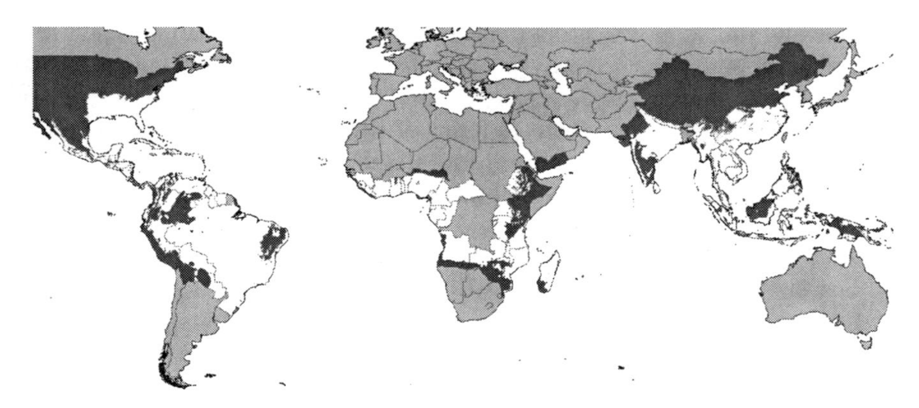

Fig. 4.2 Delineation of pseudo-target area for coffee (in white) based on FAO production statistics in 2009 (non-producing countries in light grey) in combination with topographical and climatic boundary conditions showing areas unsuitable for coffee production (in dark grey: altitude >2000 m, mean annual temperature <14°C or >28°C and mean precipitation amount <800 mm or >2800 mm). Altitude and climate maps from Heijmans *et al.* (2005).

by the Food and Agricultural Organization (FAO), can be used to define a target area. FAO administrative data is, however, often only available on a country scale.

Since the production of a certain food commodity is often restricted by climatic conditions such as minimum temperature or amount of rainfall, climatic maps do not only serve as ancillary information for interpolation, but may also help to confine the target area. An example of such a target area for coffee, which is based on FAO production data as well as altitude and climate information, is shown in Fig. 4.2.

It should be noticed that the target area as provided for coffee in Fig. 4.2 provides the area where coffee *could* be produced, but not necessarily *is* produced. Since the formal definition of the target area concerns only the area where the commodity is actually grown, reared or otherwise produced, the area in Fig. 4.2 represents a kind of pseudo-target area that is wider in terms of spatial extent than the actual ('true') target area. Despite this, such a pseudo-target area will be helpful in further defining the sampling strategy. Furthermore, it can be used in a later stage to further confine the area of possible origin (see Section 4.4).

Since sampling, together with the analysis of the marker compounds, is often the most expensive aspect of geospatial model building, another important aspect of the sampling strategy concerns the sample size. Although general guidelines are difficult to give, Webster and Oliver (1992) recommend a minimum of 100–150 observations for fitting a model of spatial structure, which can be used as a rule of thumb. The accuracy of the model can be increased significantly by increasing the sample size and it is therefore recommended to use at least several hundreds of observations.

Previous studies in which the time-averaged δ^2H and δ^{18}O composition of precipitation were modelled on a global scale were based on ~350 sample locations (see e.g., Bowen and Revenaugh, 2003; Van der Veer *et al.*, 2009). The models yielded acceptable results in terms of the uncertainty around the predicted values, although this was mostly due to the quality of the high-resolution ancillary information that was used in the interpolation and the strong relationship between these ancillary variables and the time-averaged δ^2H and δ^{18}O values.

The last aspect of the sampling strategy that should be dealt with concerns the sampling design, which defines the way that the sample locations are selected from the target area. In general, two approaches can be discerned which have different theoretical bases: random sampling versus systematic sampling. The general consensus is that systematic sampling is superior to unrestricted random sampling, and the optimal sampling network design consists of a superimposed equilateral triangular grid (see e.g., Griffith, 1996).

In practice, however, such a systematic sampling design might not be feasible because production areas of a certain food commodity within the target area are often not continuous nor are they systematically distributed. In such cases, an approximation of the systematic sampling approach could be applied, in which a sample is collected within a certain radius from the location that is provided by the design.

4.3.2 Geospatial interpolation

As mentioned in Section 4.2, the requirements for a suitable geospatial interpolation technique within this framework is that it should (I) allow assessment of the uncertainty around the predicted values and (II) allow inclusion of ancillary information. An up-to-date overview of currently available interpolation methods in space that meet these two requirements is provided in Table 4.2.

From Table 4.2 it is clear that most of these techniques are based on some form of 'kriging'. Kriging refers to a group of geostatistical interpolation methods in which the value at an unobserved location is predicted by a weighted linear combination of the values at surrounding locations (see e.g., Isaaks and Srivastava, 1989). The weights are based on a model that describes the spatial correlation structure in the data, which is referred to as the experimental variogram. For an introduction into kriging and different kriging algorithms reference is made to Isaaks and Srivastava (1989) and Goovaerts (1997).

Due to its statistical formulation, kriging allows direct quantification of the accuracy of the predicted values by means of the kriging variance, which can be regarded as a measure of the uncertainty about the predicted values. This is a distinct advantage of kriging over many of the common interpolation techniques such as inverse distance weighting (IDW), splines and triangulation.

Table 4.2 Overview of spatial interpolation techniques that provide an estimate of uncertainty and also allow inclusion of ancillary information

Method	References
Stratified kriging	Brus and Heuvelink (2007)
Cokriging	Isaaks and Srivastava (1989), Goovaerts (1997), Deutsch and Journel (1998)
Principal component kriging	Goovaerts (1997)
Kriging combined with linear regression	Delhomme (1978). Also: kriging with uncertain data
Universal kriging	Ahmed and De Marsily (1987), Goovaerts (1997), Deutsch and Journel (1998). Also: kriging with an external drift, kriging with a trend model
Regression kriging	Delhomme (1978), Ahmed and de Marsily (1987), Odeh *et al.* (1994), Odeh *et al.* (1995), Goovaerts (1997), Mardikis *et al.* (2005). Also: kriging with a guess field, residual kriging, simple kriging with varying local means
Regression methods	Odeh *et al.* (1994), Denby *et al.* (2005), Li and Heap (2008), Wilks (2008), Hastie *et al.* (2009). Includes: multiple linear regression, regression trees (CART), GAM, Ridge-regression, LASSO and artificial neural networks
Classification	Li and Heap (2008)
Bayesian Maximum Entropy	Christakos (2000), Brus and Heuvelink (2007), Brus *et al.* (2008)
Markov random fields	Norberg *et al.* (2002), Wu *et al.* (2004), Kasetkasem *et al.* (2005), Hartman (2006), Brus and Heuvelink (2007)

Source: Knotters *et al.*, 2011.

Although kriging is commonly used in the environmental sciences, its application in the field of geographical provenancing is so far limited. Van der Veer *et al.* (2009) used a form of regression kriging, so-called simple kriging with local means, to interpolate δ^2H and δ^{18}O in global precipitation using the temperature during the coldest quarter as ancillary variable. One of the main features of this approach is that it is relatively simple and intuitive. It consists of four steps:

1. Quantify the relationship between the geographical marker (δ^2H and δ^{18}O in this case) and the ancillary variable by means of regression.
2. Use the fitted function to provide an initial estimate for each grid value of the map with the ancillary variable.
3. Provide an optimized and fitted experimental variogram of the residuals of the regression.
4. Interpolate the residuals by simple kriging using the fitted variogram and add them to the map with initial estimates from step 2.

This regression approach works well when the ancillary variable is available as a continuous numerical variable such as the mean temperature or the amount of precipitation. Geological and pedological maps are of a categorical nature and in these cases another type of interpolation should be used such as stratified kriging or Bayesian Maximum Entropy (see Table 4.2).

Before interpolation using categorical information as ancillary variable, one could decide to reclassify the geological or pedological maps in order to improve the performance of the final model. Optimizing reclassification schemes can be done by hand, for example by looking at histograms and boxplots. A more elaborate way would be to use unsupervised chemometric techniques such as cluster analysis to guide optimal reclassification of categories.

Although kriging provides a direct estimate of the model variance, stochastic modelling techniques based on resampling techniques such as Monte Carlo simulations and bootstrapping provide a powerful alternative for estimating model uncertainties. For example, Bowen and Revenaugh (2003) used the related jack knife method to provide the uncertainty of the predicted δ^2H and δ^{18}O in global precipitation using the relation with latitude and altitude as ancillary variables.

The Monte Carlo technique in its simplest form uses simple random sampling of the input probability distributions to estimate the uncertainty of the model at each location within the model extent. Monte Carlo simulation and related resampling techniques like bootstrapping can in principle be used in conjunction with any interpolation technique.

Despite its heavy computational requirements, stochastic modelling therefore provides an interesting alternative to the aforementioned kriging techniques, especially for data that has been sampled according to a non-systematic (e.g., random, pseudo-random, haphazard) sampling design.

4.4 Inference on geospatial models

Statistical inference – either classical or Bayesian – is the principal statistical approach to assign measures of 'confidence' or 'certainty' to statements about one or more populations. Although the field of statistical inference has been successfully adapted for all sorts of non-spatial questions, the subject of statistical inference in a spatial context is not without controversy. Most of these issues arise because important assumptions for inference, such as probabilistic independence of observations, are likely to be violated for spatial data which often has a clear spatial correlation structure. In addition, the distribution of observations may well be conditional on their geographical locations (e.g., Brunsdon, 2009).

The spatial models described in this framework are subject to the problems pointed out above. A heuristic approach was used here by simply 'borrowing' some standard statistical tests to make informative statements about the possible origin of samples. It should be emphasized that this approach is currently used to exemplify the, as yet meagrely addressed, issue of inference on spatial models.

As mentioned in the introduction, the geospatial modelling approach for food origin authentication can provide answers to two basic questions:

1. Does the measured composition of the sample comply with the prediction interval for that area?
2. Where could the sample possibly come from according to its measured composition?

Suppose, for example, that one has five bottles of a commercial mineral water brand that were representatively sampled from a large batch of bottles of which the acclaimed origin needs to be verified. Would their (mean) composition comply with the predicted intervals? And when it turns out that the samples indicate that the batch of mineral water is possibly not from the acclaimed origin; from where could the water possibly be coming and does that include the tap around the corner?

To answer the first question at a certain level of confidence one needs to construct the appropriate prediction intervals. In statistical inference, prediction intervals for a future mean value involves predicting (or containing) the results of a future sample from a previously sampled population (e.g., Hahn and Meeker, 1991). To do so, one firstly needs an estimate of the predicted value as well as the (kriging) variance for the raster cell that covers the acclaimed origin of the sample(s). Next, a two-sided $100(1 - \alpha)\%$ prediction interval to contain a single future observation, or the mean of m future observations, based on a previous independent random sample of size n from the same population (or process), can be estimated (Hahn and Meeker, 1991):

$$[lpi, upi] = x_{avg} \pm t_{(1-\alpha/2;n-1)} s_n \sqrt{\left(\frac{1}{m} + \frac{1}{n}\right)} \qquad [4.1]$$

where lpi and upi are the lower and upper prediction limit, x_{avg} is the predicted model value at the acclaimed origin (x, y), $t_{(1-\alpha/2;\, n-1)}$ is the critical value of the t distribution with $n - 1$ degrees of freedom and s_n is the standard deviation as calculated from the (kriging) variance.

An interesting issue that arises in a spatial context is to determine the appropriate value for n. Since the predicted values and their variances in kriging are based on weighted linear combinations of the values at surrounding locations, it seems intuitive to use the number of observations from the surrounding locations as a measure of n. A possible approach would be to use the number of sample locations that fall within a certain distance from the relevant raster cell representing the acclaimed origin. Since in kriging the semivariance provides a measure of the correlation of sample values as a function of the distance between them, the variogram could moreover be used to determine the appropriate distance for sample locations to be included.

The choice of an appropriate n in Equation 4.1 is more straightforward in cases where the spatial models are based on resampling techniques such as bootstrapping and Monte Carlo simulation. Here the size of the previous independent random sample n could simply be set equal to the number of resampling runs. Since the number of runs is determined by the operator, this approach offers the opportunity to optimize the width of the prediction intervals by taking a large number of runs, which provides an additional advantage of the resampling approach.

Once an appropriate prediction interval for the predicted value of some marker compound at raster cell (x, y) has been established, the origin of a sample claimed to be of some specific area can be verified by comparison of the measured value of the marker compound in the sample (or the mean of a set of samples) to the prediction limits at (x, y). When the value measured in the sample falls within the [*lpi, upi*] interval it can be stated that there is no reason to believe that the sample does not originate from that area. However we still do not know what the chance is of including a sample from another area (Type II error). In a case where the value falls outside the [*lpi, upi*] interval it can be concluded that, although it seems unlikely, there still is a $100\alpha\%$ chance that a sample that is truly from that area will be erroneously rejected (Type I error).

Another approach would be to ask 'is the mean of the sample values significantly different from the predicted mean value at the claimed area of origin (x, y)?' In this case a t test for two independent samples could be used. In cases where $n \neq m$, the appropriate t value is given by:

$$t_{(1-\alpha/2;n+m-2)} = \frac{x_{\text{avg},n} - x_{\text{avg},m}}{\sqrt{\frac{\left[(n-1)s_n^2 + (m-1)s_m^2\right]}{(n+m-2)}}\sqrt{\left(\frac{1}{m}+\frac{1}{n}\right)}} \qquad [4.2]$$

where $t_{(1-\alpha/2;\, n+m-2)}$ is the critical t value for $(n + m - 2)$ degrees of freedom, $x_{\text{avg},n}$ is the predicted model value, $x_{\text{avg},m}$ is the mean of the test samples, s_n^2 is the model variance and s_m^2 is the variance of the test samples. By finding the critical t value for $(n + m - 2)$ degrees of freedom, it will be clear whether there is a difference between the two means at a certain level of confidence. Such a difference would indicate it is unlikely that the samples are indeed from the claimed area of origin, whereas the absence of any significant differences would imply that there are no reasons to suspect that the samples would not be from that origin. Whether and how relevant this finding is in a spatial context will be discussed below.

So far this discussion has dealt with comparing the (mean of the) measured value(s) in some test sample(s) to a single predicted raster value and its (kriging) variance, which is believed to represent its 'true' area of origin. When the claimed area of origin is not known beforehand, the approach needs to be adapted for a spatial framework. Since the geospatial models as described in this chapter provide an estimate of the predicted value as well as the (kriging) variance for all the raster cells in the model extent, one way would be to calculate t values for all the individual raster cells.

An example of a grid of such t values is provided in Fig. 4.3a, which reveals areas where the difference between the predicted and measured values are small (t value close to zero) or large (large positive or negative t value). This example is based on the $\delta^{18}O$ composition in mineral water shown in Fig. 4.1a and the kriging variance in Fig. 4.1b. By contouring the grid values at the critical t values the area or areas where the mean of the test samples is significantly different from the predicted grid values at various levels of confidence is revealed (Fig. 4.3b). The area in which the two means are not significantly different at a given significance level is referred to here as the 'area of possible origin (APO)'.

Although the methodology seems rather straightforward, it should be considered that the basic assumptions with regard to statistical inference are violated, at least to some extent. Furthermore, where t values or values for any other test statistic are calculated for more than one grid cell, as in the previous example, the situation inevitably arises that more than one hypothesis is being tested at the same time. This is referred to as multiple hypotheses testing, which has consequences for the overall significance level of the test. An intuitive way to understand this complication is to realize that if the individual significance level of the test (i.e., for one grid cell) is α, we would expect to find $N\alpha$ significant results by performing N individual tests on N grid cells.

A classical approach to this problem is to calculate an adjusted significance level α', often referred to as the family-wise or the experiment-wise α, using the Šidàk equation:

$$\alpha' = 1 - (1 - \alpha)^N \qquad [4.3]$$

where N is the number of grid cells involved in the test. The adjusted α now represents the probability of making at least one Type I error for the whole family of tests. So for an individual $\alpha = 0.05$ and $N = 50$ grid cells the family-wise significance level $\alpha' = 0.92$, which shows a 92% chance of making at least one Type I error (i.e., rejecting the claim of origin whilst the samples truly are from that area of origin). Equation 4.3 can be rearranged to give:

$$\alpha = 1 - (1 - \alpha')^{1/N} \qquad [4.4]$$

Using this equation, it is possible to calculate the individual significance level that is required to test at a certain family-wise α'. Using $\alpha' = 0.05$ and $N = 50$ grid cells, the individual significance level $\alpha = 0.001$. This is a rather strict criterion,

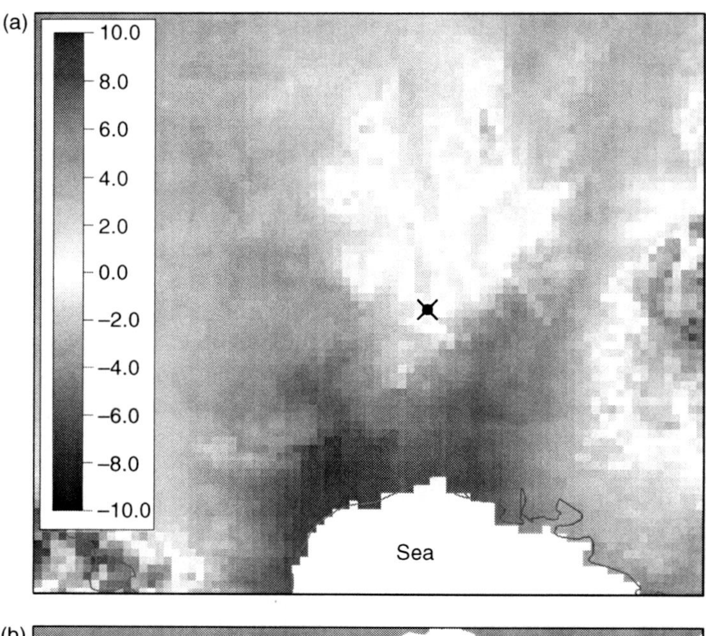

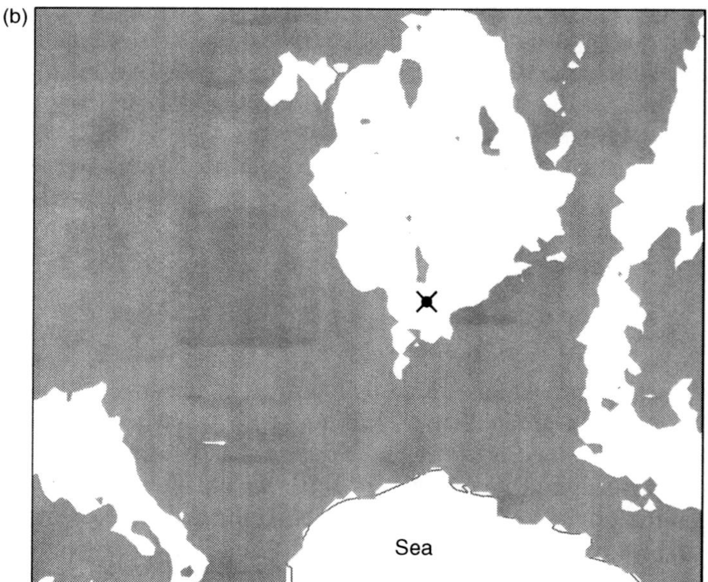

Fig. 4.3 (a) Grid of t values calculated using the predicted values for $\delta^{18}O$ in mineral water (Fig. 4.1a) and the associated uncertainty (Fig. 4.1b). The predicted mean value in the area of acclaimed origin (black cross) is −9.8‰, whereas the mean of the five test samples is −9.5‰. For n a constant value of 20 was chosen. (b) Contour map of t values where the area that has values below the critical t value for a two-sided test at a significance level of 0.05 is shown in white ($t_{crit} = 2.07$). This area represents the possible area of origin of the sample and since it includes the acclaimed origin there is no evidence to believe the five test samples are not from that origin.

especially considering the fact that most geospatial models have >>50 grid cells and this would require an even smaller value for α. This could obviously lead to unacceptably large areas of possible origin as the critical t value increases with decreasing α.

In practice the situation is less discouraging for two reasons. Firstly, the Šidàk-Bonferroni approach applies to strictly independent tests. It was previously argued that spatial data and models often exhibit a strong spatial correlation structure, so that nearby locations will be correlated making most of the tests dependent to some extent. As the Šidàk-Bonferroni approach behaves conservatively when the tests are not independent, it will give an overly pessimistic view of the actual situation.

Secondly, when the number of tests becomes very large, as is typically the case in a spatial modelling context, the approach will again behave conservatively. To provide a fair estimate of the family-wise significance level in a spatial modelling framework it is therefore important to make the correction for multiple hypothesis testing less stringent (see e.g., methods proposed by Hochberg, 1988; Shaffer, 1995). Alternatively, one could use the false discovery rate (FDR), which is defined as the ratio of the number of Type I errors by the number of significant tests (see e.g., Benjamini and Hochberg, 1995). The FDR is a widely accepted parameter for ascribing significance levels to statistical tests in large-scale testing problems.

Up to now we have only considered a situation in which statistical inference is performed on one single spatial model. As mentioned in Section 4.2, one way to increase the 'spatial discrimination power' of the geospatial modelling approach is to use multiple (independent) markers and to provide a set of geospatial models. Then, by combining the information from several independent models, one will end up with a smaller area of possible origin compared to a single model situation. The total – or family-wise – area of possible origin is then defined by:

$$APO_{total} = APO_1 \cap APO_2 \cap APO_3 \cap \ldots \qquad [4.5]$$

where APO_{total} represents the total area of possible origin as based on the intersection of the individual APOs of marker compounds 1, 2, 3,... following the procedure described above. So supposing that an appropriate definition of the areas of possible origin has been established for the individual models, and the critical t value and significance level has been chosen, a visualization of the resulting family-wise area of possible origin can be created by performing an overall intersection of the individual APOs.

This procedure again involves testing of multiple hypotheses, albeit now in terms of variables instead of spatially correlated locations. Since the correlation between the marker variables should ideally be small (see Section 4.2), the multiple hypothesis testing framework applies especially to the multiple marker/model situation. This should be taken into account when stating overall significance levels, which can be done best by using the FDR to describe the overall significance level of the test.

The topic of inference has been approached here from a classical parametric point of view, mainly because most researchers are thoroughly familiar with the theory and its applications. This approach works well assuming that the underlying distributions are normal, which might not be the case. In those cases it would be more appropriate to use non-parametric alternatives. Moreover, the Bayesian approach to spatial inference could provide a suitable alternative to the classical frequentist approach. For a general introduction see for example, Brunsdon (2009).

4.5 Conclusion and future trends

The geospatial modelling approach as described in this chapter provides an unique framework for geographical provenancing and verifying geographical origin, which plays an important role in various fields of science including food authentication, forensics, archaeology, animal migration studies, etc. So far, however, the approach has only been used in a limited number of studies, which is moreover a result of the lack of a theoretical and practical framework. This chapter tries to fill this gap by providing both theoretical concepts as well as practical guidelines for development and application of geospatial models. To bring the application of geospatial modelling in a geographical provenancing context to the next higher level, suggestions for future directions with regard to the key aspects of the geospatial modelling approach are provided here.

From Section 4.2, which deals with requirements for building a set of spatial models for food origin authentication, it can be concluded that the search for suitable geographic marker compounds should be extended in the future, as should our understanding of their relationship with ancillary variables such as climate and geology. In cases when climate data is used as an ancillary variable, using more up-to-date ancillary information instead of long term averages would moreover increase the accuracy of future geospatial models.

As discussed in Section 4.3, defining an optimal sampling strategy and using an optimal spatial interpolation technique are key aspects for building robust geospatial models. From this section it can furthermore be concluded that the general framework for sampling and spatial interpolation is currently well developed and appropriate solutions are available for most situations. Future work on geospatial modelling for geographical provenancing purposes should preferably focus on spatial interpolation methods that use resampling techniques such as bootstrapping and Monte Carlo simulations since these have several advantages over a classical approach.

The most challenging topic seems to concern statistical inference on (a set of) spatial models (Section 4.4) as it involves multiple hypothesis testing on various levels. Since this reduces the overall – or family-wise – Type I error of the test, the issue of multiple hypothesis testing deserves wider recognition in future studies. In this framework the false discovery rate (FDR) provides a suitable alternative to the much more stringent Šidàk-Bonferroni approach to control family-wise Type I errors, and should preferably be used in making uncertainty statements about tests performed on geospatial models.

4.6 References

AHMED, S. and DE MARSILY, G. (1987) 'Comparison of geostatistical methods for estimating transmissivity using data on transmissivity and specific capacity', *Water Resources Research*, **23**(9), 1717–1737.

BENJAMINI, Y. and HOCHBERG, Y. (1995) 'Controlling the false discovery rate: A practical and powerful approach to multiple testing', *Journal of the Royal Statistical Society, Serie B*, **57**, 289–300.

BOWEN, G.J. (2010) 'Isoscapes: Spatial pattern in isotopic biogeochemistry', *Annual Review of Earth and Planetary Sciences*, **38**, 161–187.

BOWEN, G.J. and REVENAUGH, J. (2003) 'Interpolating the isotopic composition of modern meteoric precipitation', *Water Resources Research*, **39**(10), 1299–1311.

BRUNSDON, C. (2009) 'Statistical inference for geographical processes', in FOTHERINGHAM, A. S, ROGERSON, P.A., *The SAGE Handbook of Spatial Analysis*, SAGE Publications Ltd, London, pp. 207–224.

BRUS, D.J., BOGAERT, P. and HEUVELINK, G.B.M. (2008) 'Bayesian maximum entropy prediction of soil categories using a traditional soil map as soft information', *European Journal of Soil Science*, **59**(2), 166–177.

BRUS, D.J. and HEUVELINK, G.B.M. (2007) *Towards a soil information system with quantified accuracy: three approaches for stochastic simulation of soil maps*, Technical Report 58, Wettelijke Onderzoekstaken Natuur en Milieu, Wageningen UR.

CHRISTAKOS, G. (2000) *Modern Spatiotemporal Geostatistics*, Oxford University Press, New York.

CRAIG, H. (1961) 'Standards for reporting concentrations of deuterium and oxygen-18 in natural waters', *Science*, **133**, 1833–1834.

DE GRUIJTER, J., BRUS, D., BIERKENS, M. and KNOTTERS, M. (2005) *Sampling for Natural Resource Monitoring*, Springer Verlag, Berlin.

DELHOMME, J.P. (1978) 'Kriging in the hydrosciences', *Advances in Water Resources*, **1**(5), 251–266.

DENBY, B., HORALEK, J., WALKER, S.E., EBEN, K. and FIALA, J. (2005) '*Interpolation and assimilation methods for European scale air quality assessment and mapping*', Technical report, European topic centre.

DEUTSCH, C.V. and JOURNEL, A.G. (1998) *GSLIB Geostatistical Software Library and User's Guide*, Oxford University Press, New York.

EHLERINGER, J.R., BOWEN, G.J., CHESSON, L.A., WEST, A.G., PODLESAK, D.W. and CERLING, T.E. (2008) 'Hydrogen and oxygen isotope ratios in human hair are related to geography', *Proc. Natl. Acad. Sci. USA*, **105**, 2788–2793.

GOOVAERTS, P. (1997) *Geostatistics for Natural Resources Evaluation*, Oxford University Press, New York.

GRIFFITH, D.A. (1996) 'Introduction: the need for spatial statistics', in ARLINGHOUS, S.L., *Practical Handbook of Spatial Statistics*, CRC Press, Boca Raton, USA.

HAHN, G.J. and MEEKER, W.Q. (1991) *Statistical Intervals: A Guide for Practitioners*, John Wiley, New York.

HARTMAN, L.W. (2006) 'Bayesian modelling of spatial data using Markov random fields, with application to elemental composition of forest soil', *Mathematical Geology*, **38**, 113–133.

HASTIE, T., TIBSHIRANI, R. and FRIEDMAN, J. (2009) *The Elements of Statistical Learning. Data Mining, Inference and Prediction*, Springer, New York.

HOCHBERG, Y. (1988) 'A sharper Bonferroni procedure for multiple tests of significance', *Biometrika*, **75**, 800–803.

ISAAKS, E.H. and SRIVASTAVA, R.M. (1989) *An Introduction to Applied Geostatistics*, Oxford University Press, New York.

KASETKASEM, T., ARORA, M.K. and VARSHNEY, P.K. (2005) 'Super-resolution land cover mapping using a markov random field based approach', *Remote Sensing of Environment*, **96**, 302–314.

KELLY, S.D., BRERETON, P., GUILLOU, C., BROIL, H., LAUBE, I., DOWNEY, G., ROSSMAN, A., HÖZL, S. and VEER, G. VAN DER (2011) 'New approaches to determining the origin of food', in HOORFAR, J., JORDAN, K., Butler, F. and PRUGGER, R., *Food Chain Integrity: A Holistic Approach to Food Traceability, Safety, Quality and Authenticity*, Woodhead Publishing Series in Food Science, Technology and Nutrition, No. 212.

KNOTTERS, M., HEUVELINK, G.B.M., HOOGLAND, T. and WALVOORT, D.J.J. (2010) *A disposition of interpolation techniques*, WOt-werkdocument 190, Statutory Research Tasks Unit for Nature and the Environment, Wageningen UR, Wageningen.

LI, J. and HEAP, A.D. (2008) *A review of spatial interpolation methods for environmental scientists*, Technical Report GeoCat # 68229, Geoscience Australia, Canberra.

MARDIKIS, M.G., KALIVAS, D.P. and KOLLIAS, V.J. (2005) 'Comparison of interpolation methods for the prediction of reference evapotranspiration – an application in Greece', *Water Resources Management*, **19**(3), 251–278.

NORBERG, T., ROSEN, L., BARAN, A. and BARAN, S. (2002) 'On modelling discrete geological structures as Markov random fields', *Mathematical Geology*, **34**, 63–77.

ODEH, I.O.A., MCBRATNEY, A.B. and CHITTLEBOROUGH, D.J. (1994) 'Spatial prediction of soil properties from landform attributes derived from a digital elevation model', *Geoderma*, **63**, 197–214.

ODEH, I.O.A., MCBRATNEY, A.B. and CHITTLEBOROUGH, D.J. (1995) 'Further results on prediction of soil properties from terrain attributes: heterotopic cokriging and regression-kriging', *Geoderma*, **67**, 215–216.

SHAFFER, J.P. (1995) 'Multiple hypothesis testing', *Annual Review of Psychology*, **46**, 561–584.

STILL, C.J., BERRY, J.A., COLLATZ, G. J. and DEFRIES, R.S. (2003) 'Global distribution of C_3 and C_4 vegetation: carbon cycle implications', *Global Biogeochemical Cycles*, **17**, 1006–1021.

VOERKELIUS, S., LORENZ, G.D., RUMMEL, S., QUÉTEL, C.R., HEISS, G., BAXTER, M., BRACH-PAPA, C., DETERS-ITZELSBERGER, P., HOELZL, S., HOOGEWERFF, J., PONZEVERA, E., VAN BOCXSTAELE, M. and UECKERMANN, H. (2010) 'Strontium isotopic signatures of natural mineral waters, the reference to a simple geological map and its potential for authentication of food', *Food Chemistry*, **118**, 933–940.

WEBSTER, R. and OLIVER, M. (1992) 'Sample adequately to estimate variograms of soil properties', *Journal of Soil Sciences*, **43**, 177–192.

WILKS, D.S. (2008) 'High-resolution spatial interpolation of weather generator parameters using local weighted regressions', *Agricultural and Forest Meteorology*, **148**(1), 111–120.

WU, K., NUNAN, N., CRAWFORD, J.W., YOUNG, I.M. and RITZ, K. (2004) 'An efficient Markov chain model for the simulation of heterogeneous soil structure', *Soil Science Society of America Journal*, **68**, 346–351.

5

New approaches for verifying food species and variety

H. Broll, Federal Institute for Risk Assessment, Germany

DOI: 10.1533/9780857097590.2.81

Abstract: Food authenticity can be verified based on modern biological methods using DNA as the target. Different approaches are available using either simple DNA hybridization or polymerase chain reaction (PCR). Today PCR is the method of choice and many species-specific systems are described in publications. The automation and independence from particular suppliers make PCR well suited for standardization. DNA sequencing (with either part or whole genome sequence determined) is a new encouraging development, making possible identification of a particular species or breed in a given food sample with exceptional precision. Instrumentation for this task is already available; however it remains expensive and so far is not applied routinely in traceability diagnostic.

Key words: PCR, species, DNA, sequencing.

5.1 Introduction: the commercial importance of food species and varieties

Consumers have the right to clear, accurate and correct information on food product packaging when selecting goods. This labelling is particularly important if the food has been processed in a way so that the individual ingredients can no longer been distinguished. In addition, so-called 'premium products' are often advertised as using a certain ingredient or variety of ingredients with specific origins in an attempt to attract consumers.

The European Union (EU) established specific legislation (Council Regulation (EC) No. 510/2006) for such products, defined as PDO (protected designation of origin), PGI (protected geographical indication) and TSG (traditional speciality guaranteed), that supports and protects the names of high quality agricultural products and foods. The aim of the regulation is to protect the standard of the regional foods, promote rural and agricultural activity, help producers to achieve a premium price for their products, and eliminate unfair competition and misleading of consumers by non-authentic products. It protects the names of certain wines, cheeses, olive oils, hams, sausages, seafood, beers, Balsamic vinegar and also regional breads, fruits, raw meats and vegetables. Examples of PDOs and PGIs in which a variety or species play the main role are the Italian wheat '*Farro della Garfagnana*' and the '*Vitellone dell'Appennino Centrale*' beef. In both cases, the key ingredient for production is only allowed to be sourced from well-defined species or breeds. In the case of '*Farro della Garfagnana*', the species *Triticum turgidum* subsp. *dicoccum* must be exclusively used if the product is to be allowed to use the name, whilst for the latter mentioned product, meat of only three cattle breeds (Chianina, Romagnola and Marchigiana) are allowed.

Another well-described PDO is '*Miel de Corse*' (Corsican honey), produced only on the island of Corsica. Corsica represents a specific flora environment, which is also represented as pollen in the honey and can be used as a marker to distinguish Corsican honey from other varieties.

In a study concerning all 820 products listed in the European Register of PDOs and PGIs (excluding spirits and wines), the European Commission's Directorate-General for Agriculture and Rural Development found that PDO and PGI agricultural products had an estimated value of €14.2 billion in 2007. It was also estimated that the 30% of PDO and PGI exported outside the EU had an equivalent value of around €700 million. In 2008 all producers together in the 27 EU member states generated €751 billion of added value in the food sector, making the market share of name-protected quality agricultural products and foods about 2% in total.

Legislation is also in place in the EU aimed at achieving a high level of health protection for consumers with food allergies. Directives 2000/13/EC and 2003/89/EC require a mandatory declaration of foods potentially causing allergenic reaction, including plant and animal species such as celery, mustard, sesame seeds, lupin and molluscs. Such food labelling legislation is only dedicated to allergenic ingredients that are deliberately introduced into food products, in contrast to traces of such allergens that have entered the food product as a result of, for example, contamination during the production process. The presence of such 'hidden allergens' can affect the safety of the food product, since it can cause a threat to the health of consumers. The presence of non-labelled allergens in a food product could therefore be interpreted as non-compliance with legislation concerning food safety.

In addition to tools like allergen management plans implemented by the food industry, detection methods are used which have been designed to identify traces of allergens on food production equipment and in food products. Combined, these methods define the essential framework used to convey the correct information regarding food product allergen content to the consumer, via appropriate use of precautionary labelling.

Another example of particular labelling needs is in the field of genetically modified (GM) foods and feeds. According to Regulation (EC) No 1829/2003 GM food and feed need to be labelled. This labelling can only be avoided if the presence of the genetically modified material is below 0.9% in relation to the individual ingredient of a food product, and if it is unintentionally present due to commingling.

In the light of rapidly evolving globalization of food production, distribution and consumption, the trade in fishery products has dramatically increased over the last decades. In many countries the situation is characterized by a decrease in traditional fish species, introduction of new species and the substitution of high-prized species with less expensive species. It has become increasingly difficult to fulfil the demand for high quality fish and products thereof, whilst at the same time guaranteeing safety and authenticity. Fraud, particularly in the field of fish products, has been reported all over the world (Rehbein, 2009). In response to this, the EU implemented a legal framework which specifies that fishery and aquacultural products may not be offered for retail use unless they are labelled with the commercial name of the species, the production method and the capture zone.

Game meat is also well known to be a target of fraud due to its limited availability and consequently relatively high price in comparison to the respective domestic species. The final consumer cannot detect the substitution in a processed product, consequently the substitution with domestic species is resulting in a large profit for the producer. In particular, the partial substitution of wild boar meat with domestic pig meat is an attractive fraud for unscrupulous manufacturers.

In general, the addition or substitution of ingredients derived from animal or plant species could be of high commercial interest. In a recent scandal in Europe, beef meat was substituted up to a considerable amount of 60% in some cases. Those cases are particularly complicated to be identified by the analyst, as he needs to know what he is looking for when applying molecular detection methods. Common to all the examples mentioned above are the need for methodologies that can distinguish and/or identify species of either plant or animal origin. Using traditional detection methods, it is possible to measure the various constituents of a certain product using classical wet chemistry techniques. If the analytical technique identifies a component which should not be present, the sample is determined not to be in compliance with the specification. But it is much more complicated if attributes such as species substitution or geographical origin are involved in the adulteration of a food product. Separation techniques like spectroscopic methods have facilitated significant improvement in the identification of fraud in foods, particularly in the determination of food product origins. However, it still has not fulfilled the potential of species-specific identification.

In the past, analytical methods to identify species specificity and authenticity mostly relied on the detection of proteins, either by electrophoresis and subsequently its comparison with a reference sample, or via the application of immunological methods using antibodies directed to proteins of the species under investigation. In general these techniques are very sensitive, accurate and easy-to-apply, but they are only applicable if the proteins are still present and not degraded, or if the proteins are in the same processing stage as they were when antibodies were developed against the protein of interest (Bonwick and Smith, 2004; Poms *et al.*, 2004). If, for example, the raw, non-degraded protein was used to develop specific antibodies, those antibodies are usually unable to detect the specific protein in a processed food product after cooking. Therefore protein electrophoresis and immunological techniques are usually only applicable to raw, unprocessed foods.

5.2 DNA-based methods for verifying food species and variety

In 1993, Allmann and co-workers demonstrated for the first time the potential of DNA-based methods for ensuring the integrity, safety and quality of food (Allmann *et al.*, 1993). The polymerase chain reaction (PCR) was used to identify the genetic material of wheat in a non-wheat food product.

The genetic material (deoxyribonucleic acid, DNA) is present in almost every cell of a plant or animal and can serve as a useful target in unambiguously identifying the source of a specific ingredient, if derived from a plant or animal species. DNA is highly species-specific. Furthermore, the same DNA is present in the majority of the cells of an organism, potentially enabling identical information to be obtained from any appropriate sample from the same source, regardless of the tissue of origin.

Through the acquisition of sequence data, DNA can potentially provide more information than protein, due to the degeneracy of the genetic code and the presence of many non-coding regions. DNA currently offers the highest specificity available for detection and identification. The probability of a given DNA sequence of 20 base pairs (bp) being present in another species is only 1×10^{-12}. Taking into consideration that the haploid genome of an averaged eukaryotic organism has a size between 10^8 and 10^{10} bp, it is statistically unlikely that the same DNA sequence will be found a second time in the same organism or in any another organism, making it particularly interesting for commercial usage.

Different species show different DNA sequences depending on the evolutionary distance between them. The further away the species are in terms of evolution, the more differences exist at the DNA level. In consequence, for two close species or analytical samples, it might hardly be possible to identify those small differences in the DNA sequence. However, sufficient differences for a DNA-based analysis generally exist at the very least between species, making it a key tool in species identification and verification.

DNA is a very stable and long-lived biological molecule. For forensic purposes, DNA has been successfully extracted from ancient mummies (Pääbo *et al.*, 1989a,b). Several other studies have also demonstrated the integrity of DNA after thorough heat and pressure treatment. Ebbehøj and Thomson (1991) investigated the quality of DNA from meat during storage and heat treatment. High-molecular weight DNA (>20 kilobase pairs [kbp]) could still be isolated from fresh meat, but storage of the unprocessed meat for a couple of days resulted in degradation of the DNA to average lengths of 15 to 20 kbp. Heating of the meat before extraction led to a further degradation of the DNA, which exhibited an average length of 300 bp when the meat was heated for 10 min at 121°C.

The influence of processing on the samples was also examined by Laube *et al.* (2007). These tests analysed sausages, home-canned foods, normal cans, cans for use under tropical conditions and ultra-high heat-treated cans of the same composition (F = 31 [ultra-high heat-treated cans] (F value 1.0 indicates a lethality of twelve decimal reductions for *Clostridium botulinum*, the most dangerous of all toxin-producing organisms, after 1 min at 121.1°C). It has been clearly shown that the quantity of extracted DNA decreased under an increasing degree of processing (Buntjer *et al.*, 1999).

The degree of DNA degradation is an important aspect when calculating the amount of an individual animal species in foods. The application of DNA-based methods facilitated even a quantitative statement in low-processed meat products, as well as in normal canned foods down to 0.1% (w/w), and in cans for use under tropical conditions, down to 1% (w/w) if a respective reference material is available which is very similar to the sample under investigation. The identification of species could even be performed successfully in ultra-high heat-treated cans (Laube *et al.*, 2007). It was found that DNA isolated from bread was around 300 bp long (Allmann *et al.*, 1992). Such highly degraded DNA still serves perfectly well for subsequent DNA-based amplification methods.

Due to their intrinsic characteristics, DNA-based methods can be standardized, making them particularly useful for routine and enforcement purposes. The result is dependent on the staff running the analysis to only a very limited extent. Above the limit of detection (LOD) a signal is either detected by the instrument or no signal is shown, and no interpretation by the analyst is necessary. Moreover, individual reagents for running the experiment are available from different suppliers, making the analysis independent from a particular provider, unlike in the case of immunological assays for which a specific antibody supplier is needed. Antibodies from different suppliers result in different assay performance.

Standardization of immunological methods usually has another disadvantage due to a batch-to-batch variation of the antibody used for the detection. Most reaction kits for food authenticity applications available on the market are based on the use of polyclonal antisera. If the antiserum is used, a new serum needs to be produced resulting in a different performance and ultimately most probably in a different scope of the whole procedure. This problem does not exist for DNA-based methods; individual components usually show the same performance from different suppliers.

For routine analysis it is extremely important that the work is organized in a way that allows as many samples as possible to be analysed at the same time, in order to save time and reduce costs. Thus high throughput is a very important factor when deciding which technique will be used for the intended analysis. DNA-based methods are well suited for high-throughput applications due to the possibility of automating the individual steps of the analysis, from extraction through to the final detection and interpretation of the results. Robotic techniques are available for the crucial pipetting steps of the DNA-based analysis, resulting in hundreds of tested samples per day. Even individual detection methods (individual primer pairs) can be combined in a single reaction, resulting in so-called multiplexing. Such multiplexing can be used either to screen for certain taxa or for the specific identification of individual species. Multiplexing can also result in the reduction of individual assays, consequently reducing work and costs.

The identification of suitable markers specific to individual species or taxa is the most common challenge faced in the application of DNA-based methods. The selected marker should distinguish between different species, but should be a feature of all races/breeds of the same species. Ideally authentic samples should be available for all breeds, which should be covered in order to demonstrate that the developed method is detecting all of them in an equal manner. However, practically this is impossible as not all breeds of a certain species are listed in a database, nor available as sample material. Thus it should be sufficient to get as many samples as possible; the author recommends analysis of approximately 20 individuals to demonstrate that the method is fit for its intended use.

In principle there are two different DNA targets available. Chromosomal DNA is present in small numbers of copies, but is usually diploid, meaning that two copies of the same DNA are present in the nucleus of the cell. In contrast, the haploid genome contains each DNA target only once, with the exception of repetitive DNA sequences present in tandem-repeats or clusters. Examples of chromosomally determined DNA targets are the intron and exon sequences of genes, ribosomal RNA genes (e.g., 18S gene sequence), ITS (internal transcribed spacer) of ribosomal coded genes and microsatellite sequences, which are usually present at both ends of the chromosomes. RNA genes in particular are present in multiple copies per haploid genome, which needs to be kept in mind when choosing such sequences as a target. Microsatellites, short sequence motifs of only two to four nucleotides, which are repeated up to hundreds of times, are more often used for the discrimination of races than for species identification.

The vast majority of species-specific identification methods target the extra-chromosomal, maternally inherited mitochondrial DNA present in various copies (usually from 10 to 500 copies). However, depending on the tissue used, the number can increase drastically. In contrast to the chromosomal genome, the exact number is usually unknown, meaning such sequences are not suitable as a target for quantitative purposes. A significant advantage of mitochondrial DNA is the fast evolution, making it simpler to discriminate between closely related species or even breeds/races (Maede, 2006; Meyer and Candrian, 1996). The mitochondrial coded *cytochrome b* (*cytb*) gene in particular is often used as a marker (Irwin *et al.*, 1991; Mafra *et al.*, 2008; Meyer *et al.*, 1994). Within the *cytb* gene, two conserved regions flank a variable DNA sequence, making it well suited for the simultaneous detection of several distinct species (Maede, 2006; Meyer and Candrian, 1996). Besides *cytb,* the 12S ribosomal RNA gene and the D-loop region of the mitochondrial genome are also commonly used as target DNA sequences for species differentiation.

5.3 DNA extraction from food samples

In order to run a PCR analysis, DNA from the sample under investigation needs to be isolated to make it appropriate for use in all subsequent analytical steps. Non-processed fresh food samples are perfectly suitable for DNA-based analyses. The DNA extracted out of such samples is of high-molecular weight and usually free of any inhibitory substances interfering with subsequent enzymatic reactions. Unfortunately, because most food products are processed, DNA is usually altered due to the processing. This is extensively reviewed by Gryson (2010).

During the production process, food might be subject to thermal treatments (cooking, pasteurization, etc.), high pressure, drying, pH modifications and a range of further crucial steps. Depending on the intensity of processing, the DNA is often degraded down to only a few hundred base pairs in length. In addition, the spectrum of foodstuffs to be analysed comprises many different food matrices which must be considered, including products which are high in fats, oils, vegetable material and animal tissues. Taking both aspects into consideration, it is obvious that not just one extraction method but several different types of extraction procedures can be applied, depending on, and optimized for, the food matrix and level of processing.

Totally different separation principles are described in the literature and used by laboratories. The most commonly used method is based on the use of the cationic detergent Cetyltrimethylammonium bromide (CTAB), which binds DNA and favours the extraction of DNA-CTAB complexes from proteins in aqueous solutions by using a surfactant (Corbisier *et al.*, 2007). However, whilst the procedure has been widely used for extraction of DNA from leaves, seeds, grains and processed food and feed stuffs, it is time-consuming and uses hazardous chemicals such as chloroform.

Alternatively, the anionic detergent sodium dodecyl sulphate (SDS) can be used to carry out the cellular lysis. Subsequently purification of the released DNA needs to be carried out, either with a combination of phenol/chloroform or with respective commercially available kits described below. Many commercially available kits also utilize detergents for the disruption of the cell walls, as an initial step for extracting DNA from plant material. Subsequently, to eliminate contaminating RNA and proteins, RNase A treatment and proteinase K digestion are usually carried out.

Some commercial kits use DNA binding to silica-based columns or magnetic beads, followed by the elution of the bound DNA with aqueous solutions in order to avoid exposure to organic solvents such as chloroform. Other methods selectively isolate DNA from cellular debris using high salt concentrations and alcohol precipitation, followed by further washing steps. Examples of commercially available DNA extraction kits are the Wizard® magnetic DNA purification system from Promega, the NucleoSpin® food kit from Macherey–Nagel or the QIAamp DNA tool kit from Qiagen. However many more kits are commercially available and it would be impossible to mention them all here.

While extraction with SDS or CTAB can be up-scaled to 1 g or even more if necessary, commercially available kits are restricted to 20–200 mg of sample material. Those 'simple' methods using SDS or CTAB are also much less expensive and applicable to a wide range of food matrices (Pirondini *et al.*, 2010). It has been calculated that the cost per sample is up to ten times higher using commercially available kits (Demeke and Jenkins, 2010). As compensation, commercially available kits used for the extraction procedure produce results in a very short time, saving labour-cost and increasing the throughput of many samples per day. Generally it could be stated that CTAB- and SDS-based extraction methods result in a higher yield of DNA, while the application of commercially available kits gives DNA of better quality for subsequent detection steps.

A general problem related to the food matrix and the respective extraction procedure is the co-purification of so-called inhibitors interfering with subsequently performed detection steps. In these detection steps, the inhibitors interact with the enzyme and result in its failure to carry out the necessary amplification process, which might subsequently lead to a negative interpretation in terms of the species under investigation; in the author's experience, the total amount of DNA extracted from the food sample is rarely the reason for the negative result. The active inhibitors including organic and phenolic compounds, polysaccharides (or products of the Maillard reaction), glycogen, fats, milk proteins, collagen, iron, cobalt or fulvic acids are much more likely to be the cause for the failure of amplification than a very low amount of DNA extracted from the food sample (reviewed in Scholz *et al.*, 1998 and Wilson, 1997). Further, even more widespread inhibitors include bacterial cells, non-target DNA and exogenous contaminants. In a few cases it is also possible to neutralize some of these inhibitors by the addition of bovine serum albumin (BSA) or spermidine to the PCR master mix. Otherwise a further purification step using phenol/chloroform or a commercially available kit is necessary to remove the inhibiting substances from the DNA solution in order to get correct (non-false negative) and reliable PCR results.

5.4 Developments in DNA-based methods

In general DNA-based methods can be divided in two categories: those which detect directly the targeted DNA and those which are amplifying the targeted DNA prior to or in parallel to the detection. All methods have in common that a quantitative determination is almost impossible as the exact relation between the number of target molecules and the amount of sample in weight-by-weight is not known. Therefore, the percentage on DNA level will not necessarily reflect the correct percentage of the targeted species of the whole food.

5.4.1 Non-amplification methods for detection – DNA hybridization

Early attempts at using DNA to detect species used quite simple methods, in which labelled DNA probes were hybridized to food samples of genomic DNA covalently linked to membranes made out of nylon; as a format slot- or dot-blot was chosen (Baur *et al.*, 1987; Ebbehøj and Thomson, 1991). It was demonstrated that under stringent conditions and for a number of species, probes comprising labelled total genomic DNA from a given species hybridize to DNA from the same species with low background. The development of such probes was simple and did not require any specific knowledge. Species-specific binding of the probes to targets resulted from hybridization of complementary repetitive sequences (Chikuni *et al.*, 1990). These repetitive sequences are arranged in tandem, found throughout the whole genome and are called 'satellite' DNA. By applying hybridization methods it was possible to unambiguously identify samples of chicken and pork DNA extracted from cooked meats and commercial products (Baur *et al.*, 1987; Chikuni *et al.*, 1990; Ebbehøj and Thomson, 1991), and pork in pork/beef admixtures (Winterø *et al.*, 1990).

The use of further species-specific satellite DNA probes has subsequently been described and identification of meat from cattle, deer, pig, chicken, turkey, rabbit, sheep and goat has been carried out successfully in raw products (Buntjer *et al.*, 1995, 1999; Hunt *et al.*, 1997). The presence of different species in mixtures and in a wide range of commercially processed, heated or canned products has also been shown (Buntjer *et al.*, 1999; Hunt *et al.*, 1997). Additionally, DNA hybridization has also been used to identify different cattle breeds (Matsunaga *et al.*, 1998). As species-specific probes are usually relatively short oligonucleotides smaller than 100 bases, hybridization is possible even after DNA degradation has been observed due to processing.

Although many protocols exist, methods based on simple hybridization have not been introduced in routine diagnostics as the time and the associated workload seems to be the limiting factor. Furthermore, methods based on amplification have been widely developed and thoroughly validated via collaborative studies, resulting in highly standardized methods. These are applicable in enforcement laboratories accredited according to ISO 17025.

5.4.2 Methods based on DNA amplification

Today the most common approach for species or taxon-specific identification relies on the use of the PCR; a process originally developed to amplify short segments of a longer DNA molecule (Mullis and Faloona, 1987; Saiki *et al.*, 1985, 1988).

A typical amplification reaction includes the target DNA isolated from the sample to be analysed, a thermostable DNA polymerase enzyme, two primers (short, synthetic oligonucleotide molecules used as initial starting points), deoxynucleotide triphosphates (dNTPs) as single bricks of the DNA chain, a reaction buffer and magnesium chloride. The primers in particular are responsible for the specificity of the individual method by annealing to the complementary strands of the DNA duplex. The primers are also responsible for the length of the amplified DNA fragments by defining the ends of the fragment. Once combined, the reaction is placed in a thermal cycler, an instrument that subjects the reaction to a series of different temperatures for varying amounts of time. These series of temperature and time adjustments is referred to as one cycle of amplification. Theoretically, each PCR cycle doubles the amount of targeted sequence (today the term 'amplicon' for the amplified target DNA is widely used; this term has been used since 1984 in conjunction with plasmids containing the origin of replication (*ori*) and the cleavage/packaging site of *herpes* virus genomes) in the reaction. Ten cycles theoretically multiply the amount of amplified products by a factor of about one thousand identical copies of the same DNA fragment; 20 cycles, by a factor of more than a million in a time period of less than two hours.

Unfortunately, this doubling of target DNA per cycle is not true in reality. A number of factors are responsible for the fact that the amplification efficiency is less than 2 per PCR cycle. In particular, DNA polymerase-inhibiting pyrophosphates are created close to the end of PCR cycles (>30) as a result of DNA synthesis, using triphosphates as single bricks. In addition, limiting primers and the loss of enzyme activities due to cycles of heating and cooling also reduce efficiency.

The pre-requisite for running PCR is a specifically equipped laboratory with trained staff, run under a quality assurance system in accordance with ISO 17025 or an equivalent accreditation scheme based on that standard. At the time of writing, all European Commission (EC) Member States are equipped with enforcement laboratories fulfilling these quality assurance standards and levels of accreditation (Zel *et al.*, 2006, 2008). The International Organization for Standardization (ISO) and the European Committee for Standardization (CEN) developed standards which describe the establishment of laboratories applying DNA-based methods. Several standards have also been established on specific methods for the purpose of identifying pathogens, plant species and GM foods, including specific methods and performance criteria (ISO 21569, 21570 and 2476). The determination of the measurement uncertainty (MU) associated with the entire procedure, which is described in a document prepared by the Joint Research Centre of the European Commission (JRC-EC) for the purpose of genetically modified food detection, can be also applied to the methods developed for species identification (European Commission, DG Health and Consumer Protection, 2004; Gustavo González and Ángeles Herrador, 2007; Macarthur *et al.*, 2010; Trapmann *et al.*, 2009).

5.4.3 Methods based on whole genomic approaches

DNA sequencing has become an affordable, popular tool for species identification due to new developments in DNA sequencing technologies and decreasing prices for service analysis. In general, DNA sequencing is used to determine individual DNA patterns base by base, before the generated DNA sequence is subsequently matched to publicly available databases in order to determine the species identified in the sample. In contrast to PCR, the DNA sequence is directly determined not merely amplified, resulting in much more comprehensive information about the sample. Using this technique, all species from which material was present in a food sample will be identified.

DNA sequencing has become a new gold standard in microbial species identification and resistance detection. The sequencing of well-defined hypervariable regions of genes such as 16S rRNA means sequence data, together with supporting biochemical data, is capable of providing unambiguous information for microbial identification.

High throughput sequencing, such as the Pyrosequencing® technology, is well suited to provide sequence data in about the same time it takes to run a PCR. Pyrosequencing® is based on the 'sequencing by synthesis' principle (Nyrén, 2007). It takes a single strand of the DNA to be sequenced and then synthesizes its complementary strand by enzymatic polymerization. Pyrosequencing® technology reads a discriminatory stretch of DNA of up to 96 samples in parallel in a very short period of time. It generates up to 400 megabase pairs in a ten hour run on an individual machine, costing around €4000–6500; a cost 1000 times less than in comparison to ten years ago. Consequently, traceability problems are within the area of application. Whilst instruments necessary to apply this technology are still quite expensive at present, the technology holds interesting potential for the near future.

5.5 Conclusion

As the global food market continues to grow and develop, it is essential that new methods for verifying food species and varieties continue to progress. At the moment PCR is the most suited technique for the detection of animal and plant species in foods. However, it is not possible to determine the exact amount of a species present, if no suitable reference material is available. While the PCR technique is the current method of choice for biological verification, DNA sequencing, where either parts or the whole genome sequence is determined, is an encouraging new development, making it possible to identify a particular species or breed in a given food sample with exceptional precision. Whilst instruments for carrying out this task are currently too expensive for routine application, the future development of this technology is sure to enhance the use of modern biological methods for the verification of food origins.

5.6 References

ALLMANN, M., CANDRIAN, U. and LÜTHY, J. (1992). Detection of wheat contamination in dietary non-wheat products by PCR. *Lancet*, **339**: 309.

ALLMANN, M., CANDRIAN, U. and LÜTHY, J. (1993). Polymerase chain reaction (PCR): a possible alternative to immunochemical methods assuring safety and quality of food – Detection of wheat contamination in dietary non-wheat products by PCR. *Zeitschrift für Lebensmittel-Untersuchung und Forschung*, **196**: 248–251.

BAUR, C., TEIFELGREDING, J. and LIEBHARDT, E. (1987). Identification of Heat Processed Meat by DNA Analysis. *Archiv für Lebensmittelhygiene*, **38**: 172–174.

BONWICK, G.A. and SMITH, C.J. (2004). Immunoassays: their history, development and current place in food science and technology. *International Journal of Food Science and Technology*, **39**: 817–827.

BUNTJER, J. B., LAMINE, A., HAAGSMA, N. and LENSTRA, J. A. (1999). Species identification by oligonucleotide hybridisation: the influence of processing of meat products. *Journal of the Science of Food and Agriculture*, **79**: 53–57.

BUNTJER, J. B., LENSTRA, J. A. and HAAGSMA, N. (1995). Rapid species identification by using satellite DNA probes. *Zeitschrift für Lebensmittel-Untersuchung und -Forschung*, **201**: 577–582.

CHIKUNI, K., OZUTSUMI, K., KOISHIKAWA, T. and KATO, S. (1990). Species Identification of Cooked Meats by DNA Hybridization Assay. *Meat Science*, **27**: 119–128.

CORBISIER, P., BROOTHAERTS, W., GIORIA, S., SCHIMMEL, H., BURNS, M., BAOUTINA, A., EMSLIE, K. R., FURUI, S., KUROSAWA, Y., HOLDEN, M. J., HYONG- HA, K., LEE, Y. - M., KAWAHARASAKI, M., DELLA SIN, X. and WANG, J. (2007). Toward metrological traceability for DNA fragment ratios in GM quantification. 1. Effect of DNA extraction methods on the quantitative determination of Bt176 corn by Real-Time PCR. *Journal of Agriculture and Food Chemistry*, **55**: 3249–3257.

DEMEKE, T. and JENKINS, G. R. (2010). Influence of DNA extraction methods, PCR inhibitors and quantification methods on real-time PCR assay of biotechnology-derived traits. *Analytical and Bioanalytical Chemistry*, **396**: 1977–1990.

EBBEH Ø J, K. F. and THOMSON, P. D. (1991). Species differentiation of heated meat products by DNA hybridisation. *Meat Science*, **30**: 221–234.

European Commission. DG Health and consumer Protection (2004). Report to the Standing Committee of the Food Chain and Animal Health on the relationship between analytical results, the measurement uncertainty, recovery factors and the provisions in the EU food and feed legislation.

GRYSON, N. (2010). Effect of food processing on plant DNA degradation and PCR-based GMO analysis: a review. *Analytical and Bioanalytical Chemistry*, **396**: 2003–2022.

GUSTAVO GONZÁLEZ, A. and ÁNGELES HERRADOR, M. (2007). A practical guide to analytical method validation, including measurement uncertainty and accuracy profiles. *Trends in Analytical Chemistry*, **26**: 227–238.

HUNT, D. J., PARKES, H. C. and LUMLEY, I. D. (1997). Identification of the species of origin of raw and cooked meat products Using Oligonucleotide Probes. *Food Chemistry*, **60**: 437–442.

IRWIN, D. M., KOCHER, T.D. and WILSON, A.C. (1991). Evolution of the cytochrome b gene of mammals. *Journal of Molecular Evolution*, **32**: 128–144.

ISO 21569, Foodstuffs – Methods of analysis for the detection of genetically modified organisms and derived products – Qualitative nucleic acid based methods.

ISO 21570, Foodstuffs – Methods of analysis for the detection of genetically modified organisms and derived products – Quantitative nucleic acid based methods.

ISO 24276, Foodstuffs – Nucleic acid based methods of analysis for the detection of genetically modified organisms and derived products – General requirements and definitions.

LAUBE, I., ZAGON, J. and BROLL, H. (2007). Quantitative determination of commercially relevant species in foods by real-time PCR. *International Journal of Food Science and Technology*, **42**: 336–341.

MACARTHUR, R., FEINBERG, M. and BERTHEAU, Y. (2010). Construction of measurement uncertainty profiles for quantitative analysis of genetically modified organisms based on interlaboratory validation studies. *Journal of the Association of Official Analytical Chemists International*, **93**: 1–11.

MAEDE, D. (2006) A strategy for molecular species detection in meat and meat products by PCR-RFLP and DNA sequencing using mitochondrial and chromosomal genetic sequences. *European Food Research and Technology*, **224**: 209–217.

MAFRA, I. FERREIRA, I. M. P. L. V. O., BEATRIZ, M. and OLIVEIRA, P. P. (2008). Food authentication by PCR-based methods. *European Food Research and Technology*, **227**: 649–665.

MATSUNAGA, T., SATO, T. and TSUJI, K. (1998). Differentiation of Japanese Black Cattle and Other Cattle Breeds by DNA Hybridisation. *Journal of Japanese Society of Food Science and Technology*, **45**: 155–157.

MEYER, R. and CANDRIAN, U. (1996). PCR-based DNA analysis for the identification and characterization of food components. *Lebensmittel-Wissenschaften und Technologie*, **29**: 1–9.

MEYER, R., CANDRIAN, U. and LÜTHY, J. (1994). Detection of pork in heated meat products by the polymerase chain reaction. *Journal of the Association of Official Analytical Chemists International*, **77**: 617–622.

MULLIS, K. B. and FALOONA, F. A. (1987). Specific synthesis of DNA in vitro via polymerase-catalyzed chain reaction. *Methods in Enzymology*, **155**: 335–350.

NYRÉN, P. (2007). 'The History of Pyrosequencing'. *Methods in Molecular Biology* **373**: 1–14.

PÄÄBO, S., HIGUCHI, R.G. and WILSON, A.C. (1989a). Ancient DNA and the polymerase chain reaction. *Journal of Biological Chemistry*, **264**: 9709–9712.

PÄÄBO, S. (1989b). Ancient DNA: extraction, characterization, molecular cloning and enzymatic amplification. *Proceedings of the National Academy of the Science of the USA*, **86**: 1939–1943.

PIRONDINI, A., BONAS, U., MAESTRI, E., VISIOLI, G., MARMIROLI, M. and MARMIROLI, N. (2010). Yield and amplificability of different DNA extraction procedures for traceability in the dairy food chain. *Food Control*, **21**: 663–668.

POMS, R.E., KLEIN, C.L. and ANKLAM, E. (2004). Methods for allergen analysis in food: a review. *Food additives and Contaminants*, **21**(1): 1–31.

REHBEIN, H. (2009). DNA-based methods. In: Fishery products: quality, safety and authenticity (ed.: Rehbein, H. and Oehlenschläger, J.), Wiley-Blackwell, Chichester, 363–387.

SAIKI, R. K., GELFAND, D. H., STOFFEL, S., SCHARF, S. J., HIGUCHI, R., HORN, G. T., MULLIS, K. B. and ERLICH, H. A. (1988). Primer-directed enzymatic amplification of DNA with a thermostable DNA polymerase. *Science*, **239**: 487–491.

SAIKI, R. K., SCHARF, S., FALOONA, F., MULLIS, K. B., HORN, G. T., ERLICH, H. A. and ARNHEIM, N. (1985). Enzymatic amplification of beta-globin genomic sequences and restriction site analysis for diagnosis of sickle cell anemia. *Science*, **230**: 1350–1354.

SCHOLZ, M., GIDDINGS I. and PUSCH C. M. (1998). A polymerase chain reaction inhibitor of ancient hard and soft tissue DNA extract is determined as human collagen type I. *Analytical Biochemistry*, **259**: 283–286.

TRAPMANN, S., BURNS, M., BROLL, H., MACARTHUR, R., WOOD, R. and ZEL, J. (2009). Guidance Document on Measurement Uncertainty for GMO Testing Laboratories. European Commission Joint Research Centre Institute for Reference Materials and Measurements. ISBN 978–92–79–11228–7.

WILSON, I.G. (1997). Inhibition and facilitation of nucleic acid amplification. *Applications in Environmental Microbiology*, **63**: 3741–3751.

WINTERØ, A.K., THOMSEN, P.D. and DAVIS, W. (1990). A comparison of DNA-hybridization, immunodiffusion, countercurrent immunoelectrophoresis and isoelectric focusing for detecting the admixture of pork to beef. *Meat Science*, **27**: 75–85.

ZEL, J., CANKAR, K., RAVNIKAR, M., CAMLOH, M. and GRUDEN, K. (2006). Accreditation of GMO detection laboratories: Improving the reliability of GMO detection. *Accreditation and Quality Assurance*, **10**: 531–536.

ZEL, J., MAZZARA, M., SAVINI, C., CORDEIL, S., CAMLOH, M., STEBIH, D., CANKAR, K., GRUDEN, K., MORISSET, D. and Van den EEDE, G. (2008). Method validation and quality management in the flexible scope of accreditation: an example of laboratories testing for genetically modified organisms. *Food Analytical Methods*, **1**: 61–72.

6

Vibrational spectroscopy in studies of food origin

G. Downey, Teagasc, Ireland

DOI: 10.1533/9780857097590.2.94

Abstract: Vibrational spectroscopy (mid-infrared, near infrared and Raman) has been applied to the study of authentication problems in foods. This chapter outlines the principles of each, the main methods of their deployment and the philosophy of their application in authentication problems. A review of recently published material describing applications for each technique is included.

Key words: vibrational spectroscopy, near infrared, mid-infrared, Raman, food authenticity, chemometrics, applications.

6.1 Introduction

Vibrational spectroscopy is the term used to describe studies of the interaction between electromagnetic radiation and the vibrational states of atomic nuclei within their respective molecules. Under normal conditions, these nuclei are in constant motion, bending and vibrating (Workman Jr. and Weyer, 2007). These vibrational states may be altered in precise ways by the absorption of radiation; given that only certain vibrational alterations are permitted by quantum theory, the frequencies or wavelengths at which any such absorptions occur are characteristic of various bond types although they are also dependent on the specific molecular environment in which a molecule is located. Therefore, measurement of the absorption of radiation over the appropriate wavelength (or frequency) range can potentially provide very specific information about the types and indeed numbers of molecules present in any given sample of material.

Difficulties arise from a number of sources. Firstly, in a biological sample such as food, the number and types of compounds containing, for example,

C=O bonds is large and variable. Therefore, spectra are usually very complex, containing information about most if not all of the molecular absorbers present in a given sample. When dealing with food samples, for example, it is often not possible to definitively relate absorbance at any specific frequency or wavelength to an individual molecular species; rather, the approach has been to accept that some or all of the recorded spectra include information about the complete molecular composition of the material under test and may be considered to be a fingerprint of it. Secondly, there are effects arising from the fact that the molecular absorbers are not harmonic (ideal) so that in addition to the fundamental (strongest) absorption, there will also be other absorptions arising from combinations and overtones, Fermi resonances and combinations. Overtone and combination band absorbances are particularly important in near infrared spectra and are found at 1690–1755 nm (first overtone), 1127–1170 nm (second overtone) and 845–878 nm (third overtone) (Workman Jr. and Weyer, 2007). The phenomena involved in Raman spectroscopy are somewhat different. In this case, a small fraction of incident light (usually from a laser) impinging on a sample is scattered and this scattered radiation contains information about the vibrational states of the molecules contained in the sample. These factors complicate the extraction of useful information from vibrational spectra of biological material and progress in their analysis has only really been possible following the development and application of suitable multivariate mathematical tools designed to isolate relevant information (chemometrics).

Among analytical techniques, vibrational spectroscopies possess a number of major advantages; chief among these are speed of analysis (generally of the order of seconds); low running costs (although the purchase cost of research-grade instruments is not insignificant); the fact that the sample is generally not destroyed; and that little or no sample preparation is needed before analysis. In addition, once a method has been developed, it is often the case, especially for near infrared (NIR), that it can be moved close to the production line and operated by production staff without any analytical training. Recent progress in the development of hand-held, portable equipment for mid-IR, NIR and Raman techniques will inevitably lead to deployment of these techniques at optimum locations either during the production process, retail distribution or at point-of-sale. One disadvantage associated with their use has been the requirement for chemometric expertise in the development and initial implementation phases of any given method; another is the fact that, since these are generally secondary methods, the collection and analysis of appropriate sample training sets is time-consuming and therefore requires significant expenditure of time and money for a lengthy period before any model is developed. Additionally, especially in the case of natural products, models must be continually monitored for performance and, should any degradation be detected, remedial action involving model adjustment or re-building is required. Despite these negative features, vibrational spectroscopy methods continue to be developed for food and other applications because of the significant advantages that a successful model offers over other candidate analytical methods.

Food authentication is an exceedingly complex analytical task since the issues which it addresses (geographic origin, identity of animal or plant species, production process, etc.) usually require information on a wide range of food constituents (often physical as well as chemical) to confirm a labelling claim. Underlying the application of vibrational spectroscopy in such tasks is the hypothesis that the spectroscopic fingerprint of a food sample contains information which is characteristic of the food and which, therefore, may be compared to other fingerprints of authentic material and thereby used to confirm or refute qualitative claims such as those relevant to authentication (Downey, 1996). This chapter will provide an overview of the application of vibrational spectroscopic techniques in the field of food authentication. It will deal initially with the specifics of each technique, follow up with a general discussion of the principles behind their use in food authentication and finish with some recent examples of reported applications.

6.2 Types of vibrational spectroscopy

Each of the vibrational spectroscopic techniques (near infrared, mid-infrared and Raman) will be dealt with separately in the following sections. Items covered will include principles, practice and recent applications.

6.2.1 Near infrared (NIR) spectroscopy

Near infrared (NIR) spectroscopy involves radiation in the wavelength range 780–2500 nm (12 820 to 4000 cm^{-1}) and, although discovered by Herschel in the early 1800s, it was generally unused until the work of Norris and co-workers in the 1950s revealed its potential in the food and agricultural sectors (Day and Fearn, 1982). While the initial and continuing major focus of the technique lies with the quantitative determination of key components of food and food ingredients (Benson, 2003; Wehling, 2003; Cen and He, 2007; McClure, 2007), its application to qualitative issues has been growing (Reid *et al.*, 2006; Manley *et al.*, 2008; Woodcock *et al.*, 2008a).

Near infrared spectra may be generally described as lacking in fine detail and comprising broad, overlapped peaks which arise chiefly from overtones and combination bands of absorptions which have their fundamental vibrations in the mid-infrared region. Strongest absorbing bonds in the near infrared are those containing atoms of carbon (C), hydrogen (H), nitrogen (N) and oxygen (O), the most common molecules present in food (Norris, 1989). Paradoxically, therefore, despite their appearance, NIR spectra are rich in information about the detailed molecular composition of a sample (Workman Jr. and Shenk, 2004). Additionally, however, recorded spectra often contain information about factors which may not be related directly to composition or functional properties. For example, spectra of powdered samples are often measured by diffuse reflectance and contain strong interferences from scattered light which are affected by mean particle size, particle size distribution, packing density, etc. This scattering interference is best removed before quantitative or qualitative modelling is attempted and a range of pre-processing mathematical treatments (such as derivatisation, multiplicative scatter correction and standard normal variate transformation among others) have been developed for this purpose (Duckworth, 2004). Following such pre-processing, a significant challenge has been the extraction of meaningful data from spectral collections and its use to generate robust and accurate predictive models. Developments in the field of chemometrics have made this possible but this is a specialist subject in its own right and will not be dealt with here; the interested reader is referred elsewhere in this book for further information while several relatively recent general publications are also recommended (Naes *et al.*, 2002; Berrueta *et al.*, 2007; Kramer *et al.*, 2007; Ozaki *et al.*, 2007; Westerhaus *et al.*, 2007).

A wide range of instruments and sampling options are now available for the collection of NIR spectra, enabling almost any sample to be scanned without special processing. Instruments exist in a number of configurations and include discrete filter, scanning monochromator, photodiode array, Fourier-transform interferometer (FT-NIR) and acousto-optical tunable filter (AOTF) systems. Variety exists in the type of illumination sources used (halogen lamp, Nernst filament, light emitting diodes), sample presentation options (reflectance, transmittance, transflectance, fibre optics) and detector technologies (PbS, InGaAs or charge-coupled devices) (Manley *et al.*, 2008). A summary diagram illustrating these combinations and variations is shown in Fig. 6.1. For a full discussion of the options available and their relative advantages and disadvantages, the interested reader is referred to other publications (McClure, 2001; Workman, 2004; McClure and Tsuchikawa, 2007; Subramanian and Rodriguez-Saona, 2009). More recently, the drive for smaller instrumentation has resulted in the development of several portable or hand-held devices which aim to bring the benefits of NIR technology out of the laboratory and into the wider world. The potential of such instruments for food authentication remains largely unconfirmed at the present and merits investigation.

6.2.2 Mid-infrared (MIR) spectroscopy

The mid-infrared (MIR) region extends from 2500 to 25 000 nm (4000 to 400 cm^{-1}) and absorption of radiation in this region is associated with fundamental molecular stretching and bending vibrations. Spectral acquisition from the original dispersive instruments was limited by their inability to scan at speed due to the use of a monochromator; nowadays, most MIR instruments use Fourier-transform (FT) spectroscopy which does not suffer from this scan speed limitation. Because fundamental absorptions are

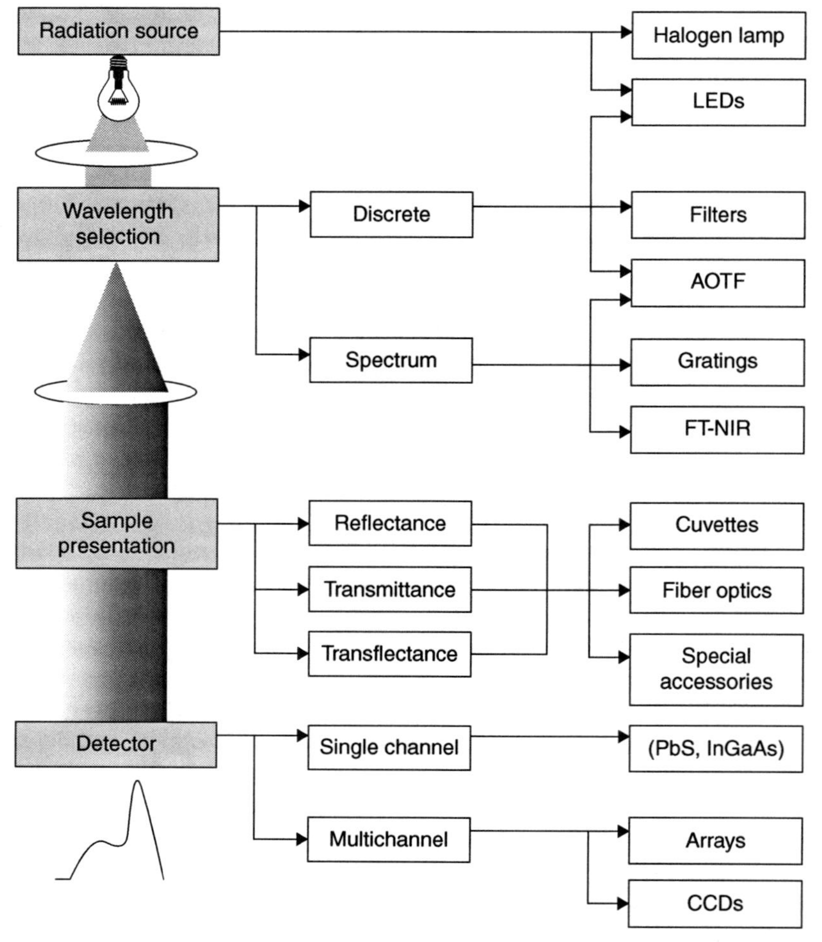

Fig. 6.1 Main features of NIR spectroscopic instrumentation. (Reproduced with permission from Blanco and Villarroya (2002); © Elsevier Ltd. 2002).

being measured and signal-to-noise ratios of modern Fourier-transform MIR instruments are very high, the detection of low concentrations of analytes is possible by this technique (Stuart, 2004; Karoui *et al.*, 2008). Advantages of MIR spectroscopy include the fact that it deals with fundamental vibrations (which are the strongest) and that there are characteristic and well-defined bands for many organic functional groups; disadvantages include the fact that transmitting materials are expensive, only short effective pathlengths may be used because most materials absorb in the MIR region, and water is a strong absorber in certain regions in the MIR spectrum which may mask contributions from other species. Sample presentation can take a wide variety of forms but the most common have been transmission, coated or smeared films, hot-pressed films or alkali halide

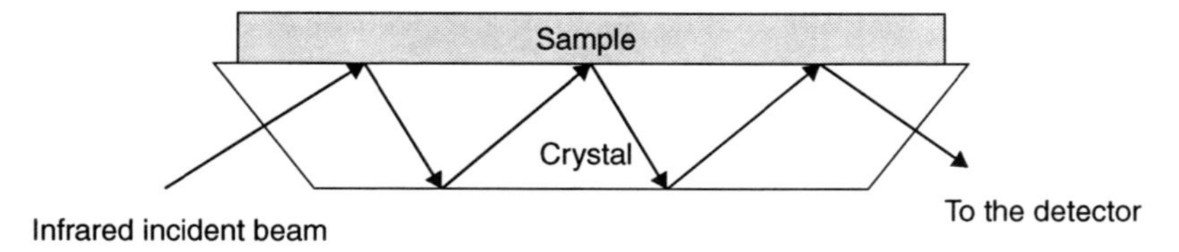

Fig. 6.2 Schematic drawing of attenuated total reflectance (ATR) sample presentation. (Reproduced by permission of ACS.)

pellets such as KBr. However, all of these methods involve only small sample sizes (of the order of milligrams) and this caused problems with biological materials given the normal levels of inhomogeneity associated with them. A major advance in sample presentation which has significantly aided analysis of such samples has been attenuated total reflectance (ATR) (Fig. 6.2). In this approach, the sample is placed in close contact with a horizontally aligned crystal of high refractive index such as ZnSe, Ge, ZnS, Si or diamond. Radiation entering one end of the crystal from the beam-splitter undergoes multiple total internal reflections at the crystal top and bottom surfaces and emerges at the other end. At each point of reflection at the top of the crystal, an evanescent standing wave is produced which may penetrate an applied sample, interact with it and produce a spectrum. This approach has enabled the use of larger samples and is especially suitable for solids, liquids, pastes and slurries although powdered material may be studied, especially if a uniform pressure can be applied to aid compaction and sample contact with the crystal surface. Most of the available crystals do have pH limitations however and may be impaired by surface scratching unless care is exercised. For these reasons, diamond is emerging as the material of choice. For a full discussion of the theory and practice of FT-IR spectroscopy, the reader is referred to other texts (Nishikida *et al.*, 1995; Stuart, 2004; Karoui *et al.*, 2008; Subramanian and Rodriguez-Saona, 2009).

FT-IR spectra of biological materials such as foods are complex but do not suffer from some of the complicating phenomena noted above for NIR spectra, for example, light scattering effects. Within the FT-IR spectral range, four regions may be generally considered; the X-H stretch region (4000–2500 cm^{-1}), triple bond region (2500–2000 cm^{-1}), the double bond region (2000–1500 cm^{-1}) and the fingerprint region (1500–600 cm^{-1}). Absorbance bands in the first three regions are generally stable and can be identified rather easily. In the fingerprint region, however, this is not the case and most of the bending and skeletal vibrations occurring in this region are subject to significant wavenumber shifts arising from relatively small electronic or steric effects. Generally, therefore, no attempt is made to identify specific peaks in this region but rather the absorbance of the complete region is used as a fingerprint of the material under study (Stuart, 2004).

Analysis of both quantitative and qualitative aspects of biological materials using this fingerprint region therefore relies on the application of the same chemometric methods as those mentioned above for NIR spectra (Karoui *et al.*, 2010).

6.2.3 Raman spectroscopy

The phenomena involved in Raman spectroscopy are somewhat different from mid- or near infrared: while IR absorbances are strongest for vibrations involving a large change in dipole moment, Raman bands are strongest for large alterations in polarisability. Hence, the information from IR and Raman spectroscopy tends to be complementary (Leugers and Lipp, 2000). In Raman measurements, a small fraction of incident light (usually from a high-energy laser source) impinging on a sample exchanges energy with the sample and is scattered; this Raman-scattered radiation (accounting for only 10^{-6} to 10^{-8} of incident light intensity) typically contains information about the vibrational states of the molecules present in the sample. This information is represented by shifts in energy of the scattered radiation from the incident radiation. Intensity of the scattered light is related to the number of molecules with which the incident light interacts, the incident light intensity and the fourth power of the difference between the frequencies of the laser and the associated molecular vibration (Kizil and Irudayaraj, 2008). Raman spectrometers may use a visible-range laser, UV-Raman instruments incorporate a UV laser light source while FT-Raman equipment utilises a near infrared laser (Nd:YAG, 1064 nm); this latter type is most commonly used because, among other benefits, it generally induces only low levels of fluorescence in samples, although the lower Raman signal intensities produced is a disadvantage (Chase, 1987). Until recently, Raman spectroscopy was a very difficult technique to use even within a laboratory, especially in the case of biological samples such as food or food ingredients. Significant improvements in instrument design, detector sensitivity, laser construction and the availability of fibre-optic sampling probes has led to an expansion of applications in this and many other fields and even into process environments. Advantages of Raman over MIR and NIR spectroscopy include its content of high-level spectral information (which can be collected easily by fibre-optic probes), fast scan speed and ruggedness due to the absence of any moving part other than the shutter in front of the detector. A significant disadvantage relates to sample fluorescence: when significant, this may be attenuated by changing to a longer wavelength laser but at the expense of sensitivity. Possible thermal degradation of biological samples is always an issue but detection limits as low as 100 ppm for specific analytes have been reported. For a fuller description of the theory and practice of Raman spectroscopy, the reader is referred to other texts (Leugers and Lipp, 2000; Lewis and Edwards, 2001; Ferraro, 2002).

As a result of the various advantages outlined above, vibrational spectroscopic methods (mid-infrared, near infrared and Raman) are now commonly applied for food analysis and it is expected that this usage will continue to grow.

6.3 Use of infrared (IR) spectroscopy in food authentication

Given that an authentic food is defined as being exactly what it claims to be, it follows that the philosophy underpinning the use of spectroscopy or any other analytical procedure for confirming the authenticity of a food or food ingredient is to establish that the material under test conforms to an appropriate specification. It is rarely the case that a quantitative specification exists which can be used for this purpose so the specification becomes a collection of reference samples which are known to be authentic examples of the food or ingredient. Therefore, the process of demonstrating consistency with such a collection is generally based on one or more mathematical operations designed to demonstrate conformity of an unknown sample to it. Any such conformity may only be demonstrated within defined statistical limits and absolute certainty is not possible; for this reason vibrational spectroscopy techniques serve best as screening methods in which the trade-off between speed, cost and accuracy are appropriately balanced. What is clear is that their proper use does not involve answering a question such as 'Where does this sample originate?' but rather 'Are the spectral properties of this sample consistent with authentic material which makes the same origin claims?'.

There are a number of mathematical ways in which answers to this question may be achieved. At the outset, these divide into approaches which (a) focus on the establishment of models for a class of authentic food or foods (class-modelling), or (b) measure differences between authentic and non-authentic foods (classification methods). Both have their particular advantages and disadvantages but classification methods have tended to be more popular, perhaps on account of their inclusion in most of the commercially available software packages. However, it is arguable that the class-modelling approach is to be preferred (Forina *et al.*, 2008, 2009; Oliveri *et al.*, 2011) although both may in fact be effectively applied in certain applications.

Class-modelling methods focus exclusively on describing known authentic material or materials; thus the emphasis is on the samples in which the analyst is primarily interested, for example, honey from Corsica or Spain or Turkey. Such models are then applied to an unknown sample spectrum and three possible outcomes are possible. In the first, the unknown will be identified as being consistent with one particular model; in the second, it may be identified as belonging to none of the models while in the third outcome, it may be identified as belonging to more than one model (Fig. 6.3).

Three added strengths of such an approach are (1) that probabilities may be assigned to the prediction outcomes, (2) they can deal successfully with

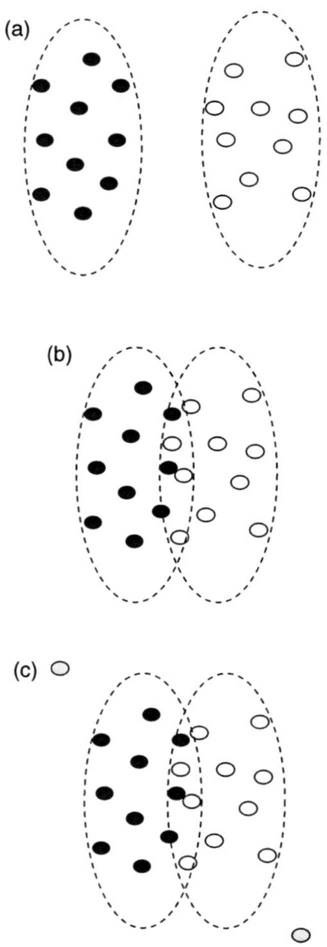

Fig. 6.3 Class-models showing (a) completely separated classes, (b) partially overlapped classes and (c) partially overlapped classes with two samples (grey filled circles) which do not fit inside either model.

complex sample distributions in hyperspace and (3) in addition to correct classification outcomes, values for sensitivity and specificity of each model are calculated. These refer respectively to the percentage of samples known to belong to a model which are correctly identified by the model and the percentage of samples known not to belong to a model which are correctly rejected by it. Recently described class-modelling approaches include potential functions

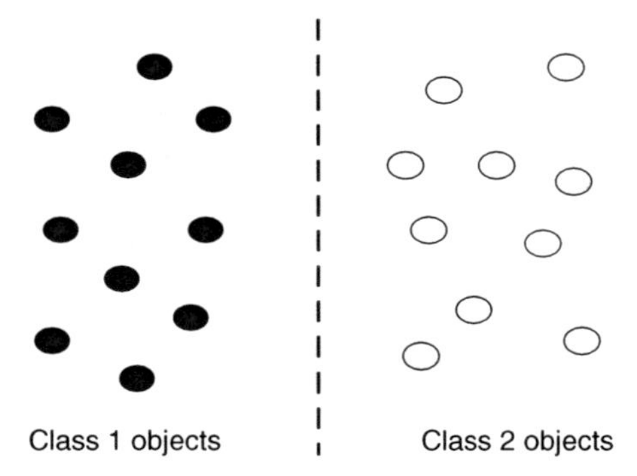

Fig. 6.4 Representation of a two-class discriminant model with the model boundary shown (broken line).

techniques (POTFUN) (Forina *et al.*, 1991), methods using unequal class spaces (UNEQ) (Derde and Massart, 1986) while a more established technique is SIMCA (soft independent modelling of class analogy) (Wold, 1976).

Classification approaches focus on the differences between groups of samples and may therefore depend significantly on those samples at the periphery of the groups involved (Fig. 6.4). While linear discriminant analysis (LDA), factorial discriminant analysis (FDA), k-nearest neighbours (KNN), support vector machines (SVMs) and artificial neural networks (ANNs) have been deployed as classification methods, perhaps the most popular approach has involved discriminant partial least-squares regression (D-PLS) (Berrueta *et al.*, 2007; Woodcock *et al.*, 2008a; Karoui *et al.*, 2010). Results of classification methods are normally restricted to a correct classification value and unknown samples are usually classed as belonging to at least one of the sample classes under study given the absence of any statistically based distance cut-off for defining group membership. Classification methods do however have a role to play in problems involving a limited number of well-defined sample classes but vigilance is required for unusual or atypical samples.

Adulteration is a problem which often arises with premium quality foods. Vibrational spectroscopy has been applied both to the detection and quantification of adulteration; this has normally involved a sequential approach in which the adulteration is detected in a first step after which a regression model is deployed to quantify the adulterant concentration (Kasemsumran *et al.*, 2005; Kelly and Downey, 2005; Kelly *et al.*, 2006a,b; Pizarro *et al.*, 2007). The general limitation of this approach however is that it assumes knowledge of the likely types of adulteration which a food may undergo (which is not always obvious) and requires the development of models for all of these. It may of course deal successfully with the most common commercial adulterations.

6.4 Applications

A number of comprehensive reviews of the application of MIR, NIR and Raman spectroscopy to food authentication problems have appeared in the recent past (Baeten *et al.*, 2008; Karoui *et al.*, 2008; Kizil and Irudayaraj, 2008; Manley *et al.*, 2008); the following section of this chapter will therefore focus on selected applications published since then (the period 2008 to 2011).

6.4.1 NIR applications

Authentication of olive oil has continued to be a significant issue addressed by a number of research groups. Casale *et al.* (2008) reported a study of 195 extra virgin olive oils (EVOOs) from Italy of which 126 were from Liguria and the remainder (69) commercial samples. A number of mathematical pre-treatments were applied to the spectra before class-modelling procedures, that is, SIMCA (soft independent modelling of class analogy), UNEQ-QDA (unequal class modelling-quadratic discriminant analysis) and MRM (multivariate range modelling) were investigated for their ability to identify Ligurian EVOO samples. According to the authors, best results were obtained using MRM on a subset of eight spectral variables; Ligurian oils were clearly differentiated from the others but there was some mis-identification of non-Ligurian oils as Ligurian. Weng *et al.* (2009) reported the use of backward propagation artificial neural networks (BP-ANNs) for discriminating between authentic and adulterated virgin olive oil and PLS for quantifying adulteration by sesame oil, soybean oil and sunflower oil in mixtures containing 5–50% of virgin olive oil by weight. Models were developed and evaluated using 100 samples as a calibration set and 52 as a separate test set. These authors stated that all of the test samples were correctly identified; in predicting the adulterant concentration, root mean square errors of cross-validation (RMSECV) equal to 1.3, 1.1 and 1.04% were achieved for sesame oil, soybean oil and sunflower oil, respectively. In a publication chiefly devoted to a novel experimental technique, Mignani *et al.* (2011) described the use of an integrating sphere and fibre optics for scatter-free absorption measurements (400–1700 nm) on a series of authentic EVOOs from the Tuscany region of Italy and the same samples adulterated by different concentrations of olive-pomace oil, refined olive oil, deodorised olive oil and refined olive-pomace oil. Due to the optical simplicity of this arrangement, data analysis by principal component analysis (PCA) and LDA was capable of predicting the fraction of adulterant in the mixture and of discriminating its type. With regard to the former, the authors claim that the results obtained by means of optical spectroscopy, without any sample treatment, are comparable with those obtained using the standard method (gas chromatography). A more complex sample collection comprising olive oil and olive oil adulterated by soybean, canola, cotton, corn and sunflower oils in binary, ternary and quaternary mixtures was described by Öztürk *et al.* (2010). These workers used a genetic inverse least-squares (GILS) procedure to develop quantitative models for each of the oils in these complex mixtures and claim to have produced models with useful accuracy although, given the fact that standard error of prediction (SEP) values were all in excess of 2% (w/w) with the exception of soybean oil, the reliable analysis of adulterants at or below 5% (w/w) would not be possible. Finally, Zhuang *et al.* (2010) reported a small study comprising only 20 samples which they claimed to effectively identify as extra virgin or ordinary olive oils; these authors also developed PLS models to predict adulteration of these oils by colza, corn, peanut, camellia, sunflower and poppy seed oils but since only an abstract of the paper is available in English, it is difficult to interpret or evaluate their conclusions properly.

Fruit has been the subject of several recent studies involving NIR spectroscopy (Paz *et al.*, 2009a,b; Kurz *et al.*, 2010; Luo *et al.*, 2011; Pérez-Marín *et al.*, 2011). Kurz *et al.* (2010) evaluated the potential of NIR for authenticity control and determination of fruit content by measuring cell wall components in a study of pure apricot purées and the same products adulterated by peaches and pumpkins; these latter were selected as likely industrial adulterants on the basis that peaches are cheaper than apricots and pumpkins usually contain relatively large amounts of carotenoids and may improve the appearance of apricot products and feign higher fruit contents. Classification studies were performed by predicting the hemicellulose (HC) and alcohol-insoluble residue (AIR) contents of samples and comparing these to pre-existing concentrations of each in pure fruits and also by applying SIMCA to spectral data (650–2500 nm) from the HC and AIR fractions extracted from pure fruits. While it was claimed that the method worked satisfactorily, the fact that HC and AIR must be prepared from test samples does reduce the analytical speed achievable. Using a total of 334 apples (*Malus domestica* Borkh.), cvs Fuji and Golden Delicious, Paz *et al.* (2009a) reported the prediction of fruit storage duration (0, 8 and 14 days) by PLS modelling; correct classification rates of 86.1% (mixed-cultivar samples) and 86.6% (single-cultivar groups) were reported in a comparison of three different instruments. Xie *et al.* (2009) described the use of NIR spectroscopy to the classification of tomatoes with different genotypes while Luo *et al.* (2011) investigated the use of this technique for apple classification. In this study, KNN, D-PLS and moving window partial least-squares discriminant analysis (MWPLSDA) were used to classify apple samples of different geographical origins, grades and varieties. FT-NIR spectra (4000–10 000 cm^{-1}) of 500 apples comprising four varieties and a number of geographical origins were collected at five equally spaced points on the circumference of the whole fruit using a fibre-optic accessory. Results indicated that MWPLSDA produced the most accurate classification models with correct classification rates between 96% and 98.6%.

Honey authentication was studied by Chen *et al.* (2008a), Woodcock *et al.* (2009) and Chen *et al.* (2011). In the former work, 71 samples (27 authentic and 44 adulterated) were scanned (4000–1100 cm^{-1}) using an FT-NIR instrument and a fibre-optic attachment. Five different calibration models were developed (details unavailable in English) and correct classification rates (D-PLS) for validation samples of between 89.3% and 93.8% were reported. No details of the types or degree of adulteration were included in the English abstract. In a more thorough study, Woodcock *et al.* (2009) set out to investigate the use of NIR spectroscopy for confirming claimed geographic origin in honeys from the island of Corsica. Authentic honeys ($n = 373$; 219 from Corsica and 154 non-Corsican) from two harvest years were scanned (1100–2498 nm) in transflectance mode (0.1 mm sample thickness) and D-PLS modelling performed on separate calibration and validation sample sets and after a variable reduction procedure (Martens' Uncertainty Test). Best results were obtained using both the latter approaches, producing correct classification rates of 90.4 and 86.3 for Corsican and non-Corsican honeys, respectively. These authors concluded that this approach had significant potential for confirming claims of geographic origin of honey. Chen *et al.* (2011) studied methods for detecting honey adulteration by high-fructose corn syrup (HFCS) using 70 unadulterated samples and 74 purchased samples which were shown analytically to contain HFCS (7–59% w/w) by isotope ratio mass spectrometry (IRMS). The best reported D-PLS model used 8 PLS loadings and produced correct classification rates of 97.9% and 95.8% for authentic and adulterated samples, respectively, in the calibration sample set ($n = 96$); 100% correct classification rates were reported for the validation sample set ($n = 48$).

Other papers published between 2008 and 2011 included the discrimination between meat from pasture- and concentrate-fed lamb (Dian *et al.*, 2008), ground meat from adult steers and young cattle (Prieto *et al.*, 2008) and detection of adulteration in pork (Fan *et al.*, 2010). Miscellaneous applications included those dealing with discrimination issues in tea (Li and He, 2008; Liu *et al.*, 2009), wine and spirits (Ferrari *et al.*, 2011; Kolomiets *et al.*, 2010; Liu *et al.*, 2008a), vinegar (Liu *et al.*, 2008b), rice (Liang *et al.*, 2009), essential oil (Bombarda *et al.*, 2008) and various herbal medicines (Chen *et al.*, 2008b; Lai *et al.*, 2011).

6.4.2 MIR applications

A number of FT-IR studies on confirming the provenance of honey have been reported. Etzold and Lichtenberg (2008) devised a study to develop calibrations for authentication of the main regional honey types of Germany and Switzerland. In this large study (1075 honey samples from the eastern region of Germany collected from 1999 to 2005 together with 131 honeydew honey samples from Switzerland and the southern part of Germany), the authors collected FT-IR spectra using a MilkoScan FT 120 (Foss Electric, Hillerod, Denmark) and used these spectra plus separate sensory and chemical data to generate predictive models for eight unifloral honeys and their blends plus a multifloral honey class. By way of conclusion, the authors reported that the most common local unifloral honey types could be correctly classified but that the decision between a unifloral honey and its blended counterpart was problematic. Another group (Hennessy *et al.*, 2008) analysed filtered and unfiltered honey samples from Mexico, Argentina, Ireland, Hungary and the

Czech Republic over the 6800–11500 nm (1470–870 cm^{-1}) range. SIMCA, FDA and D-PLS were studied to develop classification and class-models; overall correct classification rates of 93.3% (D-PLS), 94.7% (FDA) and up to 100% (SIMCA) were obtained, although models describing some classes exhibited very high false positive percentages. Samples from Hungary had the highest rate of correct identification by SIMCA while Czech and Mexican samples had the lowest; FDA and D-PLS gave similar results. These authors suggested that FT-IR data could be used to develop a screening procedure for honeys from known origins. Gallardo-Velázquez *et al.* (2009) explored FT-IR ATR spectroscopy to detect and quantify three sugar adulterants (corn syrup, high fructose corn syrup and invert sugar) in honeys from four regions of Mexico. For pure honeys, SIMCA correctly identified the geographic region of all samples in a separate validation set using spectra in the wavenumber range 1500–800 cm^{-1} although sample numbers were small ($n = 5$ of each type in validation); for the adulterated honeys, commercially useful standard errors of prediction were obtained for corn syrup (1.50–2.17%), HFCS (2.19–2.93%) and for invert sugar (1.38–2.59%). These authors did not, however, address the problem of detecting adulteration in the Mexican honey samples. In a parallel study to one reported above using NIR spectroscopy, Hennessy *et al.* (2010) described the use of ATR sample presentation following standardisation of honey samples ($n = 219$) to 70° Brix to confirm claimed Corsican origin for honeys from Corsica and several other countries (Ireland, Germany, Austria, Italy and mainland France). Both FDA and D-PLS chemometric methods were explored and the best models (2500–12 500 nm; 1470–870 cm^{-1}) produced correct classification rates of 82% (FDA) and 87% (D-PLS) when separate calibration and validation sample sets were used.

Olive oil authentication has been the subject of several reports (Groselj *et al.*, 2008; Gurdeniz *et al.*, 2008; Gurdeniz and Ozen, 2009; Hennessy *et al.*, 2009). In investigating problems of adulteration in Turkish olive oils, Gurdeniz *et al.* (2009) applied a wavelet transform to FT-IR data before modelling in a study designed to detect and quantify adulteration of EVOO with vegetable oils (rapeseed, cottonseed and a corn–sunflower binary mixture) by constructing (1) three independent models for olive oil samples adulterated with cottonseed, rapeseed and sunflower–corn binary oil mixture and (2) an overall model including all adulterated samples. In distinguishing between authentic and adulterated oil samples, oils blended with 2% and 5% adulterants could not be separated from authentic olive oils but such separation was possible for most of the oils adulterated at higher levels. Adulterant contents were predicted using PLS regression after orthogonal signal correction (OSC) pre-treatment; standard errors of prediction of between 1 and 1.4% were obtained for the three adulterant oils. Hennessy *et al.* (2009) carried out an extensive study which addressed the issue of confirming that a sample of olive oil claiming to originate in Liguria, Italy actually did so. Authentic EVOO samples ($n = 913$) from three harvests (2004–2007) were collected from Italy, France, Spain, Greece, Cyprus and Turkey; attenuated total reflectance spectra were subjected to derivative and standard normal variate data pre-treatments before analysis by PCA, FDA, and D-PLS. D-PLS was slightly more successful than FDA in classifying this data set although overall correct classification rates of 80 and 70% respectively were not high enough to recommend the procedure for commercial use. FT-IR analysis of other oils recently reported included studies on virgin coconut oil adulteration (Rohman and Che Man, 2009; Rohman and Che Man, 2011), adulteration issues in cod liver oil (Rohman and Man, 2009) and pumpkin seed oil classification on the basis of both species and genetic variety (Saucedo-Hernández *et al.*, 2011).

Other publications have described FT-IR use in authentication studies of meat (Juárez *et al.*, 2008; Rohman *et al.*, 2011), dairy products (Koca *et al.*, 2010; Kocaoglu-Vurma *et al.*, 2009; Pappas *et al.*, 2008; Woodcock *et al.*, 2008b), alcoholic drinks (Cozzolino *et al.*, 2009; Ioannou-Papayianni *et al.*, 2011; McIntyre *et al.*, 2011; Polshin *et al.*, 2011; Tarantilis *et al.*, 2008), coffee (Wang *et al.*, 2009, 2011), fruit juice (Jha and Gunasekaran, 2010; Vardin *et al.*, 2008), Chinese medicinal herbs (Zhang and Nie, 2010) and Moroccan olives (Terouzi *et al.*, 2011).

6.4.3 Raman applications

The number of published Raman applications in food authenticity studies has been very limited over this period, probably reflecting the fact that this technique is still only beginning to be explored for food classification problems. Herero *et al.* (2008) published a wide-ranging review of applications for quality assessment in meat and fish while Zou *et al.* (2009) reported on olive oil authentication. In this interesting study, a method based on the intensity ratio of the cis (=C–H) and cis (C=C) bonds normalised by the band at 1441 cm^{-1} (CH_2) was established to identify genuine and fake olive oil. Spectra of oils from China, USA, Italy, Greece and Spain were collected using a portable Raman spectrometer and the method was reportedly able to reliably distinguish genuine olive oils from olive oils containing 5% (volume percentage) or more of soybean, rapeseed, sunflower seed or corn oil. The authors suggest that this was due to the lower content of polyunsaturated fatty acids in olive oil. They also concluded that, because the test data are easy to collect and process since no chemometric operations are needed, this method is well-suited for field identification of genuine and fake olive oils. Also dealing with olive oil adulteration, El-Abassy *et al.* (2009) used Raman spectroscopy with excitation in the visible spectral range (VIS Raman). These authors claimed success in discriminating between EVOO and sunflower oil on the basis of principal component analysis using the first two principal components without mentioning that PCA is a descriptive and not a modelling technique. They also investigated the quantification of sunflower oil at concentrations between 5% and 100% (v/v) in 5% increments using partial least-squares regression analysis and claimed that a quantitative detection limit of 500 ppm (0.05%) could be achieved. In another study focussing on olive oils from PDO regions in France, Korifi *et al.* (2011) evaluated the potential of confocal Raman spectroscopy and chemometrics for authenticating the claimed PDO origins of six French virgin olive oils. D-PLS correctly classified 92.3% of French PDOs and 100% of PDO samples produced from a single olive cultivar.

Abbas *et al.* (2009) described the use of Raman spectroscopy coupled with D-PLS to discriminate between various animal fats and fats used in feedstock and production technology processes. Using samples of bovine, poultry and pig fat, D-PLS models produced high specificity values and low classification errors using spectral data in the ranges 3100–2650 cm^{-1} and 1800–1200 cm^{-1}. Eight fat types from different processing operations were also studied and similar high sensitivity and specificity values were reported. Pierna *et al.* (2011) applied Raman spectroscopy to the large Corsican honey collection referred to above (Woodcock *et al.*, 2009; Hennessy *et al.*, 2010) which was collected during the EU-funded TRACE project (www.trace.eu). A two-step mathematical analysis was employed, the first involved determination of the Fisher criterion which indicated which of the original spectral variables had important discriminating power in the Corsican/non-Corsican origin issue. Different models incorporating Corsican samples and the rest (including all the other French honeys, Italian, Austrian, German and Irish honeys) were constructed using PLS-DA and SVM as follows: (a) models for each year individually; (b) models including samples of both years, and (c) models using only the 15 variables selected by the Fisher criterion using leave-one-out cross-validation in all cases. SVM models produced the best results, with 85.5% and 94.6% correct classification for Corsican and non-Corsican honeys respectively using all variables and 82.3% and 94.5% with the reduced set of 15 variables identified as significant by Fisher's criterion.

6.5 Future trends

The main focus of research to date has undoubtedly been on laboratory studies using research-grade instrumentation. More recently, however, the commercial availability of hand-held equipment has become a reality which holds out the promise for application deployment outside of traditional laboratory, in places such as warehouses, in distribution systems and, finally, at retail sites. Challenges faced during their design and manufacture included development and construction of appropriate light sources and detectors together with the availability of the necessary built-in computer processing power. Some of this equipment is now in place in the pharmaceutical and fine chemical industries, demonstrating the capability of small hand-held devices to address issues of product identification in controlled environments and with limited reference libraries. The current research and demonstration issue is to deploy this type of miniaturised equipment for the type of complex identification challenges posed by food authenticity requirements; questions to be answered relate initially to the spectral sensitivity of the equipment and the available computing power.

With regard to on-line applications, these have already been well-established in NIR spectroscopy although the initial tendency was to use fixed-wavelength instruments for stability and reliability reasons. Full spectral equipment is now regularly used so that its transfer to food production lines for qualitative measurements should not pose any engineering or data analysis issues. Raman and mid-infrared applications are now a reality in non-food industries such as fine chemicals, plastics and pharmaceuticals using fibre-optic interfaces for spectral data collection. Applications in the food industry are not so numerous but neither engineering nor equipment issues are preventing the deployment of on-line authenticity solutions; what is lacking is simply the realisation that it can be done and, in many cases, should be done. It seems likely that increased focus on the prevention of food contamination, both accidental and intentional, to avoid public health or terrorist incidents will drive developments in this area in the foreseeable future.

6.6 Conclusions

There can be little doubt that vibrational spectroscopy has much to offer regarding food authenticity problems particularly on account of its ease of use, speed, low cost per analysis and absence of any requirement for chemical reagent use or disposal. This chapter has highlighted the main operational characteristics of each of the three methods (MIR, NIR and Raman spectroscopy) together with a discussion of some of the more recently published applications. While considerable progress has been made in the development and deployment of better chemometric data analysis techniques, in many cases it is clear that these spectroscopic models serve best as preliminary analytical measures and that in very few, if any, scenarios can 100% confidence be placed in an individual analytical result. More work continues to be required in mathematical information extraction techniques while the field of data fusion in which spectra from, for example, Raman and NIR may be concatenated and analysed simultaneously still needs thorough study. Finally, the recent availability of hand-held instruments for each of these techniques brings the prospect of their field deployment ever closer to being a reality.

6.7 References

ABBAS, O., PIERNA, J. A. F., CODONY, R., VON HOLST, C. and BAETEN, V. (2009) Assessment of the discrimination of animal fat by FT-Raman spectroscopy. *Journal of Molecular Structure*, **40**, 1284.

BAETEN, V., MANLEY, M., PIERNA, J. A. F., DOWNEY, G. and DARDENNE, P. (2008) Spectroscopic technique: Fourier transform near-infrared (FT-NIR) spectroscopy. In SUN, D.-W. (Ed.) *Modern Techniques for Food Authentication*. London, Elsevier.

BENSON, I. B. (2003) Near infra-red absorption technology for analysing food composition. IN Lees, M. (Ed.) *Food Authenticity and Traceability*. Cambridge, Woodhead Publishing, pp. 101–130.

BERRUETA, L. A., ALONSO-SALCES, R. M. and HÉBERGER, K. (2007) Supervised pattern recognition in food analysis. *Journal of Chromatography A*, **1158**, 196–214.

BLANCO, M. and VILLARROYA, I. (2002) NIR spectroscopy: a rapid-response analytical tool. *Trends in Analytical Chemistry*, **21**, 240–250.

BOMBARDA, I., DUPUY, N., DA, J. P. L. V. and GAYDOU, E. M. (2008) Comparative chemometric analyses of geographic origins and compositions of lavandin var. Grosso essential oils by mid infrared spectroscopy and gas chromatography. *Analytica Chimica Acta*, **613**, 31–39.

CASALE, M., CASOLINO, C., FERRARI, G. and FORINA, M. (2008) Near infrared spectroscopy and class modelling techniques for the geographical authentication of Ligurian extra virgin olive oil. *Journal of Near Infrared Spectroscopy*, **16**, 39–47.

CEN, H. and HE, Y. (2007) Theory and application of near infrared reflectance spectroscopy in determination of food quality. *Trends in Food Science and Technology*, **18**, 72–83.

CHASE, B. (1987) FT-Raman spectroscopy as an analytical tool. *Applied Spectroscopy*, **40**, 133–137.

CHEN, L., XUE, X., YE, Z., ZHOU, J., CHEN, F. and ZHAO, J. (2011) Determination of Chinese honey adulterated with high fructose corn syrup by near infrared spectroscopy. *Food Chemistry*, **128**, 1110–1114.

CHEN, L. Z., ZHAO, J., YE, Z. H. and ZHONG, Y. P. (2008a) Determination of adulteration in honey using near-infrared spectroscopy. *Spectroscopy and Spectral Analysis*, **28**, 2565–2568.

CHEN, Y., XIE, M. - Y., YAN, Y., ZHU, S. - B., NIE, S. - P., LI, C., WANG, Y. - X. and GONG, X. - F. (2008b) Discrimination of Ganoderma lucidum according to geographical origin with near infrared diffuse reflectance spectroscopy and pattern recognition techniques. *Analytica Chimica Acta*, **618**, 121–130.

COZZOLINO, D., HOLDSTOCK, M., DAMBERGS, R. G., CYNKAR, W. U. and SMITH, P. A. (2009) Mid infrared spectroscopy and multivariate analysis: a tool to discriminate between organic and non-organic wines grown in Australia. *Food Chemistry*, **116**, 761–765.

DAY, M. and FEARN, F. R. B. (1982) Near infra-red reflectance as an analytical technique. Part 1. History and development. *Laboratory Practice*, **31**, 328–330.

DERDE, M. P. and MASSART, D. L. (1986) UNEQ: a disjoint modelling technique for pattern recognition based on normal distribution. *Analytica Chimica Acta*, **184**, 33–51.

DIAN, P. H. M., ANDUEZA, D., JESTIN, M., PRADO, I. N. and PRACHE, S. (2008) Comparison of visible and near infrared reflectance spectroscopy to discriminate between pasture-fed and concentrate-fed lamb carcasses. *Meat Science*, **80**, 1157–1164.

DOWNEY, G. (1996) Non-invasive and non-destructive percutaneous analysis of farmed salmon flesh by near infra-red spectroscopy. *Food Chemistry*, **55**, 305–311.

DUCKWORTH, J. (2004) Mathematical data preprocessing. In ROBERTS, C. A., WORKMAN, J. and REEVES, J. B. (Eds.) *Near-Infrared Spectroscopy in Agriculture*. Madison, Wisconsin, USA, ASA,CSSA,SSSA.

EL-ABASSY, R. M., DONFACK, P. and MATERNY, A. (2009) Visible Raman spectroscopy for the discrimination of olive oils from different vegetable oils and the detection of adulteration. *Journal of Raman Spectroscopy*, **40**, 1284–1289.

ETZOLD, E. and LICHTENBERG, K. B. (2008) Determination of the botanical origin of honey by Fourier-transformed infrared spectroscopy: an approach for routine analysis. *European Food Research and Technology*, **227**, 579–586.

FAN, Y. X., CHENG, F. and XIE, L. J. (2010) Quantitative analysis and detection of adulteration in pork using near-infrared spectroscopy. IN KIM, M. S., TU, I. S. and CHAO, K. (Eds.) *Sensing for Agriculture and Food Quality and Safety II*. Proceedings of SPIE-The International Society for Optical Engineering, vol. 7676 Article Number: 7676–34 Publ. SPIE, Bellingham, WA, USA.

FERRARI, E., FOCA, G., VIGNALI, M., TASSI, L. and ULRICI, A. (2011) Adulteration of the anthocyanin content of red wines: perspectives for authentication by Fourier Transform-Near InfraRed and 1H NMR spectroscopies. *Analytica Chimica Acta*, **701**, 139–151.

FERRARO, J. (2002) *Introductory Raman Spectroscopy*. New York, Academic Press.

FORINA, M., ARMANINO, C., LEARDI, R. and DRAVA, G. (1991) A class-modelling technique based on potential functions. *Journal of Chemometrics*, **5**, 435–453.

FORINA, M., CASALE, M. and OLIVERI, P. (2009) Application of chemometrics to food chemistry. In BROWN, S. D., TAULER, R. and WALCZAK, B. (Eds.) *Comprehensive Chemometrics*. Amsterdam, Elsevier, pp. 75–128.

FORINA, M., OLIVERI, P., LANTERI, S. and CASALE, M. (2008) Class-modeling techniques, classic and new, for old and new problems. *Chemometrics and Intelligent Laboratory Systems*, **93**, 132–148.

GALLARDO-VELÁZQUEZ, T., OSORIO-REVILLA, G., LOA, M. Z.-D. and RIVERA-ESPINOZA, Y. (2009) Application of FTIR-HATR spectroscopy and multivariate analysis to the quantification of adulterants in Mexican honeys. *Food Research International*, **42**, 313–318.

GROSELJ, N., VRACKO, M., PIERNA, J. A. F., BAETEN, V. and NOVIC, M. (2008) The use of FT-MIR spectroscopy and counter-propagation artificial neural networks for tracing the adulteration of olive oil. *Acta Chimica Slovenica*, **55**, 935–941.

GURDENIZ, G. and OZEN, B. (2009) Detection of adulteration of extra-virgin olive oil by chemometric analysis of mid-infrared spectral data. *Food Chemistry*, **116**, 519–525.

GURDENIZ, G., OZEN, B. and TOKATLI, F. (2008) Classification of Turkish olive oils with respect to cultivar, geographic origin and harvest year, using fatty acid profile and mid-IR spectroscopy. *European Food Research and Technology*, **227**, 1275–1281.

HENNESSY, S., DOWNEY, G. and O'DONNELL, C. (2008) Multivariate analysis of attenuated total reflection-Fourier transform infrared spectroscopic data to confirm the origin of honeys. *Applied Spectroscopy*, **62**, 1115–1123.

HENNESSY, S., DOWNEY, G. and O'DONNELL, C. P. (2010) Attempted confirmation of the provenance of Corsican PDO honey using FT-IR spectroscopy and multivariate data analysis. *Journal of Agricultural and Food Chemistry*, **58**, 9401–9406.

HENNESSY, S. N., DOWNEY, G. and O'DONNELL, C. P. (2009) Confirmation of food origin claims by fourier transform infrared spectroscopy and chemometrics: Extra virgin olive oil from Liguria. *Journal of Agricultural and Food Chemistry*, **57**, 1735–1741.

HERRERO, A. M. (2008) Raman spectroscopy a promising technique for quality assessment of meat and fish: A review. *Food Chemistry*, **107**, 1642–1651.

IOANNOU-PAPAYIANNI, E., KOKKINOFTA, R. I. and THEOCHARIS, C. R. (2011) Authenticity of Cypriot sweet wine commandaria using FT-IR and chemometrics. *Journal of Food Science*, **76**, C420–C427.

JHA, S. N. and GUNASEKARAN, S. (2010) Authentication of sweetness of mango juice using Fourier transform infrared-attenuated total reflection spectroscopy. *Journal of Food Engineering*, **101**, 337–342.

JUÁREZ, M., ALCALDE, M. J., HORCADA, A. and MOLINA, A. (2008) Southern Spain lamb types discrimination by using visible spectroscopy and basic physicochemical traits. *Meat Science*, **80**, 1249–1253.

KAROUI, R., DOWNEY, G. and BLECKER, C. (2010) Mid-infrared spectroscopy coupled with chemometrics: a tool for the analysis of intact food systems and the exploration of their molecular structure-quality relationships – a review. *Chemical Reviews*, **110**, 6144–6168.

KAROUI, R., PIERNA, J. A. F. and DUFOUR, E. (2008) Spectroscopic technique: mid-infrared (MIR) and Fourier transform mid-infrared (FT-MIR) spectroscopies. In SUN, D.-W. (Ed.) *Modern Techniques for Food Authentication*. London, Elsevier, pp. 27–64.

KASEMSUMRAN, S., KANG, N., CHRISTY, A. A. and OZAKI, Y. (2005) Partial least square processing of near-infrared spectra for discrimination and quantification of adulterated olive oils. *Spectroscopy Letters*, **38**, 839–851.

KELLY, J. D., PETISCO, C. and DOWNEY, G. (2006a) Application of Fourier transform midinfrared spectroscopy to the discrimination between Irish artisanal honey and such honey adulterated with various sugar syrups. *Journal of Agricultural and Food Chemistry*, **54**, 6166–6171.

KELLY, J. D., PETISCO, C. and DOWNEY, G. (2006b) Potential of near infrared transflectance spectroscopy to detect adulteration of Irish honey by beet invert syrup and high fructose corn syrup. *Journal of Near Infrared Spectroscopy*, **14**, 139–146.

KELLY, J. F. D. and DOWNEY, G. (2005) Detection of sugar adulterants in apple juice using Fourier transform infrared spectroscopy and chemometrics. *Journal of Agricultural and Food Chemistry*, **53**, 3281–3286.

KIZIL, R. and IRUDAYARAJ, J. (2008) Spectroscopic technique: Fourier transform raman (FT_Raman) spectroscopy. In SUN, D. - W. (Ed.) *Modern Methods for Food Authentication*. Amsterdam, Elsevier.

KOCA, N., KOCAOGLU-VURMA, N. A., HARPER, W. J. and RODRIGUEZ-SAONA, L. E. (2010) Application of temperature-controlled attenuated total reflectance-mid-infrared (ATR-MIR) spectroscopy for rapid estimation of butter adulteration. *Food Chemistry*, **121**, 778–782.

KOCAOGLU-VURMA, N. A., ELIARDI, A., DRAKE, M. A., RODRIGUEZ-SAONA, L. E. and HARPER, W. J. (2009) Rapid profiling of Swiss cheese by attenuated total reflectance (ATR) infrared spectroscopy and descriptive sensory analysis. *Journal of Food Science*, **74**, S232–S239.

KOLOMIETS, O. A., LACHENMEIER, D. W., HOFFMANN, U. and SIESLER, H. W. (2010) Quantitative determination of quality parameters and authentication of vodka using near infrared spectroscopy. *Journal of Near Infrared Spectroscopy*, **18**, 59–67.

KORIFI, R., LEDREAU, Y., MOLINET, J., ARTAUD, J. and DUPUY, N. (2011) Composition and authentication of virgin olive oil from French PDO regions by chemometric treatment of Raman spectra. *Journal of Raman Spectroscopy*, **42**, 1540–1547.

KRAMER, R., WORKMAN, J. R. and REEVES, J. B. (2007) Qualitative analysis. In ROBERTS, C. A., WORKMAN, J. and REEVES, J. B. (Eds.) *Near-Infrared Reflectance Spectroscopy*. Madison, Wisconsin, USA, ASA,CSSA,SSSA, pp. 175–206.

KURZ, C., LEITENBERGER, M., CARLE, R. and SCHIEBER, A. (2010) Evaluation of fruit authenticity and determination of the fruit content of fruit products using FT-NIR spectroscopy of cell wall components. *Food Chemistry*, **119**, 806–812.

LAI, Y., NI, Y. and KOKOT, S. (2011) Discrimination of Rhizoma Corydalis from two sources by near-infrared spectroscopy supported by the wavelet transform and least-squares support vector machine methods. *Vibrational Spectroscopy*, **56**, 154–160.

LEUGERS, M. A. and LIPP, E. D. (2000) Raman spectroscopy in chemical process analysis. In CHALMERS, J. M. (Ed.) *Spectroscopy in Process Analysis*. Sheffield, UK, Sheffield Academic Press.

LEWIS, I. R. and EDWARDS, H. (2001) *Handbook of Raman Spectroscopy: From the Research Laboratory to the Process Line*. Boca Raton, FL, CRC Press.

LI, X. L. and HE, Y. (2008) Discriminating varieties of tea plant based on Vis/NIR spectral characteristics and using artificial neural networks. *Biosystems Engineering*, **99**, 313–321.

LIANG, L., LIU, Z. X., YANG, M. H., ZHANG, Y. X. and WANG, C. H. (2009) Discrimination of variety and authenticity for rice based on visual/near infrared reflection spectra. *Journal of Infrared and Millimeter Waves*, **28**, 353–356.

LIU, F., CAO, F., WANG, L. and HE, Y. (2008a) Discrimination of rice wine age using visible and near infrared spectroscopy combined with BP neural network. *Cisp 2008: First International Congress on Image and Signal Processing, Vol 5, Proceedings*, 267–271.

LIU, F., HE, Y. and WANG, L. (2008b) Determination of effective wavelengths for discrimination of fruit vinegars using near infrared spectroscopy and multivariate analysis. *Analytica Chimica Acta*, **615**, 10–17.

LIU, F., YE, X., HE, Y. and WANG, L. (2009) Application of visible/near infrared spectroscopy and chemometric calibrations for variety discrimination of instant milk teas. *Journal of Food Engineering*, **93**, 127–133.

LUO, W., HUAN, S., FU, H., WEN, G., CHENG, H., ZHOU, J., WU, H., SHEN, G. and YU, R. (2011) Preliminary study on the application of near infrared spectroscopy and pattern recognition methods to classify different types of apple samples. *Food Chemistry*, **128**, 555–561.

MANLEY, M., DOWNEY, G. and BAETEN, V. (2008) Spectroscopic technique: Near-infrared (NIR) spectroscopy. In SUN, D. - W. (Ed.) *Modern Techniques for Food Authentication*. London, Elsevier, pp. 65–115.

MCCLURE, W. F. (2001) Near-infrared instrumentation. In WILLIAMS, P. and NORRIS, K. (Eds.) *Near-Infrared Technology in the Agricultural and Food Industries*. St. Paul, Minnesota, USA., AACC Inc., pp. 89–106.

MCCLURE, W. F. (2007) Introduction. In OZAKI, Y., MCCLURE, W. F. and CHRISTY, A. A. (Eds.) *Near-Infrared Spectroscopy in Food Science and Technology*. USA, Wiley, pp. 1–10.

MCCLURE, W. F. and TSUCHIKAWA, S. (2007) Instruments. In OZAKI, Y., MCCLURE, W. F. and CHRISTY, A. A. (Eds.) *Near-Infrared Spectroscopy in Food Science and Technology*. Hoboken, USA, John Wiley & Sons, Inc., pp. 75–107.

MCINTYRE, A. C., BILYK, M. L., NORDON, A., COLQUHOUN, G. and LITTLEJOHN, D. (2011) Detection of counterfeit Scotch whisky samples using mid-infrared spectrometry with an attenuated total reflectance probe incorporating polycrystalline silver halide fibres. *Analytica Chimica Acta*, **690**, 228–233.

MIGNANI, A. G., CIACCHERI, L., OTTEVAERE, H., THIENPONT, H., CONTE, L., MAREGA, M., CICHELLI, A., ATTILIO, C. and CIMATO, A. (2011) Visible and near-infrared absorption spectroscopy by an integrating sphere and optical fibers for quantifying and discriminating the adulteration of extra virgin olive oil from Tuscany. *Analytical and Bioanalytical Chemistry*, **399**, 1315–1324.

NAES, T., ISAKSSON, T., FEARN, T. and DAVIES, T. (2002) *A User-Friendly Guide to Multivariate Calibration and Classification*. Chichester, NIR Publications.

NISHIKIDA, K., NISHIO, E. and HANNAH, R. W. (1995) *Selected Applications of Modern FT-IR Techniques*, Tokyo, Japan., Kodansha Ltd.

NORRIS, K. (1989) NIRS Instrumentation. In MARTEN, G. C., SHENK, J. S., BARTON, F. E. II. (Eds.) *Near-Infrared Reflectance Spectroscopy (NIRS): Analysis of Forage Quality*. Washington, DC, USDA, pp. 12–17.

OLIVERI, P., DI EGIDIO, V., WOODCOCK, T. and DOWNEY, G. (2011) Application of class-modelling techniques to near infrared data for food authentication purposes. *Food Chemistry*, **125**, 1450–1456.

OZAKI, Y., MORITA, S. and DU, Y. P. (2007) Spectral analysis. In OZAKI, Y., MCCLURE, W. F. and CHRISTY, A. A. (Eds.) *Near-Infrared Spectroscopy in Food Science and Technology*. USA, Wiley.

OZTURK, B., YALCIN, A. and OZDEMIR, D. (2010) Determination of olive oil adulteration with vegetable oils by near infrared spectroscopy coupled with multivariate calibration. *Journal of Near Infrared Spectroscopy*, **18**, 191–201.

PAPPAS, C. S., TARANTILIS, P. A., MOSCHOPOULOU, E., MOATSOU, G., KANDARAKIS, I. and POLISSIOU, M. G. (2008) Identification and differentiation of goat and sheep milk based on diffuse reflectance infrared Fourier transform spectroscopy (DRIFTS) using cluster analysis. *Food Chemistry*, **106**, 1271–1277.

PAZ, P., SANCHEZ, M. T., PEREZ-MARIN, D., GUERRERO, J. E. and GARRIDO-VARO, A. (2009a) Evaluating NIR instruments for quantitative and qualitative assessment of intact apple quality. *Journal of the Science of Food and Agriculture*, **89**, 781–790.

PAZ, P., SANCHEZ, M. T., PEREZ- MARIN, D., GUERRERO, J. E. and GARRIDO- VARO, A. (2009b) Instantaneous quantitative and qualitative assessment of pear quality using near infrared spectroscopy. *Computers and Electronics in Agriculture*, **69**, 24–32.

PÉREZ-MARÍN, D., SÁNCHEZ, M. - T., PAZ, P., GONZÁLEZ-DUGO, V. and SORIANO, M. - A. (2011) Postharvest shelf-life discrimination of nectarines produced under different irrigation strategies using NIR-spectroscopy. *LWT – Food Science and Technology*, **44**, 1405–1414.

PIERNA, J. A. F., ABBAS, O., DARDENNE, P. and BAETEN, V. (2011) Discrimination of Corsican honey by FT-Raman spectroscopy and chemometrics. *Biotechnologie Agronomie Societe et Environnement*, **15**, 75–84.

PIZARRO, C., ESTEBAN- DIEZ, I. and GONZALEZ- SAIZ, J. M. (2007) Mixture resolution according to the percentage of robusta variety in order to detect adulteration in roasted coffee by near infrared spectroscopy. *Analytica Chimica Acta*, **585**, 266–276.

POLSHIN, E., AERNOUTS, B., SAEYS, W., DELVAUX, F., DELVAUX, F. R., SAISON, D., HERTOG, M., NICOLAÏ, B. M. and LAMMERTYN, J. (2011) Beer quality screening by FT-IR spectrometry: impact of measurement strategies, data pre-processings and variable selection algorithms. *Journal of Food Engineering*, **106**, 188–198.

PRIETO, N., ANDRES, S., GIRALDEZ, F. J., MANTECON, A. R. and LAVIN, P. (2008) Discrimination of adult steers (oxen) and young cattle ground meat samples by near infrared reflectance spectroscopy (NIRS). *Meat Science*, **79**, 198–201.

REID, L. M., O'DONNELL, C. P. and DOWNEY, G. (2006) Recent technological advances for the determination of food authenticity. *Trends in Food Science and Technology*, **17**, 344–353.

ROHMAN, A. and CHE MAN, Y. B. (2009) Monitoring of Virgin Coconut Oil (VCO) adulteration with palm oil using Fourier transform infrared spectroscopy. *Journal of Food Lipids*, **16**, 618–628.

ROHMAN, A. and CHE MAN, Y. B. (2011) The use of Fourier transform mid infrared (FT-MIR) spectroscopy for detection and quantification of adulteration in virgin coconut oil. *Food Chemistry*, **129**, 583–588.

ROHMAN, A. and MAN, Y. B. C. (2009) Analysis of cod-liver oil adulteration using fourier transform infrared (FTIR) spectroscopy. *Journal of the American Oil Chemists Society*, **86**, 1149–1153.

ROHMAN, A., SISMINDARI, ERWANTO, Y. and MAN, Y. B. C. (2011) Analysis of pork adulteration in beef meatball using Fourier transform infrared (FTIR) spectroscopy. *Meat Science*, **88**, 91–95.

SAUCEDO-HERNÁNDEZ, Y., LERMA-GARCÍA, M. A. J. S., HERRERO-MARTÍNEZ, J. M., RAMIS-RAMOS, G., JORGE-RODRÍGUEZ, E. and SIMÓ-ALFONSO, E. F. (2011) Classification of pumpkin seed oils according to their species and genetic variety by attenuated total reflection Fourier-transform infrared spectroscopy. *Journal of Agricultural and Food Chemistry*, **59**, 4125–4129.

STUART, B. H. (2004) *Infrared Spectroscopy: Fundamentals and Applications*, Chichester, UK, John Wiley & Sons Ltd.

SUBRAMANIAN, A. and RODRIGUEZ-SAONA, L. (2009) Fourier transform infrared (FTIR) spectroscopy. In SUN, D. - W. (Ed.) *Infrared Spectroscopy for Food Quality Analysis and Control*. London, UK, Elsevier.

TARANTILIS, P. A., TROIANOU, V. E., PAPPAS, C. S., KOTSERIDIS, Y. S. and POLISSIOU, M. G. (2008) Differentiation of Greek red wines on the basis of grape variety using attenuated total reflectance Fourier transform infrared spectroscopy. *Food Chemistry*, **111**, 192–196.

TEROUZI, W., DE LUCA, M., BOLLI, A., OUSSAMA, A., PATUMI, M., IOELE, G. and RAGNO, G. (2011) A discriminant method for classification of Moroccan olive varieties by using direct FT-IR analysis of the mesocarp section. *Vibrational Spectroscopy*, **56**, 123–128.

VARDIN, H., TAY, A., OZEN, B. and MAUER, L. (2008) Authentication of pomegranate juice concentrate using FTIR spectroscopy and chemometrics. *Food Chemistry*, **108**, 742–748.

WANG, N., FU, Y. and LIM, L. - T. (2011) Feasibility study on chemometric discrimination of roasted arabica coffees by solvent extraction and Fourier transform infrared spectroscopy. *Journal of Agricultural and Food Chemistry*, **59**, 3220–3226.

WANG, J., Soojin, J., BITTENBENDER, H. C., GAUTZ, L. and QUING, X. L. (2009) Fourier transform infrared spectroscopy for Kona coffee authentication. *Journal of Food Science*, **74**, C385–C391.

WEHLING, R. (2003) Infrared spectroscopy. In NIELSEN, S. S. (Ed.) *Food Analysis* 3rd ed. LONDON, U.K., Kluwer Academic/Plenum Publishers.

WENG, X. X., LU, F., WANG, C. X. and QI, Y. P. (2009) Discriminating and quantifying potential adulteration in virgin olive oil by near infrared spectroscopy with BP-ANN and PLS. *Spectroscopy and Spectral Analysis*, **29**, 3283–3287.

WESTERHAUS, M., WORKMAN, JR J., REEVES, J. B. and MARK, H. (2007) Quantitative analysis. In ROBERTS, C. A., WORKMAN J R., J. and REEVES, J. B. (Eds.) *Near-Infrared Reflectance Spectroscopy*. Madison, Wisconsin, USA, ASA,CSSA,SSS A.

WOLD, S. (1976) Pattern recognition by means of disjoint principal components models. *Pattern Recognition*, **8**, 127–139.

WOODCOCK, T., DOWNEY, G. and O'DONNELL, C. P. (2008a) Better quality food and beverages: the role of near infrared spectroscopy. *Journal of Near Infrared Spectroscopy*, **16**, 1–29.

WOODCOCK, T., FAGAN, C., O'DONNELL, C. and DOWNEY, G. (2008b) Application of near and mid-infrared spectroscopy to determine cheese quality and authenticity. *Food and Bioprocess Technology*, **1**, 117–129.

WOODCOCK, T., DOWNEY, G. and O'DONNELL, C. P. (2009) Near infrared spectral fingerprinting for confirmation of claimed PDO provenance of honey. *Food Chemistry*, **114**, 742–746.

WORKMAN, J. J. (2004) Near-infrared spectrophotometers. In ROBERTS, C. A., WORKMAN, J R. J. and REEVES, J. B. (Eds.) *Near-Infrared Reflectance Spectroscopy*. Madison, Wisconsin, USA, ASA, CSSA, SSSA.

WORKMAN, Jr J. and SHENK, J. (2004) Understanding and using the near infra-red spectrum as an analytical method. In ROBERTS, C. A., WORKMAN, JR J. and REEVES III, J. B. (Eds.) *Near-Infrared Spectroscopy in Agriculture*. Madison, Wisconsin, USA, ASA, CSSA and SSA.

WORKMAN, Jr J. and WEYER, L. (2007) Introduction to near-infrared spectra. In Workman Jr J. and Weyer, L. (Eds.) *Practical Guide to Interpretive Near-Infrared Spectroscopy*. London, CRC Press.

XIE, L., YING, Y. and YING, T. (2009) Classification of tomatoes with different genotypes by visible and short-wave near-infrared spectroscopy with least-squares support vector machines and other chemometrics. *Journal of Food Engineering*, **94**, 34–39.

ZHANG, L. and NIE, L. (2010) Discrimination of geographical origin and adulteration of radix astragali using Fourier transform infrared spectroscopy and chemometric methods. *Phytochemical Analysis*, **21**, 609–615.

ZHUANG, X. L., XIANG, Y. H., QIANG, H., ZHANG, Z. Y., ZOU, M. Q. and ZHANG, X. F. (2010) Quality analysis of olive oil and quantification detection of adulteration in olive oil by near-infrared spectrometry and chemometrics. *Spectroscopy and Spectral Analysis*, **30**, 933–936.

ZOU, M. - Q., ZHANG, X. - F., QI, X. - H., MA, H. - L., DONG, Y., LIU, C. - W., GUO, X. and WANG, H. (2009) Rapid authentication of olive oil adulteration by Raman spectrometry. *Journal of Agricultural and Food Chemistry*, **57**, 6001–6006.

7

Chemometrics in studies of food origin

B. Vandeginste, VICIM, The Netherlands

DOI: 10.1533/9780857097590.2.117

Abstract: Chemometrics focuses on the mathematical and statistical modeling of analytical data in order to obtain relevant chemical information. It copes with the peculiarities of analytical data acquired in a wide range of domains. One of these domains is food research and, in particular, the verification of the authenticity of food products. Chemometrics offers a wide palette of tools for developing decision rules on the authenticity of the origin of food. Important aspects in food traceability are that chemometric models are easily maintainable with good updatability, simple, robust for defective data and easily convertible into a set of specifications. These aspects were particularly challenging to the WP6 team, Statistical Specifications, in the TRACE project.

Key words: chemometrics, classification, verification, traceability, food, TRACE project.

7.1 Introduction

In the early 1970s, the demand for new and improved mathematical and statistical methods to analyze the wealth of data generated by modern analytical instruments was ever growing. A distinct need to visualize large collections of (spectral) data and to relate those data to the sample property emerged. A new discipline, 'chemometrics' was born. From the earliest stage, food research benefited from the advances in modern instrumentation and chemometrics. Thirty years ago, in September 1982 an interdisciplinary IUFOST meeting 'Food research and data analysis' was organized in Oslo with the

---

Disclaimer: The publication reflects only the authors' views and the European Commission is not liable for any use that may be made of the information contained therein.

aim of exposing food researchers to new capabilities offered by computer aided analysis of multivariate food research data. At that meeting, Forina and coworkers (1983) presented new results on the classification of olive oils according to their fatty acid profiles. Typical data sets concerned the fatty acid profiles (8 acids) of 572 olive oils collected in six Italian regions. At that time no less than eight chemometric techniques were applied to this data set, including linear discriminant analysis (LDA) and soft independent modeling of class analogy (SIMCA). Classifications according to the origin of a food were not yet perfect, but the tremendous chemometric capabilities were clearly demonstrated. A year later, in September 1983, a two-week NATO Advanced Study Institute: Chemometrics – Mathematics and Statistics in Chemistry took place in Cosenza (Italy). Here, Forina and Lanteri (1984) showed that the fatty acid composition of Portuguese olive oils changed over time, indicating the need for model updating. Both studies were early seeds for the traceability of the origin of food. How far have we come since then? Well, new algorithms such as support vector machines (SVMs) (Cristianini and Shawe-Taylor, 2000; Üstün *et al.*, 2006) and artificial neural networks (ANNs) (Zupan *et al.*, 1997) cope with highly non-linear data. By applying robust techniques like Robust SIMCA (Daszykowski, 2007a) we experience fewer problems with outliers. Techniques are available to replace missing data (Walczak and Massart, 2001). Data sets with many variables and few objects can be analyzed by kernel partial least squares (PLS) (Rannar *et al.*, 1995). Better clustering techniques are available (Daszykowski *et al.*, 2002a). Probabilistic versions of classification by discriminant PLS (D-PLS) assign class probabilities (Pérez *et al.*, 2010). Decision tree-based methods put less emphasis on prior variable selection (Caetano *et al.*, 2005). And of course, modern computers allow the processing of large data sets and the application of computing-intensive procedures like genetic algorithms (GAs) (Wehrens and Buydens, 1998). In TRACE a work-package was dedicated to testing, applying and providing fit-for-purpose chemometric tools for the verification of the origin of food commodities such as olive oil, honey, meat and cereals.

7.2 Samples and data

Obviously the quality of chemometric models depends on both the representativeness of the samples taken from the object under investigation and the quality of the data. For instance one of the challenges within the TRACE project was to collect a statistically representative sample of olive oils originating from a defined growing region, for example, Liguria. This requires a 'statistical' sample to be drawn from the population of all candidate sampling locations in that region. Therefore each candidate sampling location is described by a set of quantitative and qualitative descriptors or attributes. For olive oils, attributes are, for example, farmers, producers, soil composition, manuring, harvesting year, pesticide treatments, cultivar composition, and phytosanitary conditions. These attributes position all candidate sampling locations in multivariate space. A statistical sample of n samples out of m candidates is then obtained by finding n samples that best span the multivariate space. This approach was successfully applied in a study on the authenticity of wines by applying the Kennard and Stone (1969) and potential function algorithms (Forina *et al.*, 1991; Millan *et al.*, 1998). The quality of the data is reflected by the uncertainty associated with the values or measurements, the number of missing values (holes in the data matrix) and outlying values. We cannot put a figure on the degradation level with increasing number of missing values. Eliminating missing values either by removing the column (variable) or row (sample) containing the missing value is a valid option, but there are better ones. If it is done that way, we advise removing the object, as this is less influential in the modeling step later on. A simple alternative is to replace a missing value with either the mean or median of the variable (column), certainly not with a zero. Another option is a random fill (within the range of the variable). However, principal component imputation (Stanimirova and Walczak, 2008) has proven to be a better approach. This algorithm proposes the best possible value of the missing elements in the data set by an iterative procedure that switches between the principal components space and the original space. The algorithm starts by substituting the missing elements by the corresponding column and row means. Then the data matrix is decomposed into a scores and loadings matrix by a principal components analysis (PCA). The original matrix is reconstructed by taking the product of the scores matrix and loadings matrix after deletion of the insignificant loading vectors. Note that only the missing values are replaced whereas the original data are retained in the data set. The procedure iterates by repeating the algorithm on the newly obtained matrix and is continued until the imputed values stabilize.

We distinguish two types of outliers, outlying variable values and outlying objects. Outlying objects may be visualized in a Robust PCA plot and be removed (Locantore *et al.*, 1999). Outlying variable values are identified by applying a classical statistical outlier test on each variable (Massart *et al.*, 1997). Removal of an outlying value poses a particular problem as it creates a missing value that should not be filled by imputation. Either the object in which the outlier is present should be removed or the subsequent data analysis should be robust for outliers. Obviously the simultaneous presence of outliers and missing values poses a circular problem. An outlier may be flagged by the presence of missing values and imputed values may be influenced by the presence of outliers. The measurement uncertainty affects the classification results in two ways. First it introduces an uncertainty in the model parameters (to be compared with the uncertainty of the slope and intercept in straight line calibration). Second, the uncertainty in the values of a new object presented to the model affects its class prediction. Of course, all effects depend on the magnitude of the uncertainty with respect to the class separation (see Section 7.5.1). How to take the measurement uncertainty into account when checking a product against a certain upper limit is described in a Eurachem/Citac Guide (Ellison and Williams, 2007). In the introduction we briefly touched on the verification of the origin of olive oils based on the fatty acid profile. Nowadays, such profiles can easily be manipulated by adulteration with other oils. This demonstrates that attributes providing perfect traceability may be of no value if these attributes are easily counterfeited (López-Díez *et al.*, 2003). The trace element composition and Sr isotope ratio of a food commodity are far more robust against adulteration, certainly when considered in relation to the soil composition. In many instances attributes are a spectroscopic fingerprint (e.g., near infrared spectrum) of a food commodity without allocating peaks to specific compounds or functional groups. Here, we deal with signals sometimes requiring signal improvement, correction and restoration prior to the modeling step. An issue here is whether uninformative wavelengths should be deleted (Centner *et al.*, 1996).

7.3 The modeling step

Data modeling is the core step in setting up a food traceability system. The first decision to be made is whether we want to verify the declared origin of a food product or we want to classify a product in one or more predefined categories. This can be explained by taking the example of the verification of the origin of chicken. Three possibilities are distinguished. Let us first consider the situation where we want to verify whether a chicken is European or not. A model (and specifications) are then developed for a European chicken. A chicken that fits these specifications is then decided to be European. A chicken that does not meet the specifications is then decided to be non-European (without further specification of the country). If in a second situation we want to verify whether a chicken is a Chinese chicken or not, then a model (and specifications) are developed for a Chinese chicken. A chicken that does not fit the specifications of a Chinese chicken is then judged not to be a Chinese chicken (not necessarily a European chicken). Third, we want to verify whether a chicken is Chinese or European. We thus need a model that discriminates between these two origins: either it is Europe or it is China. If the purpose of the traceability system is to protect the European farmer, then we are in the first situation. If we want to protect the Chinese farmer, then we are in the second situation. If we want to protect the European farmer from imports specifically from China then we are in the third situation. From the first two cases we observe that the modeling step in principle only requires products from the origin we want to verify. If we want to verify whether a chicken originates from Europe, we need a representative sample of European chicken and do not necessarily need to model non-European chickens. So it may be that when applying the model, Chinese chicken will fail the model but chicken from other regions, for example, South-America will not. This is the so-called asymmetric case. The class of interest is well defined, but the rest is just a heterogeneous collection outside the class. The third case is a two-class problem. Both classes are well defined. This may be expanded to the *n*-class case. However, any *n*-class problem may be reduced to a two-class case: a defined or target class and the combined other classes. In this case the origin of food may be verified by discriminative models. The *n*-class case may also be reduced to an asymmetric case with a single target class. In this case the origin of food may be verified by class modeling. The second decision to be made is whether we want to convert the chemometric model into an easily verifiable set of attribute specifications, for instance, characteristic variable ranges for a certain class. This requirement imposed some tough chemometric challenges to WP6 as excellent performance of neural networks, SVMs and, to a lesser extent, SIMCA were associated with difficult or impossible model interpretation.

7.3.1 The verification of the origin of food by class modeling

In order to build a class model, a data set is collected of objects belonging to that class. In the context of traceability, objects are food products originating from the target region. Several techniques are available to develop class models: (i) parametric techniques based on a multivariate normal distribution, (ii) non-parametric (distribution free) techniques and (iii) techniques in-between the two. Parametric techniques, such as unequal covariance matrix classification (UNEQ) (Derde and Massart, 1988) are less applicable as they require a large sample set in order to derive reliable statistical parameters. Density or potential function classifiers, such as ALLOC (Coomans *et al.*, 1981) and POTFUN (PF) (Forina *et al.*, 1991) are non-parametric methods. In PF classification, each object of the training set is surrounded by a potential field in data space. A test object is classified into the class with the highest cumulative potential at the location of the test object. This cumulative potential is the sum of all individual potentials at the position of the test object and is an estimate of the probability density function of the class. This permits the development of Bayesian classification rules. SIMCA is well studied and widely applied in food chemistry. Specifically the robust version of SIMCA (Daszykowski *et al.*, 2007b) is particularly suited in traceability studies. Recently, the performance of class modeling techniques for the verification of the origin of olive oils has been critically reviewed (Forina *et al.*, 2008). In another study the performance of UNEQ and SIMCA for the verification of Italian olive oils (Marini *et al.*, 2006) were compared. The principle of SIMCA is the development of a class model by PCA. The number of factors retained in the model is determined by cross-validation, and the model is validated by submitting a test set. Whether an object fits the class model or not is evaluated by considering two distances: (i) the Mahalanobis distance (MD) (Vandeginste *et al.*, 1998), that is, the distance of an object from the data center, corrected for the correlation and (ii) the orthogonal distance (OD) of the object from the model or residual (Vandeginste *et al.*, 1998). In fact, these two

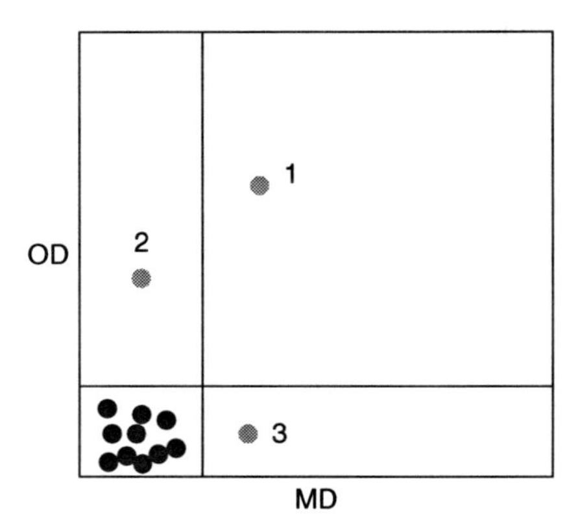

Fig. 7.1 Plot of orthogonal distance (OD) *vs* Mahalanobis distance (MD) in SIMCA class modeling. Objects in the lower left corner fit the model. Object 1 is a bad leverage object. The residual of object 2 is too large. Object 3 is a good leverage point.

distances determine whether the object is an outlier with respect to the model or not. By plotting OD *vs* MD (and including cut-off lines for both distances obtained during the modeling step, we decide whether an object belongs to the class or not (Fig. 7.1).

Objects below both cut-off lines belong to the class, for which the model is developed and the objects exceeding one of the cut-off lines are not classified as belonging to the class: object 1 in Fig. 7.1 is a bad leverage point that does not fit the model at all, the residual of object 2 is too large and object 3 is a good leverage point (helps the modeling but is located too far from the class center). In the context of traceability we may make two types of errors:

(i) we decide that a product is not authentic whereas it is. This is a false rejection or false negative. Here, we make a type I error. The probability of such an error is α;
(ii) we decide that a product is authentic whereas it is not. This is a false acceptance or false positive. Here, we make a type II error. The probability of such an error is β.

When developing a class model we optimize the probability α of a false rejection. Considering the costs associated with a wrong accusation of a farmer or producer, it is clear that α should be as small as possible. However, once we have accepted the probability α, the value of β – the risk that we accept unauthentic products – is defined as well. In some instances it may take large values. It is relatively easy to assess α, as it follows from the modeling and validation step. However, the assessment of β is not. Here, we have to make an intelligent guess on the type of falsely labeled products that will be examined in the future and have such products available in a test set for checking the false positive rate of the model. The latter is less obvious. In fact β depends completely on what we mean by 'other'. The SIMCA modeling technique does not depend on the number of classes we want to include in the traceability system. For each class a verification model can be developed independently of the objects belonging to other classes. For each model we tune α. An initial guess of β is then obtained by fitting all other objects to the class model being considered. Here, we assume that the other objects are somehow a representative sample for unauthentic products, which is not necessarily the case. A nice property of SIMCA as a classifier is that new objects may be an outlier for all models or some objects may fit several class models because the classes overlap. All three class modeling techniques, SIMCA, UNEQ and POTFUN were applied on TRACE data sets of near infrared (NIR) spectra of honeys and olive oils by Oliveri *et al.* (2011). Distinct differences in α-error and β-error were observed. Low β-errors were associated with high α-errors and *vice versa*. This trade-off phenomenon of both types of error is often observed. Instead of choosing the best technique (or one of the models) one could try and fuse the models into a single decision. This is further discussed in Section 7.5.4.

7.3.2 The verification of the origin of food by classification models

Why would classification methods be applied in addition to adequate class modeling techniques already available for the verification of food authenticity? There are a couple of reasons. First, SIMCA models are not easily converted into specifications expressed in the original variables. Second, and this is the most important reason, when the data are two- or multi-class, the emphasis may be on classification rather than on verification. This allows us to apply techniques that aim to optimize the discrimination between classes and are easily interpreted. So, next to class modeling techniques one can apply discriminating techniques such as LDA (a parametric technique), discriminant partial least squares (D-PLS) and SVMs. SVM is only applicable in the two-class situation but is particularly suited when the borderlines between classes are highly non-linear, whereas D-PLS models are more easily interpreted. D-PLS is a popular method that has been extensively applied to classify a wide variety of food commodities, including butters (Macatelli *et al.*, 2009). SVM applies complex non-linear operators (kernel transformation) to linearize non-linear between-class boundaries. Although SVM is only applicable for binary classifications (two classes), its big potential in verification studies has been demonstrated in the TRACE project by discriminating Italian from non-Italian olive oils and Ligurian from non-Ligurian olive oils analyzed by FT-IR spectroscopy (Caetano *et al.*, 2007) and the discrimination of Corsican honeys from other regions (Pierna *et al.*, 2011). A second class of classifiers to be considered are decision trees such as classification and regression trees (CART) (Breiman *et al.*, 1984) and genetic programming (GP) (Gilbert *et al.*, 1997; López-Díez *et al.*, 2003). Both, CART decision trees and GP trees are expressed in the original variable space. Therefore both are easily implemented in a traceability system. In addition the trees are relatively sparse fulfilling the requirement of model sparsity.

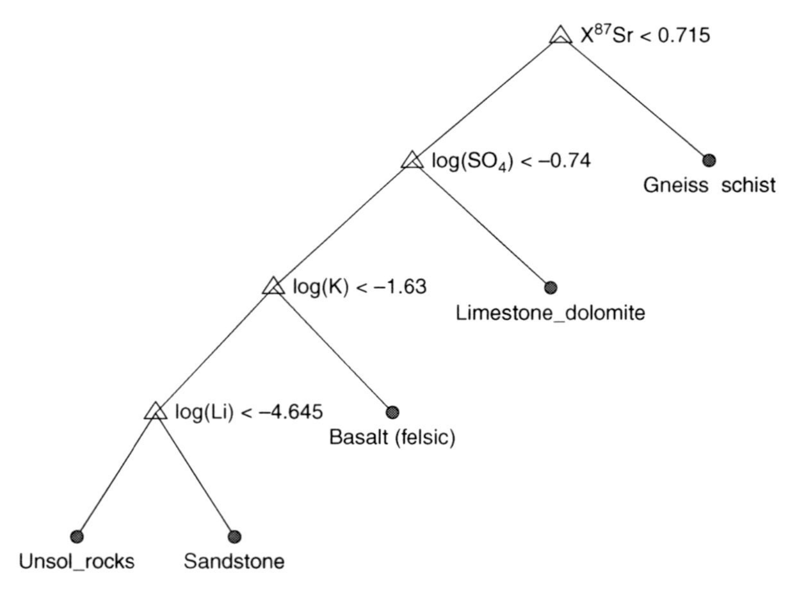

Fig. 7.2 CART decision tree of mineral waters characterized by their trace element concentrations (see text for explanation).

This is illustrated by the CART tree obtained for the classification of mineral waters based on their trace element composition (Fig. 7.2). The mineral water data set collected within TRACE consisted of 23 main and trace element concentrations measured for 331 mineral waters collected over the whole of Europe. This set was split in a modeling set (235 samples) and a test set (96 samples) using the Duplex algorithm (Snee, 1977). Mineral waters were categorized according to nine different geologies of origin. As one can see from Fig. 7.2, the tree consists of five nodes, remarkably using only three trace element concentrations and the Sr isotope ratio. The names attributed to the nodes correspond to geology of the majority of the waters classified in that node. One should bear in mind however that because class sizes were unbalanced in this study, bigger classes have a larger influence on the classification than smaller classes. Moreover, because there are fewer nodes than classes there will be some mixing up of the nodes. This node impurity is illustrated in Table 7.1 showing that 73% of the Gneiss (Gn) mineral waters are classified in node 1, 5% in node 2, and 22% in Node 5 (row 3 in Table 7.1). Such a confusion matrix is a serious complication when developing a traceability system, as it affects the reliability of the decision rules. Therefore, additional decision rules may be required that resolve specific confusions.

This may be achieved by developing discriminative models (e.g., a D-PLS model) that resolve ambiguities between two or more classes. The performance of a sparse CART model was clearly demonstrated on verification of wines from the protected designation of origin Valencia (Gonzalvez *et al.*, 2009).

Table 7.1 Distribution (%) of mineral waters over the five nodes obtained in the CART tree in Fig. 7.2

	Node 1 (Gn)	Node 2 (Ld)	Node 3 (Bf)	Node 4 (Ur)	Node 5 (S)	Total
Ur	0.00	0.04	0.00	0.83	0.13	1
Ld	0.00	0.43	0.02	0.41	0.14	1
Gn	0.73	0.05	0.00	0.00	0.23	1
Gr	0.50	0.10	0.00	0.30	0.10	1
Grs	0.07	0.04	0.00	0.30	0.59	1
S	0.11	0.19	0.08	0.22	0.39	1
M	0.00	0.50	0.00	0.33	0.17	1
Bm	0.00	0.00	0.67	0.17	0.17	1
Bf	0.00	0.00	1.00	0.00	0.00	1

Note: Symbols in the first column refer to the soil types (not further explained).

The model included the concentration of two trace elements (Li and Mg) only from a total set of 38 that were measured. Although CART can cope with outliers and/or missing values, trees are sensitive for minor changes in attribute values. Techniques to improve the CART stability are discussed in Section 7.5.5. Genetic programming (GP) has been successfully applied for the verification of Corsican honeys analyzed by Cryoprobe ^{1}H-NMR (nuclear magnetic resonance) spectroscopy (Donarski *et al.*, 2008). The decision tree linked 13 variables by the operators +, −, average, Min, > and SQRT, with an overall classification rate of 97.3%. Finally, Kohonen ANNs (Garcia-Gonzalez *et al.*, 2009; Groselj *et al.*, 2010) can be considered as an alternative to class modeling. However, classical ANN models are difficult to interpret. Bottle-neck networks are somewhat more easily interpreted as Novic and Groselj (2009) showed on the classification of honeys (collected by TRACE) according to their origin based on their trace element concentrations.

7.3.3 Model performance

Irrespective of the modeling technique, models need to be optimized and validated. Optimization, for example, by cross-validation is part of the modeling step. Model validation requires an independent test set. Therefore, a representative subset of objects has to be selected from each class. One may randomly select objects to be included in the test set. However a better approach is to apply a stratified sampling approach, for example, by selecting samples according to the Kennard and Stone (1969) algorithm. This assures that the test set spans the entire data space. If the size of the data set does not permit a test set to be extracted we may validate by G-fold cross-validation (Frank and Todescini, 1994). In non-probabilistic multi-class classifications, measures of performance of a classifier are derived from the misclassification or confusion matrix (Table 7.2) of the class assignments obtained with the test set. From that table the specificity (or class purity) and sensitivity of the classifier with respect

Table 7.2 Misclassification or confusion matrix

Predicted class\true class	Class 1	Class 2	Class 3	α-error	Specificity (or class purity)	Sensitivity
Class 1 (n_1)	n_{11}	n_{12}	n_{13}	$(n_{12} + n_{13})/n_1$	$n_{11}/(n_{11} + n_{12} + n_{13})$	n_{11}/n_1
Class 2 (n_2)	n_{21}	n_{22}	n_{23}	$(n_{21} + n_{23})/n_2$	$n_{22}/(n_{21} + n_{22} + n_{23})$	n_{22}/n_2
Class 3 (n_3)	n_{31}	n_{32}	n_{33}	$(n_{31} + n_{32})/n_3$	$n_{33}/(n_{31} + n_{32} + n_{33})$	n_{33}/n_3
β-error	$(n_{21} + n_{31})/(n_2 + n_3)$	$(n_{12} + n_{32})/(n_1 + n_3)$	$(n_{13} + n_{23})/(n_1 + n_2)$			

to a given class are obtained. The sensitivity of a classifier with respect to class g is the percentage of objects belonging to class g that are correctly classified. The specificity of a classifier with respect to class g is the percentage of objects belonging to class g found in the group of objects classified in class g. The efficiency in classifying a certain class is the mean of sensitivity and specificity. The sensitivity of a classifier is equal to $(1-\alpha)$ and its specificity is related to β.

In the asymmetric case classification performance is usually indicated by its sensitivity or probability (α) of making a type I error (false negative rate). If enough 'other' products were available and classified, a binary confusion matrix ($n_3 = 0$ in Table 7.2) is obtained from which the specificity and the probability (β) of making a type II error (false positive rate) can be calculated. We note here that the sensitivity of a class is independent of its specificity. For instance, when n_{21} and n_{31} in Table 7.2 are large and n_{12} and n_{13} are small, the sensitivity of class 1 will be high (corresponding to a small α-error), but its specificity will be low (the β-error is large). Although it would appear that the β-error is independent of the α-error, one should realize that less specific models may have a better sensitivity. The performance of CART for discriminating Italian from non-Italian olive oils and Ligurian from non-Ligurian olive oils (Caetano *et al.*, 2007) demonstrates this very well (Table 7.3). This was confirmed by a recent study by Oliveri *et al.* (2011) on the SIMCA, UNEQ and POTFUNF classifications of honeys and olive oils collected by TRACE.

The question arises whether a classification should be optimized for maximal sensitivity, maximal specificity or maximal efficiency as a compromise. For quality issues except in some extreme cases, it is desirable to have a low number of authentic products that are classed as unauthentic due to trade and litigation issues, whereas a number of unauthentic samples classed as authentic can be tolerated. The stakeholders in question will define what an acceptable tolerance is. This means that a high sensitivity is required, whereas a low specificity may be acceptable. This condition is met for the aforementioned discrimination of non-Italian from Italian olive oils, but not for the Liguran from the non-Ligurian olive oils, although the efficiency of both

Table 7.3 Sensitivity and specificity of the classification of olive oils by CART

Olive oils	CART performance	
Italian *vs* non-Italian	Sensitivity	Specificity
First harvest	90.7	13.3
Second harvest	85.7	43.6
First + second harvests	87.6	31.4
Ligurian *vs* non-Ligurian		
First harvest	34.8	81.7
Second harvest	35.3	92.7
First + second harvests	11.3	94.8

Source: Caetano *et al.*, 2007.

classifications is about the same. In fact, instead of optimizing for maximal efficiency, one could better optimize sensitivity.

For safety issues the reverse is true. Here the labeling of unsafe products as being safe is less desirable. The tolerance of mislabeling is based on risk assessment expressed in a loss matrix L (Frank *et al.*, 1994). The off-diagonal elements of that matrix represent the loss associated with assigning a product to class g' whereas it is class g.

An important aspect is the modeling and discriminatory power of a single variable or set of variables. The modeling power is related to the sensitivity and the discriminatory power is related to the specificity of the classifier. Therefore, the modeling step usually (depending on the technique) includes the selection of an optimal subset k out of K variables. One way is to select the variables sequentially (univariate method) by applying the Fisher criterion (Pierna *et al.*, 2011). However, in this way the multivariate nature of the data, that is, the correlation between the variables, is disregarded. Therefore a stepwise selection is preferred by starting the modeling with the variable with the best discriminating power. The next important variable is added to the already selected variable by cross-validation. The procedure of adding variables to the subset of variables is continued until classification results stabilize. Nowadays computing-intensive variable selection methods are available, such as genetic algorithms, to select a subset of k variables from a set of K (Leardi *et al.*, 1992). The performance of the classifier is then optimized.

7.4 Conversion of verification and classification models into specifications

Classifiers are characterized by (a set of) conditions imposed on the outcome of a model on the input of attribute values of an object (food product). In principle any classifier is applicable in a traceability system as long as all model parameters and decision rules (intervals) are stored in the system together with a reference to the underlying modeling data set. A system requirement might be that decision rules are expressed in the original variables. This imposes requirements not only on the performance of the classifier, but also on its interpretability in terms of the original (manifest) variables. Therefore a major challenge in the TRACE project was to find a compromise between model performance and easy interpretation of the model, allowing us to formulate a set of easily applicable specifications.

7.4.1 Specifications expressed in the original variables

The ideal and simplest form of a specification is an interval specification for a single discriminative attribute. It specifies the interval within which the verified attribute of the object should be found with a certain probability. Interval specifications can be expanded to several attributes. A more complex form of specifications may be assigned to a combination of variables, for example, a weighted sum or to some other mathematical expression.

Interval specifications

Let us consider the case that during the modeling step one finds that the interval of the Na concentration of a limestone mineral water is between 1.3 and 371 ppm. A simple specification (or decision rule) would be that 'when 1.3 ppm < [Na] < 371 ppm, then the water is a limestone water'. One should realize that this is not necessarily true because other waters than a limestone water may fit that interval as well. This error is expressed by the β-error. As we want to keep the β-error within some accepted limits, we look for attributes (variables) with a minimal β-error for a given α-error. In other words we look for the attribute with best discriminatory power conditional to an acceptable modeling power. This is nothing else than classical variable selection (n-class case) by the Fisher criterion that maximizes the ratio of the between-class variance and the within-class variance. One should also note that although the reverse specification 'when a water is a limestone water then 1.3 ppm < [Na] < 371 ppm' is valid, it does not make sense in the context of traceability studies as the authenticity of the water is unknown.

The discriminating power of variables is easily compared by inspecting the box-plots of all variables for each class. Figure 7.3a shows the box-plots of the Mg content in honeys collected by TRACE in 18 regions. The solid lines give tentative lower and upper specification limits for a honey from Shalkidiki. From this plot we conclude that the discriminating power of Mg is good as long as we do not consider honeys from Barcelona (Bar) and Mue. However, if there is a concern about honeys from Barcelona, the potassium content would provide a better specification for Shalkidiki honeys, as can be seen from Fig. 7.3b. This example immediately points to the question whether it is advisable to include specifications of more variables in the decision rule. Let us consider two specifications for the grey dots in Fig. 7.4, one for x_1 and one for x_2: if $a < x_1 < b$ then Object is dot; if $c < x_2 < d$ then Object is dot.

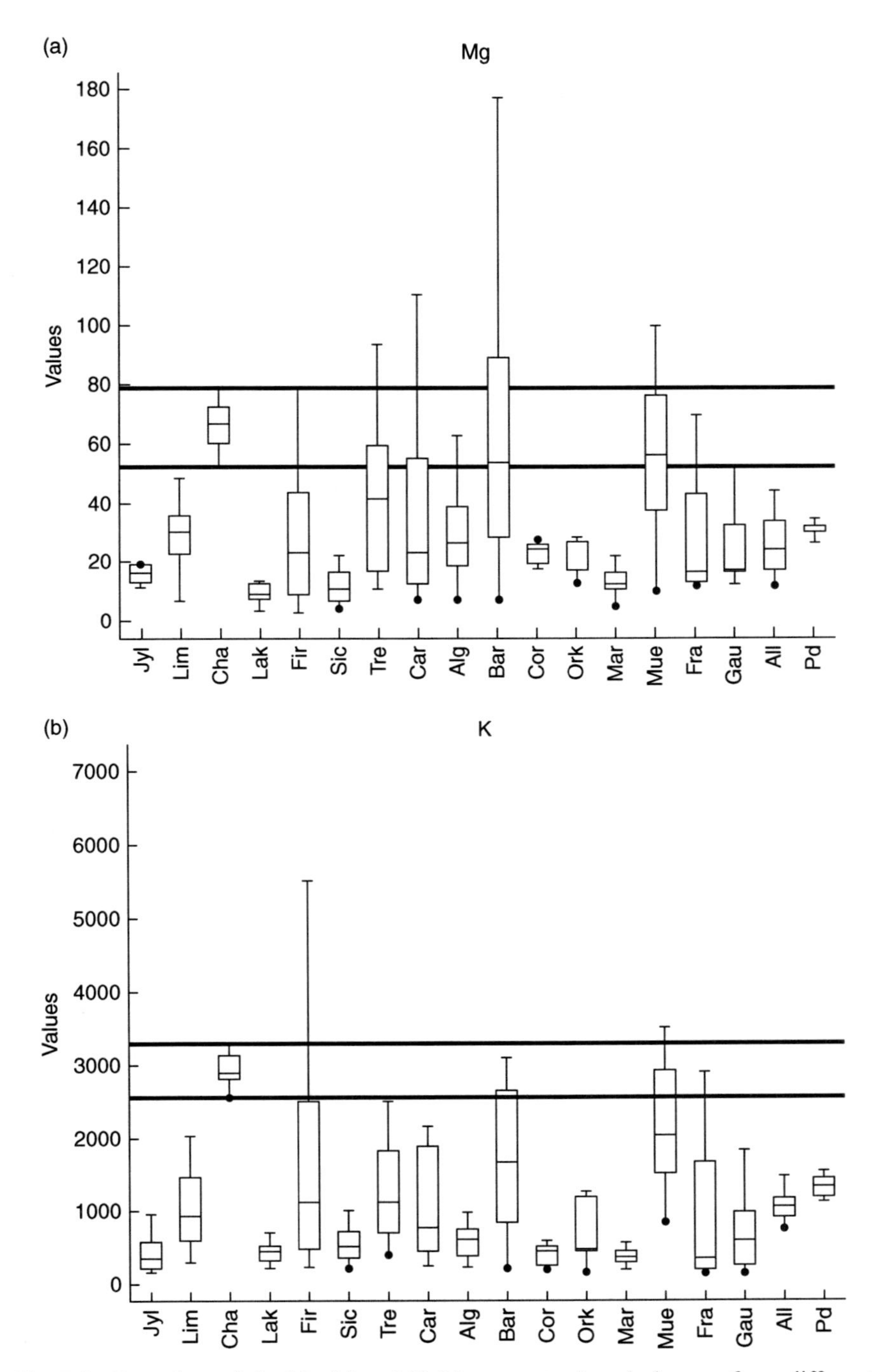

Fig. 7.3 Box-plots of the Mg (a) and K (b) concentrations in honeys from different regions.

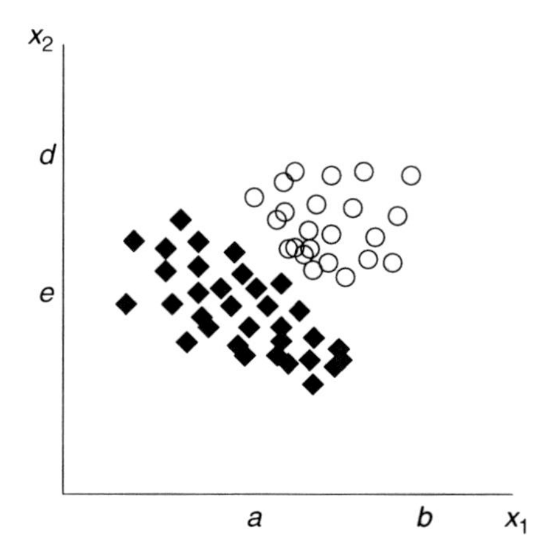

Fig. 7.4 Bivariate distributions of two classes with interval specifications for class (○).

Both specifications have a considerable β-error. However by combining both specifications in a bivariate specification 'if $a < x_1 < b$ AND $c < x_2 < d$ then object is dot' the specification is almost β-error free. Obviously, if there is still a β-error in the bivariate case, adding more variables may further reduce the β-error. One should note that this multi-univariate approach involves multiple univariate testing that requires a Bonferroni correction of α (Massart *et al.*, 1997). The example also demonstrates that a truly multivariate probability density function much better describes the data, as it takes the correlation between the variables into account. If we do so, the modeling becomes similar to the parametric modeling technique UNEQ or the Hotelling T^2 statistics (Vandeginste *et al.*, 1998, p. 228). As discussed before, parametric methods are less highly recommended as they require large data sets (the object-to-variable ratio should be greater than 3). Moreover, because in traceability studies we only accept small α-errors, tails of the distribution become important. However, this part of the probability density distribution may significantly deviate from normality. Decision rules consisting of a set of (upper and/or lower) interval specifications, as described above are directly obtained from CART decision trees. As discussed before, CART optimizes the tree efficiency. The derivation of decision rules from a CART tree is rather straightforward. This is illustrated by the classification of Italian olive oils according to their origin based on their NIR spectra (Fig. 7.5).

From the tree the following decision rules (specifications) can be formulated: (i) if signal at 1161.1 cm^{-1} > 0.00234799 then origin is Italy or (ii) if signal at 1161.1 cm^{-1} < 0.00234799 and signal at 1454.3 cm^{-1} < 0.00327995 then origin is Italy, else origin is not Italy. The purity of the end-nodes leads to a confusion matrix from which the sensitivity and specificity of the classifier are calculated (Table 7.3). We should note here that the CART tree is optimized for the best efficiency conditional on the smallest possible variable subset.

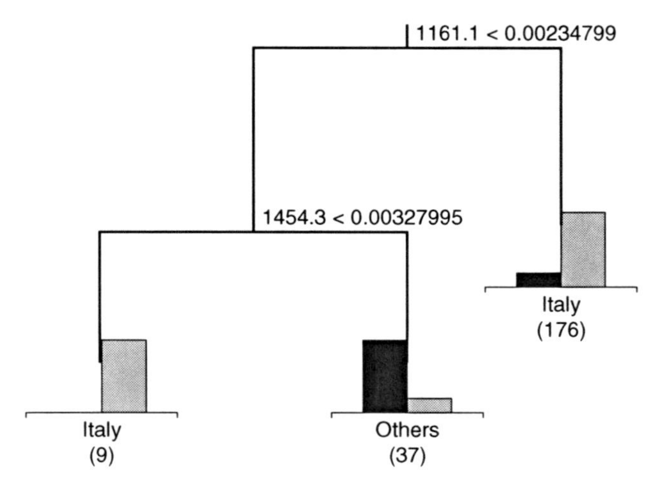

Fig. 7.5 Classification tree obtained for the discrimination of the Italian olive oils by NIR: second harvest. The values between parentheses represent the total number of samples classified in that node for the calibration set (grey – Italian samples, black – others, i.e., non-Italian samples).

Specifications in the format of a mathematical expression

As indicated before, a GP classifier generates and optimizes a desired computational function or mathematical expression that links a (minimized) subset of variables in order to produce explanatory rules. For instance, the set operators {+, –, /, *, 0.1, 1, 3, 5, rand, $\log_{10}$, 10^x, tanh} may link the variables in order to optimize the efficiency of the classifier. These computational functions or mathematical expressions directly yield specifications. For instance the tree that discriminates Corsican from non-Corsican honeys published by Donarski *et al.* (2008) is converted into the following specification (see reference for an explanation of the variables): If SQRT(V2446) > Min{[V11338 + V10010 + Min(V25180,V25340) – V13421], [AVG(Min (V8454, AVG(V24446,V22440)), V11338 + V25149 – AVG(V7003,V24435))]} then honey is Corsican. Discriminating methods, such as LDA and D-PLS yield linear boundaries between classes. These boundaries are expressed in the form of weighted sums of the original variables (Vandeginste *et al.*, 1998). A specification derived from a D-PLS model may take the following form:

if (log Mg – 1.232) * 0.329 – (log Na – 1.354) * 0.119 + (log SO_4 – 1.595) * 0.262 + (log NO_3 + 0.150) * 0.243 – (log X87Sr + 0.149) * 304.73 + (log Li + 1.83637) * 0.120 + (log Co + 3.928) * 0.403 + (log Ga + 5.0) * 0.193 + (log Rb + 2.736) * 0.305 + (log Cd + 5.252) * 0.207 – (log Ce + 4.760) * 0.254 + (log U + 3.3817) * 0.0875 > 0, then water is limestone water; else water is unsolid_rock water.

We should remark here that the variable scaling and transform are included in the model, allowing us to input untransformed concentrations in the model.

We also remark that the water is either classified as a limestone water or as an unsolid_water. Here, too, sparse models are preferred.

7.4.2 Interval specifications in latent space

We have already mentioned that by PCA a data matrix (**X**) is decomposed into a product of scores matrix (**T**) and an orthogonal loadings matrix (**V**). Each row of the scores matrix constitutes a reduced set of new attributes characterizing an object. Thus the attributes $x_1, x_2, \ldots, x_n$ are replaced by a smaller set $t_1, t_2, \ldots, t_p$ ($p < n$). For each class, we may define the range of these new attributes. Knowing these t-ranges we may derive decision rules for each class expressed in t in the same way as we did for the original variables x. Because the new attributes are a weighted sum of the original variables, these interval specifications for t may be back-transformed into intervals for the weighted sum of the manifest variables. One should realize that PCA optimizes retention of variance instead of discrimination between classes as LDA does. Therefore we recommend decomposing the data matrix by calculating the scores on canonical variates instead of PCs (Vandeginste *et al.*, 1998). Moreover, the above procedure does not take into account the distance of the objects from the PCA model and does not take correlation into account when calculating the scores ranges. In fact, PCA applied on a specific class (e.g., objects from a certain region) is the core of SIMCA modeling. In TRACE we demonstrated that it is feasible but not easy to convert SIMCA cut-off lines as shown in Fig. 7.1 into a set of specifications. Duchesne and MacGregor (2004) published a nice overview on the establishment of multivariate specification regions for incoming materials in a factory, which is a problem very similar to checking food authenticity.

7.5 Recent and future trends

Chemometrics has always suffered from the presence of defective data, outliers, missing values, censored data or various combinations of these defects. Determining how to cope with these phenomena is still high on the chemometrics agenda. Second, new challenges are imposed by the heterogeneity of data acquired from several sources requiring to combine or to fuse data and models. Finally, probabilistic (Bayesian) approaches in classification remains work in progress.

7.5.1 Probabilistic classifiers

We already mentioned that classes may overlap. Class modeling techniques may allocate a sample to a single class, to none of the classes or to several (most often two) classes. In the latter case the class of the sample remains undecided. This calls for a probabilistic approach that provides the probabilities that a given object belongs to a certain class. This probability depends on a number of uncertainties: (i) the uncertainty of the attributes characterizing the object, (ii) the uncertainty of the model parameters and (iii) the uncertainty of the decision limits (Bro *et al.*, 2005; Faber and Bro, 2002). In 1987 van der Voet and Coenegracht (1987) presented a probabilistic version of SIMCA and CLASSY that is based on the distribution of the objects within the class. Pérez *et al.* (2010) developed a probabilistic version of D-PLS that constructs Gaussian functions centered at the D-PLS model outcome for the modeling samples. The width is defined by the standard error of the predicted sample. After adding all Gaussian functions a class envelope is obtained. A Bayesian approach is then used to classify a new object. Recently a new probabilistic k-nearest neighbor algorithm was proposed to attribute class probabilities to classification results obtained for meat and olive oils (Friel and Pettitt, 2010).

7.5.2 Robust classifiers

Outliers and other defective data may reduce model quality. The classical approach is to detect outliers or rogue values. As discussed before, the elimination or substitution of outliers may in its turn affect the quality of the model. Not to say that such elimination or substitution of outliers may even lead to undesired manipulation of the data. When possible, robustified methods are preferable. Several steps in the modeling process may be robustified. The simplest robustifier is robust column centering. Instead of using the column means as in classic centering, the column medians are subtracted from each data element. Robust autoscaling is a little bit more demanding. Here we need a robust estimator of variance, the Q_n estimator. After a robust autoscaling we may calculate a robust PCA (robust variance–covariance matrix). There are several ways to make PCA robust. In spherical PCA (Locantore *et al.*, 1999), the objects are weighted according to the inverse of their Euclidean distances to the robust center (the median) of the data. The objects, which are far away from the data majority, get small weights, diminishing their influence in the final PCA model. Robust PCA is the kernel of robust-SIMCA. Since the classification rules in classic SIMCA are derived from PCA, a robust-SIMCA model can be obtained by replacing the classic PCA approach by its robust version. Similarly to the robust PCA method, the class model fit is based on a distance–distance plot, where the distances are calculated using robust scores and the residuals from the constructed robust model (Daszykowski *et al.*, 2007a). A robust-SIMCA approach can be preceded by a robust preprocessing procedure. Similar to classic SIMCA, one has to choose a method for uniform design, for example, the Kennard and Stone (1969) or Duplex (Daszykowski *et al.*, 2002b) algorithms, and the complexity of the robust-SIMCA model constructed for each individual class. Rousseeuw *et al.* (2006) robustified several chemometric techniques by replacing classic estimators for variance and covariance by the minimum covariance determinant (MCD). One of these tools is class modeling by a robust Hotelling T^2 test. A robust variance–covariance matrix is also the kernel of robust PLS (Hubert and Vanden

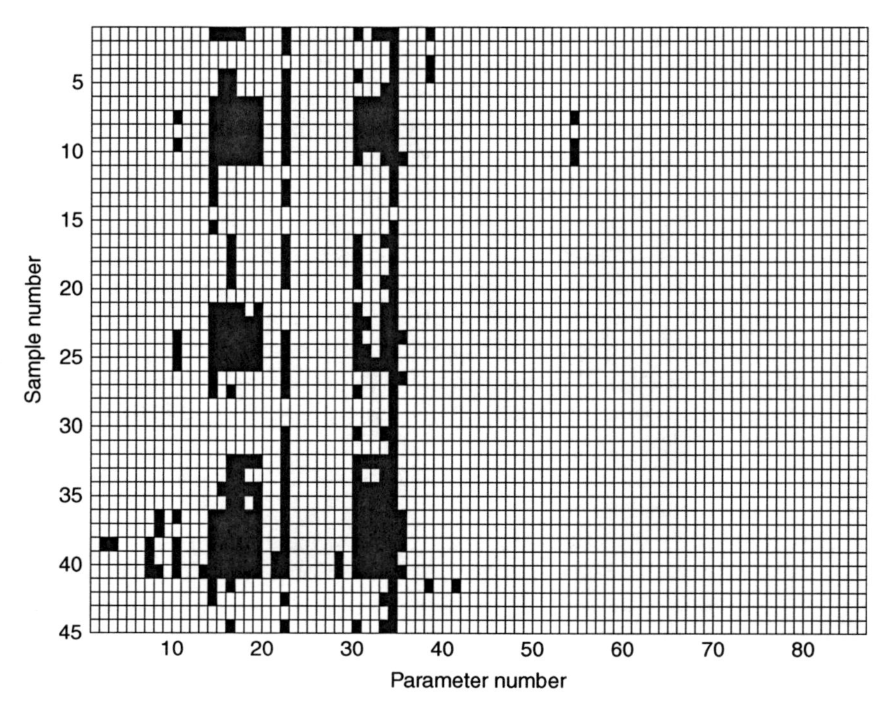

Fig. 7.6 Missing values in a typical data set measured in the TRACE project.

Branden, 2003) based on the SIMPLS algorithm from De Jong (1993). Debruyne *et al.* (2009) demonstrated that outliers added to a wine data set, were perfectly recovered by robust LS-SVM. Liebmann *et al.* (2010) compared robust and classical PLS methods and a tutorial on robust methods is available from Daszykowski *et al.* (2007b). Hubert *et al.* (2008) reviewed robust multivariate methods, including LDA, quadratic discriminant analysis (QDA), PCA, PCR and PLS.

7.5.3 Coping with corrupt data

Data cleaning and correction are important aspects. First of all the dataset has to be inspected for rogue values. These are clear blunders that should be corrected. More difficult is the detection and treatment of outliers. Some objects may be atypical and be an outlier for two reasons: either an attribute value is outlying or the assigned class is wrong. In TRACE, for example, the assignment of the geological property of a mineral water was not always straightforward. For one reason or another, some trace element concentrations may be missing. In multivariate analysis this would result in either the deletion of the object with the missing value (row of the data matrix) or the deletion of the trace element (column of the data matrix) from the matrix.

Figure 7.6 depicts a typical data set obtained in TRACE. Deletion of objects with missing values would reduce the data set to a few objects only.

Deleted variables on the other hand could be discriminating. Instead, one can replace missing values by the most likely value. We have already mentioned that principal components fill is the preferred method. However, the simultaneous occurrence of outliers and missing values poses a problem. Outlying objects cannot be detected as long as there are missing data and imputed values are influenced by the presence of outliers. In fact imputed values will move objects towards outlying objects that become less easily detectable. One particular type of data needing special attention are chromatographic data that need alignment before analysis. Several alignment procedures have been proposed (Walczak, 2005).

A particular concern arises when the origin of products is characterized by trace element profiles. Different laboratories may report different left censored (less than) values for the same analyte caused by different detection limits of their analytical method. Several methods have been proposed to substitute or to impute censored data (Jain *et al.*, 2008; Lind, 2010). Expectation maximization (EM) by treating censored data as missing values is one of these methods. The PCA iterations are continued until the imputed values are in agreement with the reported LOD.

7.5.4 Data and model fusion

In traceability studies one may collect data from various sources for the same set of objects, for example, spectroscopic data from different techniques (NMR, NIR, MS), concentration profiles, physical parameters and so on. The result is a collection of data sets. If all sets are modeled individually, we obtain as many classifiers as there are data sets, each with their own performance. In order to arrive at a final and unique decision, results from the different classifiers have to be combined. However, we cannot exclude the possibility that the global conclusion remains undecided as the classifiers may yield conflicting answers. In addition, the separate modeling of the sets does not exploit the synergy between data from different sources. We distinguish three levels of fusion (Liu and Brown, 2004). The first level, the data-level (low level) is simply the concatenation of all sets in a single set by putting all columns together. The second level is to concatenate all sets in latent space, called full hierarchical modeling. The third level is the fusion of models by hybrid modeling.

At the lowest level of fusion, variables may be selected among all sources. In this way the correlation between variables among different sources is included in the variable selection process. Concatenation of data sets may result in a landscape matrix that needs special attention during analysis (Kettaneha *et al.*, 2005). By selecting variables a landscape matrix may be converted into a portrait matrix that involves fewer difficulties when processing. At the second level multivariate analysis (e.g., PCA) models are constructed for each source separately. In the next step scores are combined in a new data set. This new scores data set is subsequently subjected to a multivariate analysis (MVA). Many papers discuss scaling issues arising at both fusion levels (Forshed *et al.*, 2007). Model fusion, the highest level of fusion is a form of hybrid modeling. It is not particularly designed to model data from several sources, but it aims to develop a hybrid model that is superior to the individual models. One of the first hybrid methods, CLASSY (Van der Voet and Coenegracht, 1987), was developed in 1987. It tries to take advantage of combining SIMCA and ALLOC in a single technique. Later on, other techniques were hybridized. For instance, instead of developing a PCR or PLS model, one may develop a hybrid PCR-PLS model that creates a combined space of PLS and PCA factors (Fang *et al.*, 2006). In pattern recognition one could fuse the results of individual classifiers obtained for different data sources by applying a Bayesian approach as Roussel *et al.* (2003) demonstrated on the classification of white grapes using a D-PLS classifier for each data source: aroma, FT-IR and UV. CART models may be combined by boosting and bagging (Cao *et al.*, 2010).

7.5.5 Model maintenance and updating

When utilizing a model over an extended period of time, it may be necessary to maintain or even to update the model. This is a well-known issue in multivariate calibration models based on NIR spectroscopy (Capron *et al.*, 2005). We all know that simple univariate calibration models need to be checked regularly by measuring check samples. In the context of traceability, there are two main reasons for model maintenance or updating. The first reason is a change in the analytical measurement process, for instance, another laboratory is involved in the analysis of products. A second reason is the occurrence of seasonal effects, for instance, one has to check whether a sample set from a new harvest is still well predicted by the model of previous harvests. Studies in TRACE clearly revealed large laboratory effects, specifically at low concentration levels of trace elements. There are two options if predictions by the new laboratory continue to fail. Either the laboratory has to take corrective actions until it is in agreement with the model, or the model has to be updated in order to include the variability of the new laboratory. Therefore, a subset of the modeling objects has to be selected and be analyzed by the new laboratory. A representative subset can be obtained by the use of the Kennard and Stone algorithm (Daszykowski *et al.*, 2002b). Studies in TRACE revealed large harvest effects in the classification of olive oils leading to completely different CART trees (Caetano *et al.*, 2007). One of the contributing causes could be the intrinsic instability of CART trees for small perturbations. A more stable model is obtained by creating a random forest (Lin *et al.*, 2010) of trees built for random sub-sets of the modeling set. Random forests (RF) are a variant of bagging. Bagging is a kind of bootstrap aggregation that builds trees from bootstrap samples. These are samples of the same size as the training set drawn from the original data matrix after replacement. All classifiers are then combined into a single one (Cao *et al.*, 2010). If there is a large harvest year effect, one has several options. Either the modeling is repeated after adding representative samples of the new year to the modeling set, or separate models are developed for each year. When sampling locations are identical over the years, three-way models (Durante *et al.*, 2011) may be considered. Which approach to adopt depends on the purpose of the traceability system being updated.

7.5.6 Bayesian approach for decision-making

ARTHUR (Harper *et al.*, 1977), the first comprehensive software package for pattern recognition introduced in chemometrics, comprised a Bayesian classifier. It calculated the *a posteriori* probability that a given object is member of class k, taking into account the *a priori* probability of class k and the risk (R_k) associated with the misclassification of an object in class k. In authenticity studies, there are two possible states: either the product is authentic (S_y) or the product is not authentic (S_n). The *a priori* probability $Pr(S_y)$ that a product is authentic depends on our trust in the producer. For a trusted producer we may judge that this probability is high, for example, equal to 0.95. For a new producer (no history records), this probability may be set at 0.5 and for a suspected producer, equal to 0.05. After checking a product against specifications, there are two possible test results: either the test result is 'authentic' (T_y) or the result is 'not authentic' (T_n). The conditional probability of a true state S_i given a test result T_j is given by:

$$\Pr\left(S_i/T_j\right)=\frac{\Pr\left(S_j\right)\Pr\left(T_j/S_i\right)}{\Sigma_i \Pr\left(S_i\right)\Pr\left(T_j/S_i\right)} \quad [7.1]$$

In classification the above probabilities may be derived from the confusion table:

$$\Pr\left(T_y \mid S_y\right)=(1-\alpha);\Pr\left(T_y \mid S_n\right)=\beta;\Pr\left(T_n \mid S_n\right)=(1-\beta);\Pr\left(T_n \mid S_y\right)=\alpha$$

By substituting these probabilities in the above equation we may calculate four likelihoods (i) the likelihood that a producer (or product) is 'authentic' conditional to an 'authentic' test result: $Pr(S_y|T_y) = (1-\alpha)Pr(S_y)/[(1-\alpha)(Pr(S_y) + \beta Pr(S_n)]$, (ii) the likelihood that a producer (or product) is 'authentic' conditional to an 'not authentic' test result: $Pr(S_y|T_n) = \alpha Pr(S_y)/[\alpha(Pr(S_y) + (1-\beta) Pr(S_n)]$ and the likelihood that a producer is 'not in compliance' (iii) conditional to an 'authentic' test result and (iv) conditional to an 'unauthentic' test result. The posterior 'probabilities' are listed in Table 7.4 as a function of producer trust, α-error and β-error.

As we can see, a high value of β (=0.9) in combination with a small α (= 0.1) does not change our prior confidence we already had in the producer, irrespective of the outcome of the test. One nice aspect is that irrespective of the level of trust in a producer, a negative test result always leads to the

Table 7.4 Posterior probability of authenticity as a function of the α and β-error of the decision rule and trust in the producer

		Posterior compliance probability					
		Trusted producer		Suspected producer		New producer	
α	β	T = y[a]	T = n[b]	T = y	T = n	T = y	T = n
0.01	0.01	1	0.08	0.92	0	0.99	0.01
0.01	0.5	0.95	0.15	0.18	0	0.66	0.02
0.01	0.9	0.91	0.47	0.11	0.01	0.52	0.09
0.1	0.01	1	0.48	0.91	0.01	0.99	0.01
0.1	0.5	0.94	0.64	0.17	0.02	0.64	0.17
0.1	0.9	0.9	0.9	0.1	0.1	0.5	0.5

[a]Test outcome is authentic.
[b]Test outcome is unauthentic.

conclusion that there is only a small chance that the product is authentic. However an 'authentic' test result for a new and suspected producer will only lead to a decision of 'authentic' if the β-error is small (=0.01). This points to the necessity of having a good estimate of the β-error when developing Bayesian decision rules.

7.6 Conclusion

A traceability system should be cost-effective, easily maintainable and accurate when non-compliance of a product or producer is concluded. It implies that the measurement process should be cheap, thus simple with a minimal number of analytes measured on a minimal number of samples. Cost of sampling depends on the region(s) of interest, defined by the scope of the traceability system. Measurement costs are reduced by using fast and cheap screening methods that should flag suspected products before proceeding to a more demanding method of analysis. Which model is developed again depends on the system requirements. A possible work flow when initiating a traceability study is provided in Table 7.5.

In the asymmetric case, only class modeling is applicable, such as SIMCA or POTFUN. If products from distinct but well-defined regions should be discriminated, several models can be developed such as LDA, CART, GP, D-PLS, ANN, SVM (only 2-class case). But SIMCA and POTFUN are still applicable. Usually one tries all methods and selects the one with the best performance or the model that is the easiest expressed in the original variables. However, one should be aware that discrimination methods optimize the global performance over all categories (expressed by sensitivity and specificity). Therefore, an unbalanced number of samples per category should be avoided. Also in the case of multi-class data one could opt for developing a class model for each

Table 7.5 Work flow of a traceability study

Start of traceability study: problem formulation	
I want to verify the authenticity of a product Select a representative sample of products from the region you want to verify. The data belong to a single category.	I want to assign a product to specified regions Select a representative sample of products from each region (~equal size). The data are categorized.
When wrong assignments of unauthentic products are an issue	
You need a reasonably large sample of unauthentic products as well. The data are categorized in two classes: authentic and unauthentic.	You may consider products from regions other than the target region being representative for unauthentic products.
Specificity of the classifier should be optimized	
Visual inspection of the data	
Variables Inspect box-plots (for all variables for each category) Inspect histograms Consider data transformation (logarithm) of some variables Visualize spatial distribution of missing values	
Objects Inspect PCA plot of all objects and identify possible outlying objects Replot PCA after deletion of outlying object(s) The PCA plot should indicate how well classes are clustered and/or separated The PCA plot should also indicate whether classes are linearly separable Consider data transformation to improve the situation (logarithm, autoscaling)	
Prepare data for analysis	
Apply statistical test for rogue values and/or outliers Replace censored values Impute missing values in the absence of outliers Apply specific preprocessing techniques designed for spectra and chromatograms Consider autoscaling when range of variables differs a lot Consider variable selection when data are categorized Split the data set in a modeling set (70%) and test set (30%) by the Kennard & Stone or Duplex algorithm (each class separately) If split is impossible prepare later on for G-fold cross-validation and/or bootstrap	
When the conversion of the classifier into specifications is not required	
Model the authentic samples by SIMCA or POTFUN. Optimize the model for high sensitivity (low α). Inspect distance–distance plot (SIMCA) of modeling set. Validate model with the test set. Inspect distance–distance plot of (SIMCA) test set.	Develop and validate SIMCA or POTFUN models for each class (see left for work flow). Validate models with test set. Evaluate sensitivity and specificity of each model. Alternatives: 1) Develop and validate a multi-class CART or GP decision tree (specifically when model sparsity is an issue).

(Continued)

Table 7.5 Continued

Inspect possible outliers in the distance–distance plots and remove their influence by robust autoscaling (when autoscaling was applied) followed by robust-SIMCA. Evaluate the classification of unauthentic samples when available and estimate specificity. Try to improve sparsity of the model by removing variables with a poor modeling power without losing sensitivity and/or specificity.	Validate models with a test set. Evaluate node impurities. Improve CART results by boosting or decompose node impurities by additional LDA or D-PLS. 2) Develop and validate binary classifiers for a selected target class against the rest by D-PLS or LDA. When sensitivity and/or specificity are insufficient, consider binary classifications by SVM or multi-category classifications by ANN. Another approach is pairwise classification, where a new classifier is created for each pair of classes. A final classification is obtained by combining the individual classifiers.
When classifiers need to be converted into a set of specifications	
Prepare the availability of unauthentic products (may be heterogeneous). Develop, optimize and validate binary classifiers for authentic *vs* unauthentic products by D-PLS or LDA. Collect validation data in confusion matrix.	Develop, optimize and validate a multi-class CART or GP decision tree. Validate models with a test set. Evaluate node impurities. Improve results by boosting. Alternative: Develop and validate binary classifiers for a selected target class against the rest by D-PLS or LDA (see work flow above).
Convert models into specifications	
Decompose the CART model into a set of < and > conditions Rewrite the mathematical expression represented by GP in an explicit way Describe the LDA or D-PLS boundary between two classes in an algebraic way	
Classifier maintenance	
Check classifier(s) with new products of trusted origin Check results of a new laboratory by analyzing a subset of the product base Check validity of classifiers for new harvests or other changes Remedial actions to be considered 1) Improve stability of, for example CART models by boosting or random forests 2) Add the new samples to the available sample base and repeat the whole modeling step 3) define a new sample base and repeat the whole modeling step 4) Consider the option of multi-way modeling	
Advanced options	
Data fusion at the 1st and 2nd level Hybrid modeling Fusing results of individual classifiers by a Bayesian approach	

group. As these models are independent from each other, an unbalanced number of samples per category is less a problem. Moreover, each model is individually maintained and updated. A typical traceability study usually involves the development of several classifiers as was the case in the study on tracing the geographical origin of honeys based on volatile compounds profiles (Stanimirova *et al.*, 2010). As one can see there is plenty of choice of techniques, but this can also be dangerous, as pointed out by Wold *et al.* (1984) who warned against applying different pattern recognition methods and then selecting the results that 'look best'. Not the best results, but the best method should be selected. On the other hand, when several classifiers have been developed, we would like to use all information that is in the models by making them acting together (or in competition) to decide on the class of a new object. This is what boosting does (Freund and Schapire, 1997) which was applied in chemometrics by Zhang *et al.* (2005). The bulk of papers on food authentication reports either the sensitivity, specificity or efficiency to indicate classification performance. It makes little sense to report efficiencies alone, as it does not discriminate type I from type II errors. In the best case sensitivities around 95% are reported being commented as (very) good. Whether this is good enough is debatable, as it would mean that there is a big chance to decide that a product (or producer) is not authentic whereas it is. Such high error will never stand in court. Therefore, to improve both analytical and chemometric capabilities in order to achieve sufficient performance is still a great challenge. It is clear that prior to the modeling, data should be visually inspected, for example, by PCA plots, box-plots or by newer visualization techniques, such as parallel coordinate plots (Yang *et al.*, 2003), before and after scaling or any other transform. Although uninformative variables usually do not deteriorate model performance, one wants to select variables from the point-of-view of cost-effectiveness. CART and GP intrinsically optimize the variable set. Inspection of D-PLS and SIMCA models reveal the modeling power of the variables. Fisher ratios are good indicators for the discriminating variables in LDA. Variable selection before SVM and ANN modeling may require a brute-force approach by, for example, a genetic algorithm.

Within TRACE a Matlab chemometrics toolbox was developed with modules for CART, R-SIMCA, SVM, Kohonen ANN, probabilistic D-PLS including algorithms to impute missing values, boosting CART trees and sub-group selection by the Kennard and Stone algorithm. This toolbox is available from the author on request.

7.7 Further reading

Michèle Lees (Ed), *Food Authenticity and Traceability*, Woodhead Publishing Ltd, 2003.

Michael Jee (Ed), *Oils and Fat Authenticity*, Blackwell Publishing, Ltd, 2002.

7.8 Acknowledgement

The author acknowledges all TRACE partners of Work-package 6, Statistical Specifications, for their contributions to the results referenced here and for their stimulating discussions during the work-package team meetings.

TRACE was a project in the Sixth Framework Programme of the European Union, Project No. FOOD-CT-2005–006942.

More information on the TRACE project is available on http://www.trace.eu.org/.

7.9 References

BREIMAN L, FRIEDMAN J H, OLSHEN R A and STONE C G (1984), *Classification and Regression Trees*, Belmont, CA: Wadsworth International Group.

BRO R, RINNAN A and FABER N M (2005), 'Standard error of prediction for multilinear PLS – 2. Practical implementation in fluorescence spectroscopy', *Chemometrics and Intelligent Laboratory Systems*, **75**, 69–76.

CAETANO S, AIRES-DE-SOUSA J, DASZYKOWSKI M and VANDER HEYDEN Y (2005), 'Prediction of enantio selectivity using chirality codes and Classification and Regression Trees', *Analytica Chimica Acta*, **544**, 315–326.

CAETANO S, USTUN B, HENNESSY S, SMEYERS-VERBEKE J, MEISSEN W, DOWNEY G, BUYDENS L and VANDER HEYDEN Y (2007), 'Geographical classification of olive oils by the application of CART and SVM to their FT-IR', *Journal of Chemometrics*, **21**, 324–334.

CAO D-S, XU Q-S, LIANG Y-Z, CHEN X and LI H-D (2010), 'Automatic feature subset selection for decision tree-based ensemble methods in the prediction of bioactivity', *Chemometrics and Intelligent Laboratory Systems*, **103**, 129–136.

CAPRON X, WALCZAK B, DE NOORD O E and MASSART D L (2005), 'Selection and weighting of samples in multivariate regression model updating', *Chemometrics and Intelligent Laboratory Systems*, **76**, 205–214.

CENTNER V, MASSART D L, DE NOORD O E, DE JONG S, VANDEGINSTE B G M and STERNA C (1996), 'Elimination of uninformative variables for multivariate calibration', *Analytical Chemistry*, **68**, 3851–3858.

COOMANS D, MASSART D L, BROECKAERT I and TASSINC A (1981), 'Potential methods in Pattern Recognition 1. Classification aspects of the supervised method ALLOC', *Analytica Chimica Acta*, **133**, 215–224.

CRISTIANINI N and SHAWE-TAYLOR J (2000), *An introduction to support vector machines and other kernel-based learning methods*, Cambridge, UK: Cambridge University Press.

DASZYKOWSKI M, KACZMAREK K, STANIMIROVA I, VANDER HEYDEN Y and WALCZAK B (2007a), 'Robust SIMCA – bounding influence of outliers', *Chemometrics and Intelligent Laboratory Systems*, **87**, 95–103.

DASZYKOWSKI M, KACZMAREK K, VANDER HEYDEN Y and WALCZAK B (2007b), 'Robust statistics in data analysis – a review. Basic concepts', *Chemometrics and Intelligent Laboratory Systems*, **85**, 203–219.

DASZYKOWSKI M, WALCZAK B and MASSART D L (2002a), 'Looking for natural patterns in analytical data. 2. Tracing local density with OPTICS', *Journal of Chemical Information and Computer Sciences*, **42**, 500–507.

DASZYKOWSKI M, WALCZAK B and MASSART D L (2002b), 'Representative subset selection', *Analytica Chimica Acta*, **468**, 91–103.

DEBRUYNE M, SERNEELS S and VERDONCK T (2009), 'Robustified least squares support vector classification', *Journal of Chemometrics*, **23**, 479–486.

DE JONG S (1993), 'SIMPLS: an alternative approach to partial least squares regression', *Chemometrics and Intelligent Laboratory Systems*, **18**, 251–263.

DERDE M P and MASSART D L (1988), 'Comparison of the performance of the class modeling techniques UNEQ, SIMCA, and PRIMA', *Chemometrics and Intelligent Laboratory Systems*, **4**, 65–93.

DONARSKI J A, JONES and CHARLTON A L (2008), 'Application of cryoprobe ^{1}H nuclear magnetic resonance spectroscopy and multivariate analysis for the verification of corsican honey', *Journal of Agricultural and Food Chemistry*, **56**, 5451–5456.

DUCHESNE C and MACGREGOR J F (2004), 'Establishing multivariate specification regions for incoming materials', *Journal of Quality Technology*, **36**, 78–94.

DURANTE C, BRO R and COCCHI M (2011), 'A classification tool for N-way array based on SIMCA methodology', *Chemometrics and Intelligent Laboratory Systems*, **106**, 73–85.

ELLISON S L R and WILLIAMS A, Ed. (2007), *Use of uncertainty information in compliance assessment*, EURACHEM/CITAC Guide.

FABER N M and BRO R (2002), 'Standard error of prediction for multiway PLS 1. Background and a simulation study', *Chemometrics and Intelligent Laboratory Systems*, **61**, 133–149.

FANG Y, CHO H and JEONG M (2006), http://citeseerx.ist.psu.edu/viewdoc/download?doi=10.1.1.123.5218&rep=rep1&type=pdf.

FORINA M, ARMANINO C, LEARDI R and DRAVA G (1991), 'A class-modeling technique based on potential functions', *Journal of Chemometrics*, **5**, 435–453.

FORINA M, ARMANIO C, LANTERI S and TISCORNIA E (1983), 'Classification of Olive Oils from their fatty acid composition', in Martens H and Russwurm H Jr (Ed.) *Food Research and Data Analysis*, London: Applied Science, 189–214.

FORINA M and LANTERI S (1984), 'Data analysis in food chemistry', in KOWALSKI B R (Ed), *Chemometrics – Mathematics and Statistics in Chemistry*, Dordrecht: D. Reidel Pu, 305–349.

FORINA M, OLIVERI P, LANTERI S and CASALE M (2008), 'Class-modeling techniques, classic and new, for old and new problems', *Chemometrics and Intelligent Laboratory Systems*, **93**, 132–148.

FORSHED J, IDBORG H and JACOBSSON S P (2007), 'Evaluation of different techniques for data fusion of LC/MS and ^{1}H-NMR', *Chemometrics and Intelligent Laboratory Systems*, **85**,102–109.

FRANK I E and TODESCINI R (1994), *The data analysis handbook*, Amsterdam: Elsevier Science.

FREUND Y and SCHAPIRE R E (1997), 'A decision-theoretic generalization of on-line learning and an application to boosting', *Journal of Computer and System Sciences*, **55**, 119–139.

FRIEL N and PETTITT A N (2010), 'Classification using distance nearest neighbours', *Statistics and Computing*', http://dx.doi.org/10.1007/s11222-010-9179-y.

GARCIA-GONZALEZ D L, LUNA G, MORALES M T and APARICIO R (2009), 'Stepwise geographical traceability of virgin olive oils by chemical profiles using artificial neural network models', *European Journal of Lipid Science and Technology*, **111**, 1003–1013.

GILBERT R J, GOODACRE R, WOODWARD A and KELL D B (1997), 'Genetic programming: a novel method for the quantitative analysis of pyrolysis mass spectral data', *Analytical Chemistry*, **69**, 4381–4389.

GONZALVEZ A, LLORENS A, CERVERA M L, ARMENTA S and DE LA GUARDIA M (2009), 'Elemental fingerprint of wines from the protected designation of origin Valencia', *Food Chemistry*, **112**, 26–34.

GROSELJ N, VAN DER VEER G, TUSAR M, VRACKO M and NOVIC M (2010), 'Verification of the geological origin of bottled mineral water using artificial neural networks', *Food Chemistry*, **118**, 941–947.

HARPER A M, DUEWER D L, KOWALSKI B R and FASCHING J L, (1977), 'ARTHUR and experimental data analysis: the heuristic use of a polyalgorithm' in KOWALSI B R (Ed.), *Chemometrics: Theory and Application*, Washington: American Chemical Society, 14–52.
HUBERT M, ROUSSEEUW P J and VAN AELST S (2008), 'High-breakdown robust multivariate methods', *Statistical Science*, **23**, 92–119.
HUBERT M and VANDEN BRANDEN K (2003), 'Robust methods for partial least squares regression', *Journal of Chemometrics*, **17**, 537–549.
JAIN R B, CAUDILL S P, WANG R Y and MONSELL E (2008), 'Evaluation of maximum likelihood procedures to estimate left censored observations', *Analytical Chemistry*, **80**, 1124–1132.
KENNARD R W and STONE L A (1969), 'Computer aided design of experiments', *Technometrics*, **11**, 137–148.
KETTANEHA N, BERGLUNDB A and Wold S (2005), 'PCA and PLS with very large data sets', *Computational Statistics and Data Analysis*, **48**, 69–85.
LEARDI R, BOGGIA R and TERRILE M (1992), 'Genetic Algorithms as a Strategy for feature selection', *Journal of Chemometrics*, **6**, 267–281.
LIEBMANN B, FILZMOSER P and VARMUZA K (2010), 'Robust and classical PLS regression compared', *Journal of Chemometrics*, **24**, 111–120.
LIN X, SUN L, LI Y, GUO Z, LI Y, ZHONG K, WANG LU X, YANG Y and Xu G (2010), 'A random forest of combined features in the classification of cut tobacco based on gas chromatography fingerprinting', *Talanta*, **82**, 1571–1575.
LIND P (2010), 'QSAR analysis involving assay results which are only known to be greaterthan, or less than some cut-off limit', *Molecular Informatics*, **29**, 845–852.
LIU Y and BROWN S (2004), 'Wavelet multiscale regression from the perspective of data fusion: new conceptual approaches', *Analytical and Bioanalytical Chemistry*, **380**, 445–452.
LOCANTORE N, MARRON J S, SIMPSON D G, TRIPOLI N, ZHANG J T and COHEN K L (1999), 'Robust principal component analysis for functional data', *Sociedad de Estadistica e Investigacion Operativa Test*, **8**, 1–74.
LÓPEZ-DÍEZ E, BIANCHI G and GOODACRE R (2003), 'Rapid quantitative assessment of the adulteration of virgin olive oils with hazelnut oils using Raman spectroscopy and chemometrics', *Journal of Agricultural and Food Chemistry*, **51**, 6145–6150.
MACATELLI M, AKKERMANS W K, BUCHGRABER M, PATERSON A and VAN RUTH S (2009), 'Verification of the geographical origin of European butters using PTR-MS', *Journal of Food Composition and Analysis*, **22**, 169–175.
MARINI F, MAGRI A L, BUCCI R, BALESTRIERI F and Marini D (2006), 'Class-modeling techniques in the authentication of Italian oils from Sicily with a Protected Denomination of Origin (PDO)', *Chemometrics and Intelligent Laboratory Systems*, **80**, 140–148.
MASSART D L, VANDEGINSTE B G M, BUYDENS L M C, de JONG S, P J Lewi and Smeyeres-Verbeke J (1997), *Handbook of Chemometrics and Qualimetrics: Part A*, Amsterdam: Elsevier Science BV.
MILLAN C P, FORINA M, CASOLINO C and Leardi R (1998), 'Extraction of representative subsets by potential functions method and genetic algorithms', *Chemometrics and Intelligent Laboratory Systems*, **40**, 33–52.
NOVIC M and GROSELJ N (2009), 'Bottle-neck type of neural network as a mapping device towards food specifications', *Analytical Chimica Acta*, **649**, 68–74.
OLIVERI P, DI EGIDIO V, WOODCOCK T and DOWNEY G (2011), 'Application of class-modeling techniques to near infrared data for food authentication purposes', *Food Chemistry*, **125**, 1450–1456.
PÉREZ N F, FERRÉ J and BOQUE R (2010), 'Multi-class classification with probabilistic discriminant partial least squares (p-DPLS)', *Analytica Chimica Acta*, **664**, 27–33.
PIERNA J F, ABBAS Q, DARDENNE P and BAETEN V (2011), 'Discrimination of Corsican honey by FT Raman spectroscopy and chemometrics', *Biotechnology Agronomy Society Environment*, **5**, 75–84.

RANNAR S, GELADI P, LINDGREN F and WOLD S (1995), 'A PLS Kernel algorithm for data sets with many variables and few objects. 2. Cross-validation, missing data and examples', *Journal of Chemometrics*, **9**, 459–470.

ROUSSEEUW P J, DEBRUYNE M, ENGELEN S and HUBERT M (2006), 'Robustness and outlier detection in chemometrics', *Critical Reviews in Analytical Chemistry*, **36**, 221–242.

ROUSSEL S, BELLON-MAUREL V, ROGER J-M AND GRENIER P (2003), 'Fusion of aroma, FT-IR and UV sensor data based on the Bayesian inference. Application to the discrimination of white grape varieties', *Chemometrics and Intelligent Laboratory Systems*, **65**, 209–219.

SNEE R D (1977), 'Validation of regression models: methods and examples', *Technometrics*, **19**, 415–428.

STANIMIROVA I, ÜSTÜN B, CAJKA T, RIDDELOVA K, HAJSLOVA J, BUYDENS L M C and WALCZAK B (2010), 'Tracing the geographical origin of honeys based on volatile compounds profiles assessment using pattern recognition techniques', *Food Chemistry*, **118**, 171–176.

STANIMIROVA I and WALCZAK B (2008), 'Classification of data with missing elements and outliers', *Talanta*, **76**, 602–609.

ÜSTÜN B, MELSSEN W J and BUYDENS L M C (2006), 'Facilitating the application of Support Vector Regression by using a universal Pearson VII function based kernel', *Chemometrics and Intelligent Laboratory Systems*, **81**, 29–40.

VANDEGINSTE B G M, MASSART D L, BUYDENS L M C, de JONG S, LEWI P J and SMEYERS-VERBEKE J (1998), *Handbook of Chemometrics and Qualimetrics: Part B*, Amsterdam: Elsevier Science.

VAN DER VOET H and COENEGRACHT P M J (1987), 'New probabilistic versions of the SIMCA and CLASSY classification methods', *Analytica Chimica Acta*, **192**, 63–75.

WALCZAK B (2005), 'Fuzzy warping of chromatograms', *Chemometrics and Intelligent Laboratory Systems* **77**, 173–180.

WALCZAK B and MASSART D L (2001), 'Tutorial. Dealing with missing data', *Chemometrics and Intelligent Laboratory Systems*, **581**, 15–127.

WEHRENS R and BUYDENS L M C (1998), 'Evolutionary optimisation: a tutorial', *TRAC-Trends in Analytical Chemistry*, **17**, 193–203.

WOLD S, ALBANO C, DUNN III W J, EDLUND U, ESBENSEN K, GELADI P, HELLBERG S, JOHANSSON E, Lindberg S and SJOSTROM M, in KOWALSKI B R (Ed), *Chemometrics, Mathematics and Statistics in Chemistry*, Reidel, Dordrecht, 1984, 74.

YANG J, WARD M O and RUNDENSTEINER E A (2003), 'Interactive hierarchical displays: a general framework for visualization and exploration of large multivariate data sets', *Computers and Graphics*, **27**, 265–283.

ZHANG M H, XUB Q S, DAEYAERT F, LEWI P J and MASSART D L (2005), 'Application of boosting to classification problems in chemometrics', *Analytica Chimica Acta*, **544**, 167–176.

ZUPAN J, NOVIC M and RUISANCHEZ I (1997), 'Kohonen and counterpropagation artificial neural networks in analytical chemistry', *Chemometrics and Intelligent Laboratory Systems*, **38**, 1–23.

Part III

Applications to food commodities

8

Using new analytical approaches to verify the origin of wine

B. Médina, M. H. Salagoïty, F. Guyon and J. Gaye, Service Commun des Laboratoires (SCL), France, P. Hubert, Centre Etudes Nucléaires de Bordeaux Gradignan, Université de Bordeaux, France and F. Guillaume, Institut des Sciences Moléculaires (ISM), Université de Bordeaux, France

DOI: 10.1533/9780857097590.3.149

Abstract: Modern methods described here, have for ultimate goal to check the main components of a wine label: geographical origin, real composition and vintage. Detecting counterfeiting, ensuring traceability, guaranteeing the authenticity of products and other investigations more in the judiciary domain have given us the opportunity to apply proven methodologies generally used in other areas (environment, geology, food-products in general).

Key words: wine, authenticity, origin, vintage, advanced analytical methods.

8.1 Introduction

There have been many examples of fraud and falsification in the wine domain. Modern verification methods designed to combat such falsification ultimately aim to check three key elements: geographical origin, real composition according to the label (grape variety for example) and vintage, all of which can be very difficult to assess without prior knowledge.

These three elements are intrinsically linked and usually inseparable; they are the main components of a wine label and the consumer is generally unaware that underlining them are very strict conditions of production, having respect of which ensures fair trade. Detecting counterfeiting, ensuring traceability, guaranteeing the authenticity of products and other investigations more in the judicial domain have given us the opportunity to apply proven methodologies generally used in other areas, such as environmental studies, geology and food product analysis.

In this chapter, methods and techniques for the verification of wine origin are discussed. The simple techniques used to produce valuable results in areas such as traceability of a given wine through 'cellaring', and grape variety recognition are considered, before more evolved isotopic techniques for assessing geographical and botanical origin, as well as compliance with production legislation, are described. The use of trace minerals, certain geochemical isotope ratios, wine tasting and contaminants are then examined alongside wine 'dating' using ultra-low radioactivity, among other methods, and the use of Raman spectrometry to investigate counterfeit wine. Finally, an unforgeable chemical identity card for wine is proposed, using NMR as the ultimate tool.

The methods discussed in this chapter are listed below and are those used to solve the toughest problems facing an official control laboratory:

- traditional methods, GC, HPLC, AA, ECHP;
- advanced mathematical and statistical techniques: classification into groups, selection of features, partial least square (PLS) regression, etc.;
- NMR, IRMS: direct analysis, stable isotopes;
- ICP/MS: mineral, isotope ratios of minerals;
- ultra-low radioactivity: radioactive isotopes;
- Raman spectroscopy; and
- organoleptic testing (wine tasting).

The use of these methods has led to many real problems being solved, and examples of their use in facilitating verification procedures are given below.

8.2 Wine authentication using wine tasting and standard analytical methods

Very often it is not necessary to jump to up to date methods to solve authentication problems and the most simple methods can be used.

8.2.1 Wine tasting

Wine tasting has been in use since man first fermented grapes,[1] in order to compare, set prices, determine provenance or detect defects. Although very subjective in principle, modern techniques permit sensory evaluation to be used in an objective way; defects are more easily detected,[2] triangular tests permit a reference sample to be compared to a wine treated with a specific oenological practice,[3] and this can be a very efficient way to detect counterfeit wine from the original. Following the appropriate Quality Assurance/Quality Control procedure (ISO 17025) is mandatory in ensuring confidence in results.

8.2.2 Traceability through cellaring and differentiation of close appellations

Cellaring follows fermentation and continues through bottling.[4,5] During this period the winemaker undertakes various clarifications which result in a clear and stable wine. On average cellaring lasts from 16 to 18 months. In terms of traceability, the challenge is to identify chemical components present at the end of the fermentation which provide a good representation of the original and are still found at the end of the cellaring process. Tasting was the method traditionally used for this purpose and it served as the traceability base for transactions. The use of simple methods such as GC, HPLC, AA and ECHP, within a strict quality assurance regime (ISO 17025) may be sufficient to ensure the authenticity of a wine, including verification of its geographical origin. It is, in the latter case, mandatory to have an authentic reference sample so that results can be compared. The most stable components of wine, including alcohols, higher alcohols, organic acids and some metals, act as robust markers which change little over time and can be representative of a production lot for a very long time. The determination of these components is used to verify that a wine tasted at different times is the same, or to demonstrate that two wines of close 'appellations' (i.e., of different geographical origin but still very close) actually differ far more than expected.

Figure 8.1 shows variations in the chosen parameters after wine has been aged for six months, during which time it has undergone numerous oenological treatments including racking, sulphiting, clarification and aeration. Logically there is an increase of ethanal and ethyl-acetate, but the other parameters do not vary. After six months the 'footprint' is still perfectly recognizable.

This method has also been put to the test to characterize two close appellations, as demonstrated in Fig. 8.2. In this example, one road divides the two vineyards, with St Emilion on one side and Pomerol on the other. These geographic boundaries were set a long time ago and have been validated by the INAO (National Institute for Appellation of Origin and quality). Once again the diagram demonstrates a clear difference; a different shape emerges for each wine which is immediately recognizable.

8.2.3 Identification of grape varieties

Recognition of varietals and deconvolution of blends is performed in the knowledge that according to European Regulation (RCE 491/2009 and 607/2009): a varietal wine must consist of at least 85% of the variety mentioned.[5] This led us to investigate blended as well as pure varietals, as the French Controlled Appellation System (AOC) allows the cultivation of specific grape varieties (positive list) according to geographical location.

Wine from white grape varieties

In testing wine from white grape varieties, two sets of criteria are used for differentiation: shikimic acid content and protein profiles.[6] The shikimic acid

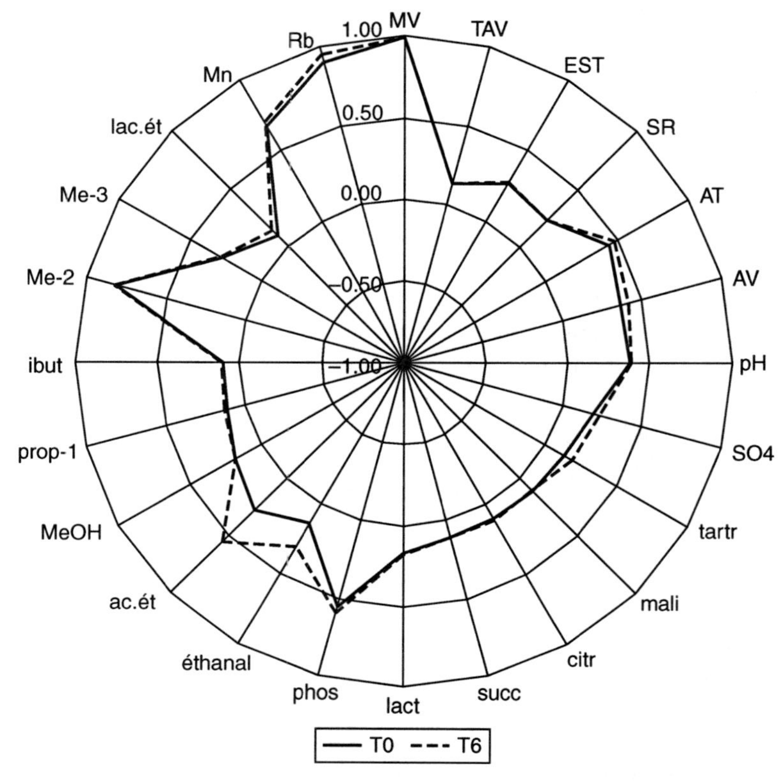

Fig. 8.1 Classical wine parameter variations during the 'cellaring' of a wine with 'appellation'. Comparison between T0 and T6 months.

is determined using HPLC and is associated with the protein profile using ECHP (Fig. 8.3a), as shikimic acid alone does not allow sufficient discrimination (Fig. 8.3b). A database was developed with six varieties studied and 90 samples of single varietal wines analysed.

Data processing using multidimensional statistical analysis (SIMCA)[7] and the above determinations allowed classification into different grape varieties and the development of a model (Fig. 8.3c). Deconvolution of blends using PLS leads to predictions fairly close to reality (Table 8.1), and the exploration of protein profiles is a promising tool for the recognition of white grapes without using PCR and DNA testing.[8]

Wine from red grape varieties

Here again two sets of criteria for differentiation are necessary: shikimic acid (Fig. 8.4a) and anthocyanin profile (HPLC), normalized to 100% (Fig. 8.4b). A database for French varietals was developed, with 11 varieties studied and 120 samples of single varietal wines analysed. In-house wine-making was

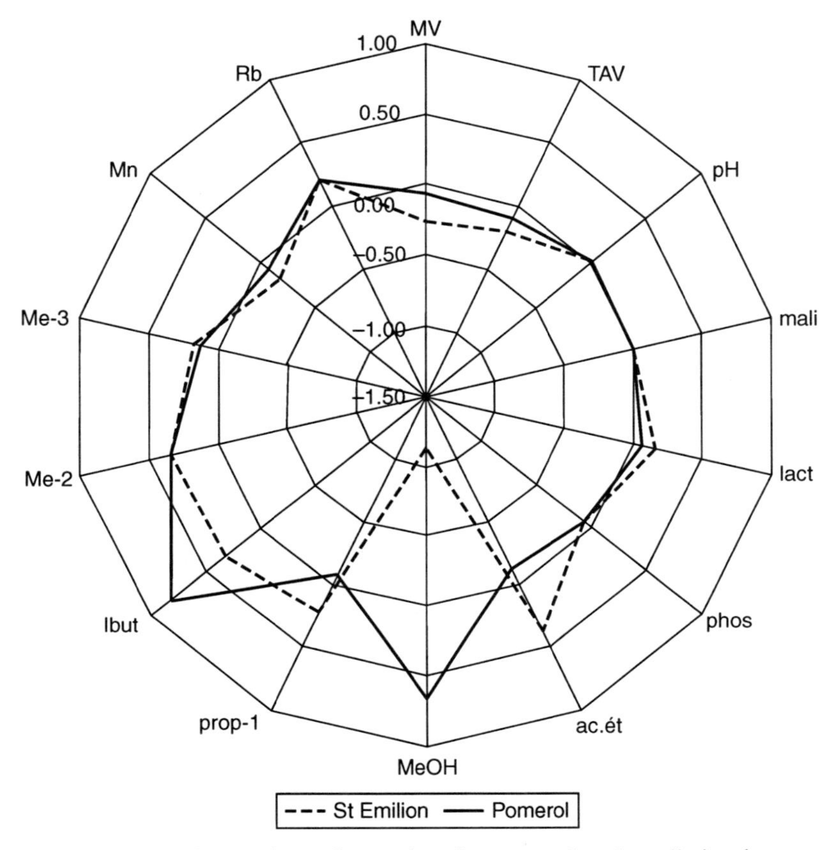

Fig. 8.2 Comparison of two wines from very close 'appellations'.

used to make the samples from authentic grapes, provided by official bodies. Classification using multidimensional statistical analysis (SIMCA) according to the main production areas in France was carried out and PLS was used to deconvolute the blends. The results were as follows.

Beaujolais and Bourgogne

For wine varietals of Beaujolais and Bourgogne, a two dimensional representation (based on shikimic acid and % free anthocyanins) is sufficient to perfectly distinguish Gamay and Pinot Noir wines (Fig. 8.4c).

Bordeaux varietals

The correct classification of Bordeaux varietals was attained up to 90%, and the prediction of blending was satisfactory. However, this method does still require some improvements (Table 8.2).

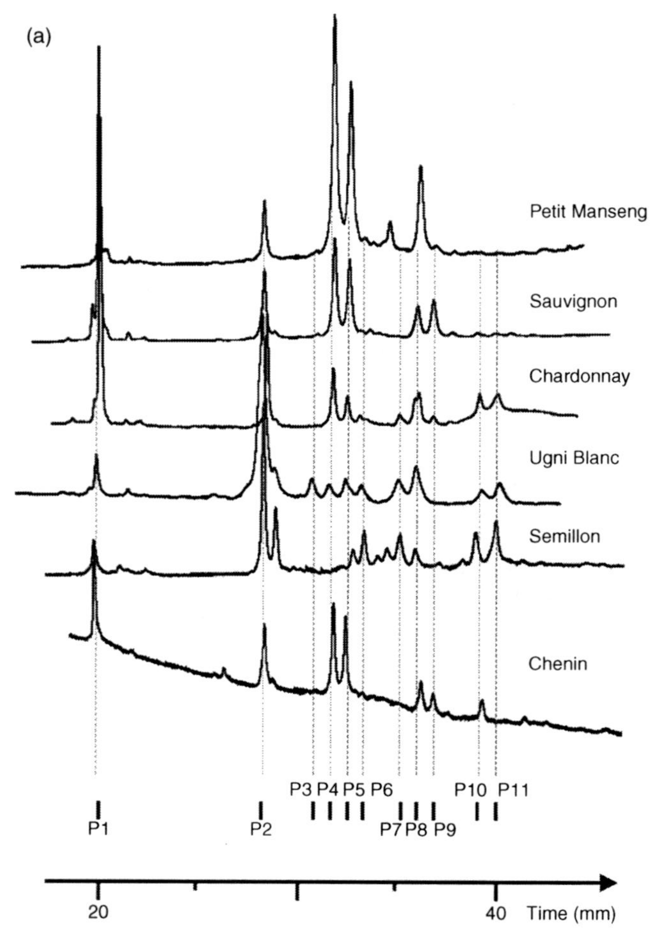

Fig. 8.3 (a) White wine varietal protein profile using ECHP; (b) Shikimic acid content of white varietal wine (mg/L); (c) White wine varietal classification using SIMCA.

Languedoc and Rhône varietals

When analysing all of the main varietals traditionally found in this region, the SIMCA algorithm does not lead to a perfect separation; nevertheless Syrah and Cinsault are easily distinguished (Fig. 8.4d).

This study shows that varieties can be correctly recognized, both in white wines and red wines using simple chemical parameters, and it highlights a promising tool for the deconvolution of mixtures. Whilst several varieties are poorly recognized, the determination of the percentages of blends of the main varietals is accurate enough to meet the requirements of European regulations. Such a path using anthocyanins and phenolic acids was taken to characterize other varieties like Malbec, popular in Argentina,[9] and Carmenère in Chile[10]

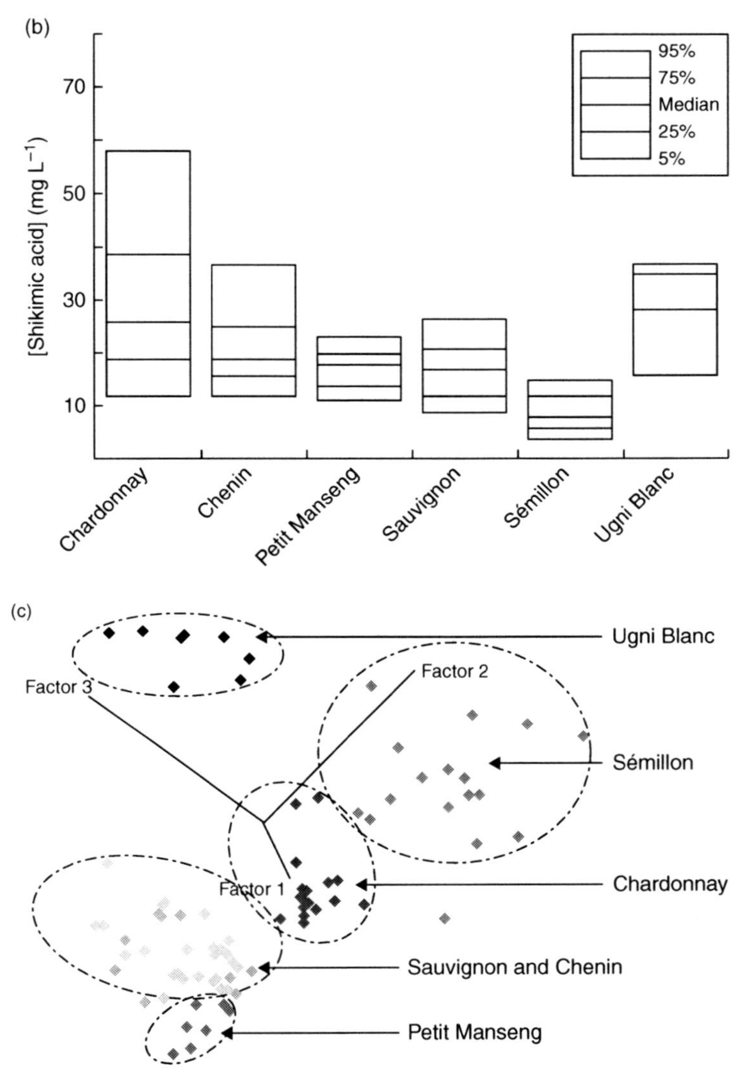

Fig. 8.3 (*Continued*)

Table 8.1 Composition prediction (%) of blended varietal white wines

	Sauvignon		Ugni Blanc		Chardonnay	
	Real	Pred	Real	Pred	Real	Pred
Mix 1	*50*	**35**	*50*	**46**	—	—
Mix 2	*90*	**81**	*10*	**16**	—	—
Mix 3	—	—	*50*	**40**	*50*	**41**

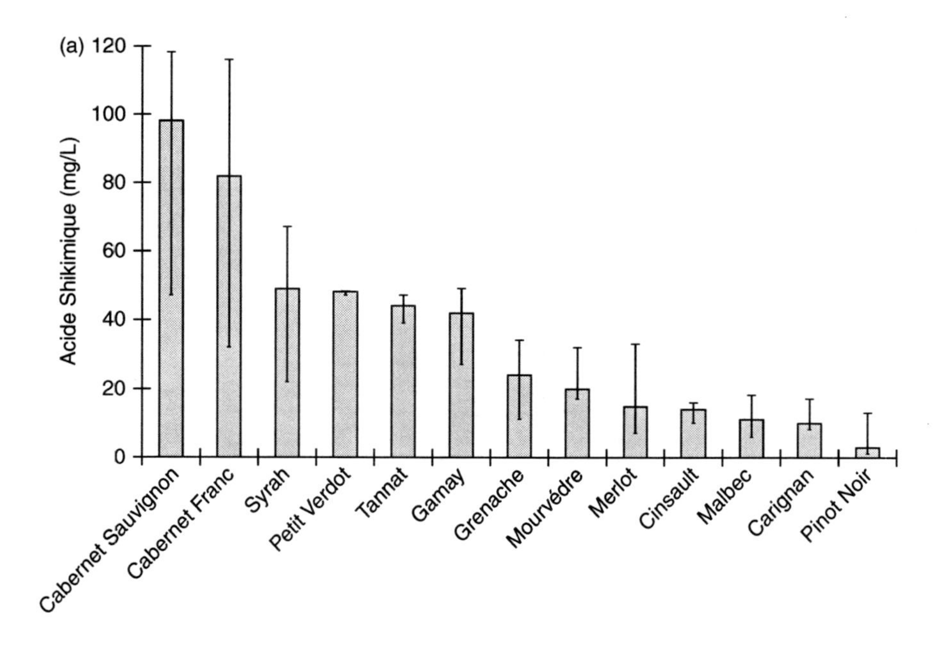

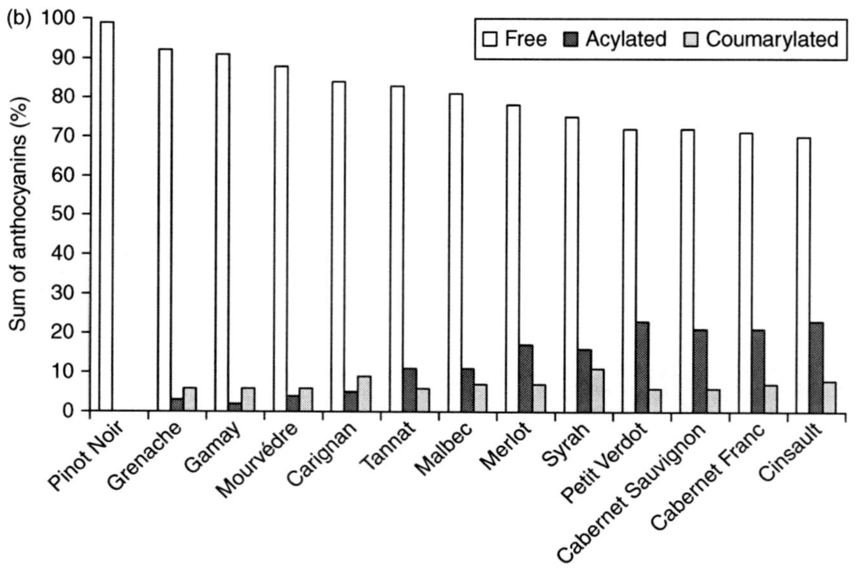

Fig. 8.4 (a) Shikimic acid content of red varietal wine (mg/L) mean and natural variation; (b) Anthocyanin content of red varietal wine (mg/L) mean and natural variation; (c) Two dimensional Bourgogne red wine varietal discrimination; (d) SIMCA plot of Languedoc-Rhône red wine varietal.

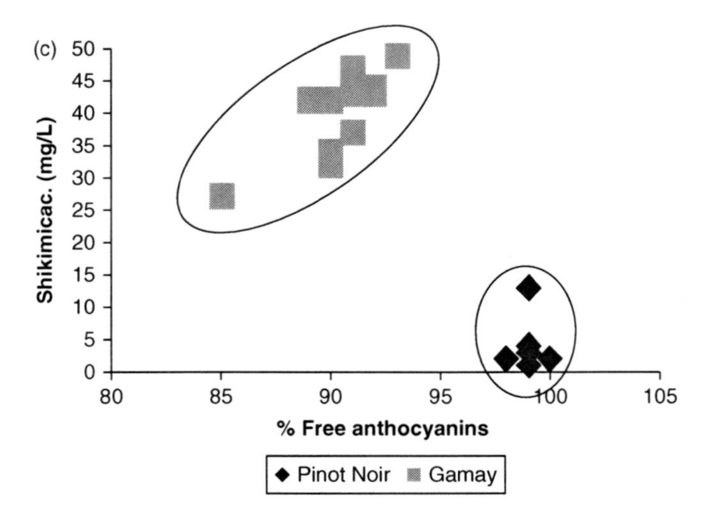

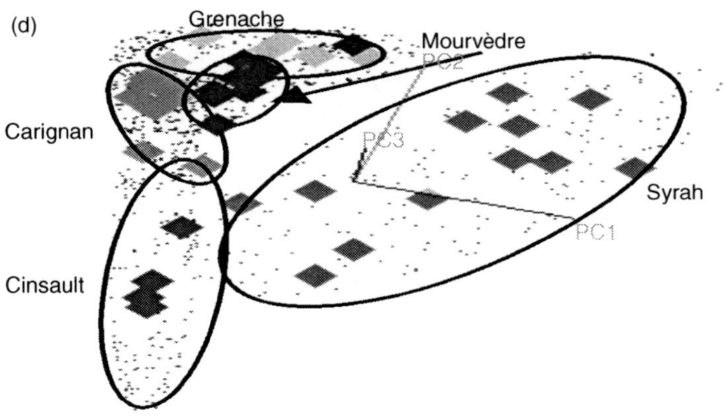

Fig. 8.4 (*Continued*)

Table 8.2 Composition prediction (%) of blended varietal red wines

	Merlot		Cab. Sauvignon		Cab. Franc	
	Real	Pred	Real	Pred	Real	Pred
Mix 1	*50*	**40**	*25*	**27**	*25*	**37**
Mix 2	*10*	**19**	*50*	**47**	*40*	**38**
Mix 3	*85*	**67**	—		*15*	**33**

and elsewhere, where the varietals are promoted before the region. The match between grape variety and geographical origin can also be highlighted easily, allowing the Controlled Appellation of Origin to be enforced. The models still need some improvement in terms of accuracy but are good and robust enough to be used in their current form, having been successfully tested in the real world.

8.3 Determining the geographical origin of wine

Wine reflect the composition of the soil and the climate where the grapes have grown; using mineral determinations and their isotopic ratios is a must for determining the geographical origin.

8.3.1 Determination of trace elements and 'heavy' isotopes using ICP/MS

Early work on minerals[11] using GFAA (Graphite Furnace Atomic Absorption) such as nickel and chromium in wine, led to surprising results indicating that, whilst stainless-steel was less of a factor in the content of those metals, the large variation in glass composition over time, especially in stabilising of colour, has led to the variation of chromium in wine over time and can be used for dating. For red wines of the Bordeaux area in green and white bottles, low to very low chromium content is found up to the 1950s, with a large increase seen which begins around 1955, with a peak around 1965 followed by a long linear reduction in a direct relationship with solubility phenomenon[12] (Fig. 8.5). It remains to be seen if this process will continue over time and in other areas with different glass processing and composition.

ICP/MS methods have been developed for wine, and ultra-trace minerals such as lanthanides were accurately determined and used successfully.[13,14] Work proved that a direct link exists with the soil where the wine is produced,[15] but for white wine, the frequent use of bentonite pollutes the signal somewhat.

Lead and strontium[16,17,18]

The following study of the geographical origin of certain wines results from the use of an ICP Mass Spectrometer (ICP/MS) (Fig. 8.6). This analytical tool can give access to the majority of the heavier elements of the periodic table and their isotopes.

The determination of lead isotope ratios in wines, using different ICP MS devices, and comparison of the precision and accuracy of these ratios highlighted the feasibility of this technique.[19] Of the four stable Pb isotopes,

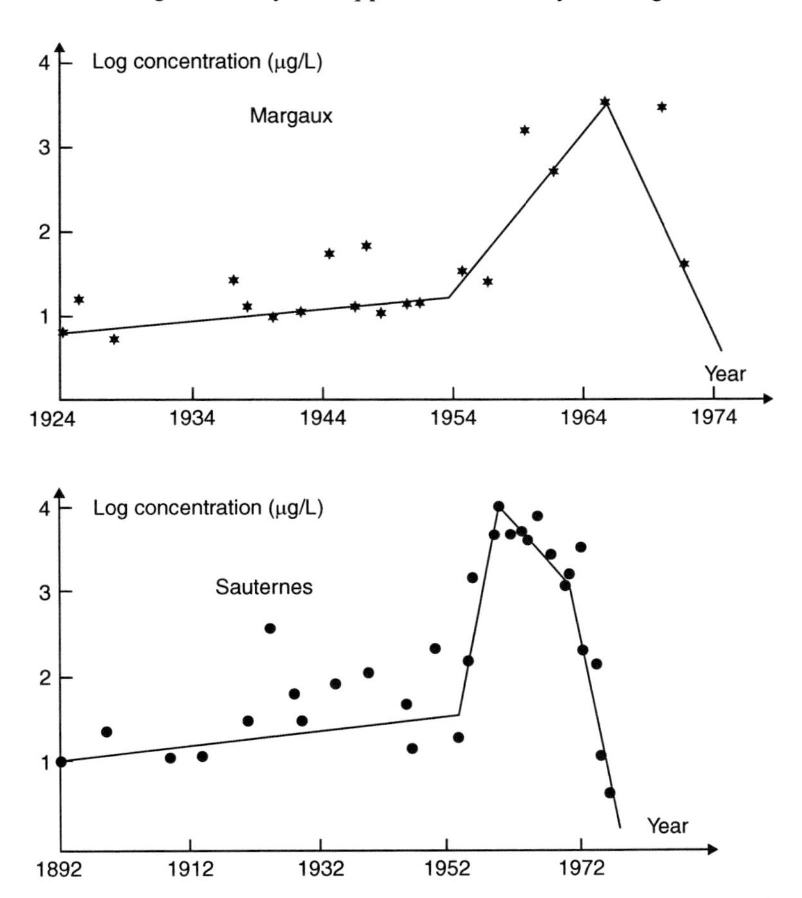

Fig. 8.5 Concentration of chromium in wine over a secular sequence of Bordeaux wines.

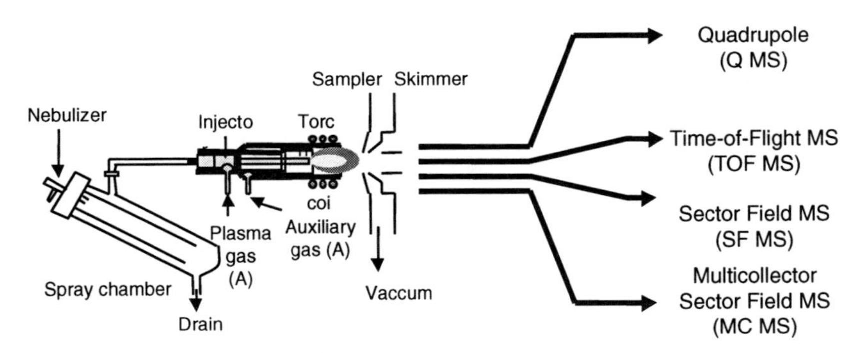

Fig. 8.6 Schematic diagram of an ICP/MS.

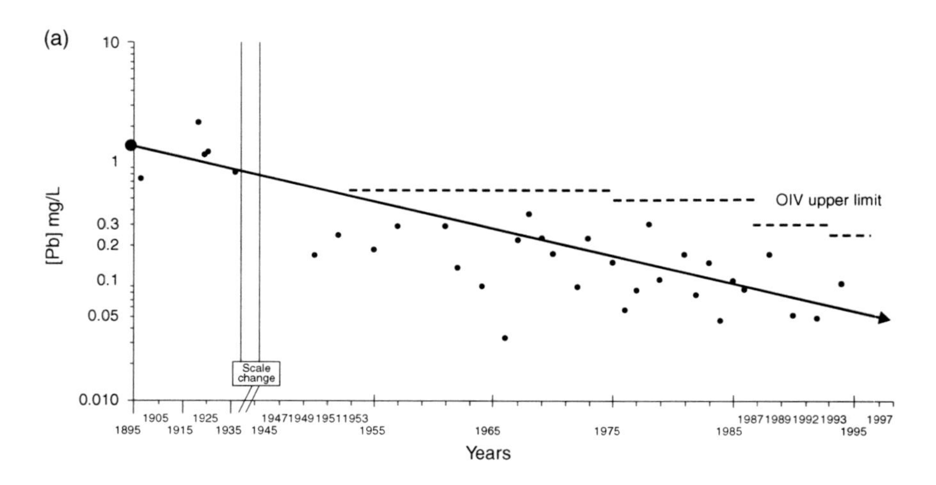

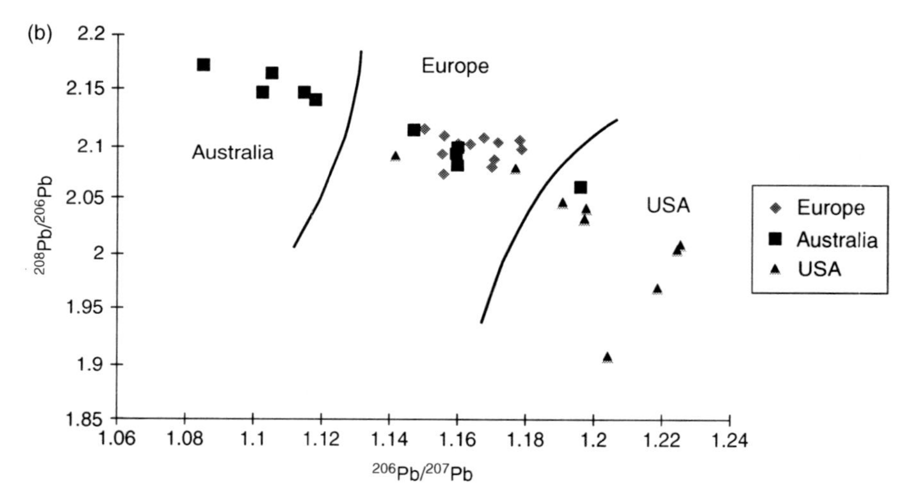

Fig. 8.7 (a) Lead concentration variation in wines from the last century to the present according to age; (b) Lead isotopic ratio and continental origin of wine; (c) Lead isotopic ratios and age of wine (same chateau).

only Pb-204 is not 'radiogenic', while the isotopes Pb-206, Pb-207 and Pb-208 are formed by the decay series of U-238, U-235 and Th-232, respectively. Lead in wines (leaving aside pollution due to equipment, which is rare) used to result almost exclusively from air pollution.[16] Lead isotopic ratios vary from continent to continent (mainly due to the antiknock agent added to petrol) and are transmitted to wine. This phenomenon in turn allows the classification of wine into three classes according to their continent of origin (Fig. 8.7b). However it should be noted that as levels of lead in wine decline due to international efforts to reduce anthropogenic lead levels in the environment,

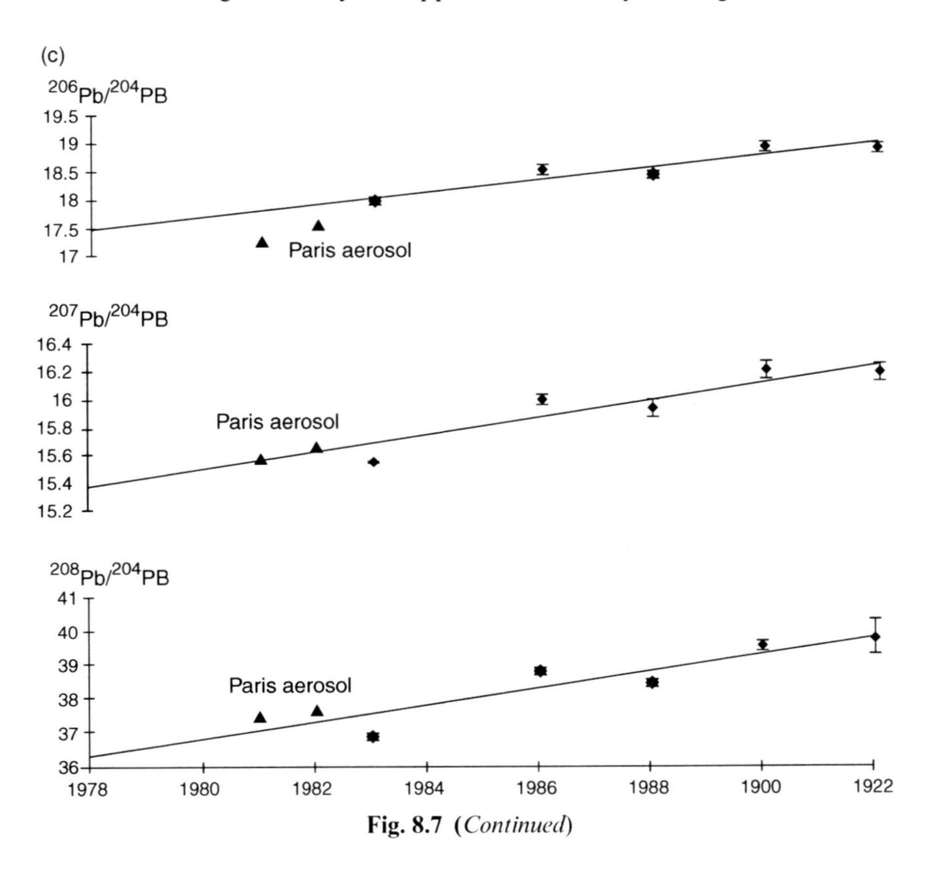

Fig. 8.7 (*Continued*)

(e.g., lead in petrol) lead isotope ratios are gradually becoming less relevant markers of origin.

Strontium has four stable isotopes, with a varied abundance of Sr-87 due to the beta-decay of Rb-87. Strontium isotope ratios (also determined by ICP /MS) are a soil characteristic and, being in equilibrium in the water supply of the vine, are found in identical form in water, grapes and later in the wine.[20] Due to the specificity of the geological formation, $^{87}Sr/^{86}Sr$ ratio varies, as demonstrated in the following examples:

Bourgogne region on Carbonaceus Jurassic 0.708–0.709
Bordeaux region on sedimentary soils: 0.711–0.712

Isotope ratios of strontium can be precisely determined in wine using ICP/ MS.[21] Considering lead and strontium, we potentially have a simple characterization of wine with two parameters, one from the air and another from the ground. The classification resulting from this combination is good and seems universally applicable (Fig. 8.8).

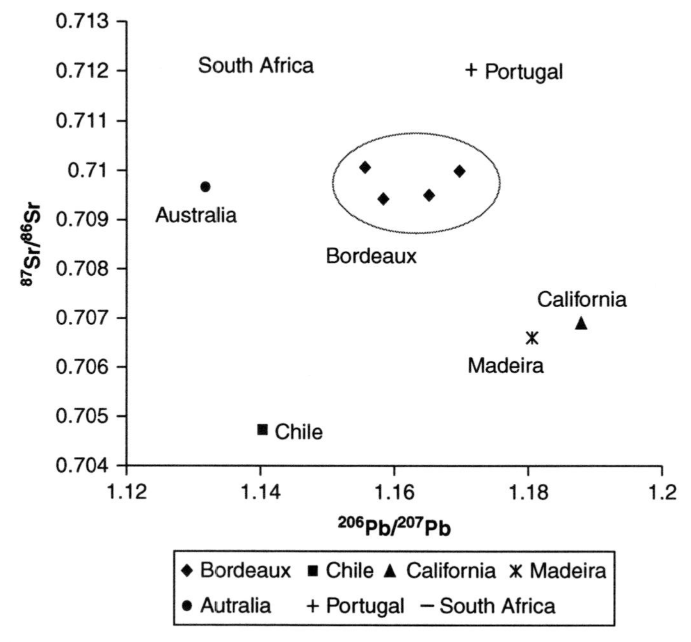

Fig. 8.8 Lead isotopic ratio *vs* strontium isotopic ratios and country origin of wines.

Neodymium, Samarium, Uranium and Thorium
The use of very precise multicollector ICP/MS permits other geochemical parameters and ratios to be determined, such as the lanthanides Nd, Sm and also U. Other ratios may also be useful, such as Li[22] and B.

Neodymium shows 7 stable isotopes and its Nd-143 isotope is enriched by the alpha-decay of Samarium 147 with a period of 106 million years.

Nd, Sm are in the sub ppb level
Neodymium and samarium can be used to determine both the age of a soil and a probable location where the vine was grown. Table 8.3 shows the results for 268 wines from all over the world (from Reference [18]), with Sm/Nd ratio showing high variation, implying very different ages of the initial rocks. This ratio is probably more meaningful at the local level within a region and representative of the 'terroir' effect. The results of Sm/Nd ratios obtained in this case with a conventional quadrupole ICP/MS are unlikely to be accurate enough for a final conclusion and those results need to be confirmed on a high resolution instrument. The value of the crustal Pb/Nd ratio is close to 1 (0.77), whilst for wine it is always higher, indicating a strong anthropogenic influence (varying from 8 in Australia to 300 in Europe).

Table 8.3 Geochemical parameters: neodymium, samarium, lead, uranium and thorium values (in µg/L) and ratios according to wines from different types and geographical origin

	n values	Nd	Sm	Pb	Th	U	Pb/Nd	Sm/Nd
White	73	0.92	0.23	29.9	0.17	0.50	56.41	0.272
Red	173	0.13	0.06	12.7	0.56	0.66	98.30	0.462
Country								
Argentina	4	0.38	0.09	17.7	0.09	0.23	47.27	0.242
Australia	16	0.73	0.16	5.1	0.13	0.69	7.88	0.267
Chile	10	0.14	0.05	12.8	0.07	0.24	163.23	0.495
France	140	0.89	0.66	31.9	0.66	0.74	311.36	0.552
Germany	11	2.15	0.58	44.3	0.25	0.85	27.79	0.269
Italia	10	0.41	0.13	22.3	0.08	0.20	90.33	0.387
New Zealand	8	0.38	0.10	18.1	0.13	0.40	75.22	0.267
Portugal	10	0.70	0.16	40.6	0.10	0.31	134.57	0.269
South Africa	8	0.47	0.12	28.2	0.13	0.41	69.04	0.296
Spain	11	0.33	0.12	17.7	0.26	0.46	250.47	0.842
USA	10	0.30	0.09	35.3	0.05	0.23	199.58	0.394

Source: From Reference 18.

In the bedrock, the isotopic ratio $^{234}U/^{238}U$ generally reflects the balance of the radioactive chain. Because in its oxidized state it is relatively soluble in water, U-234 is preferentially leached, and the activity ratio $^{234}U/^{238}U$ in soils is usually less than one (below equilibrium). In contrast, groundwater may have very high compositions up to 10 and it is probably higher in wine. Uranium-238 and thorium-232 were determined in wines (Table 8.3) with a quadrupole ICP/MS.[18]

To conclude this section on verifying the geographical origin of wine using stable isotopes, lead with 4 usable isotopes, Sr-87/Sr-86, Nd-143/Nd-144, Sm, Th, U and their combinations permit a very rugged approach to the authentication of wine linked to environmental conditions, combining all influences of potential variation in composition: anthropogenic, geographical and process. Already used in archaeological applications[23] and for food products such as cocoa,[24] this multi-geochemical tracers approach can be considered as a 'watermark' of a wine.

8.3.2 The use of ratios of light isotopes for determining geographical origin

The distribution of oxygen-18 is not uniform across the globe (Fig. 8.9), and this property has been used to find the geographical origin of wines, with particularly good results obtained for the classification of wines from the Southern hemisphere.

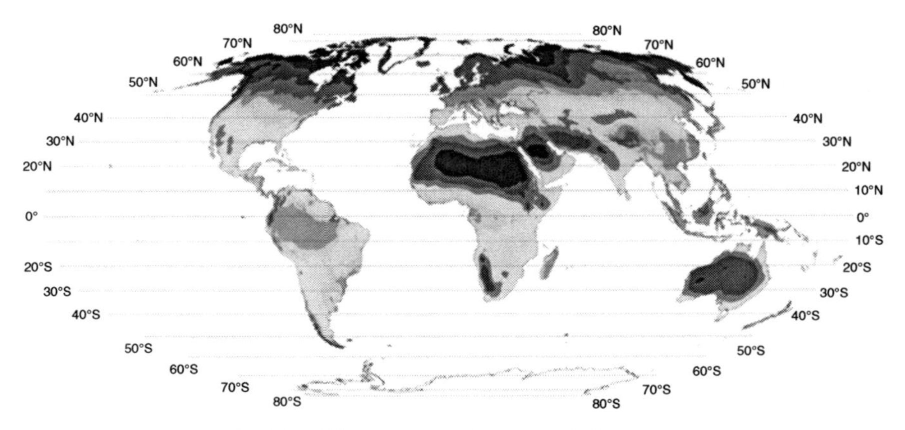

Fig. 8.9 ^{18}O concentration around the world.

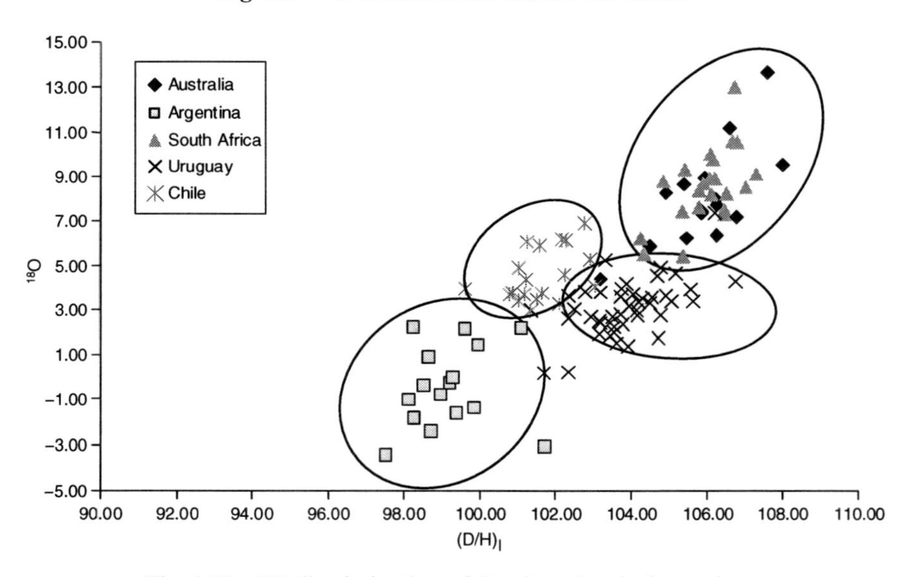

Fig. 8.10 2D discrimination of Southern hemisphere wines.

Using a two-dimensional (2D) representation with ^{18}O and $(D/H)_I$ a good separation is visually observed between four different countries (Fig. 8.10). The combination of isotopic values with other parameters such as minerals is worth considering in relation to classification. For optimal use of all of these values, multidimensional analysis becomes an indispensable tool. A very good classification is obtained using the method 'SIMCA' for ^{13}C, ^{18}O, $(D/H)_1$, $(D/H)_2$, and 10 minerals. Table 8.4 gives the result of this classification and only three wines from South Africa are misclassified out of 174 wines from the Southern hemisphere. Selecting the best predictors for a classification, only three: ^{18}O, $(D/H)_I$, lithium can be retained for an optimal classification. The resulting three-dimensional (3D) graph is shown in Fig. 8.11.

Table 8.4 SIMCA classification of Southern hemisphere wines

SIMCA classification	Chile predict	South Africa predict	Uruguay predict	Australia predict	Argentina predict	% predict
Chile	47	0	0	0	0	100.0
South Africa	0	32	0	3	0	91.4
Uruguay	0	0	44	0	0	100.0
Australia	0	0	0	26	0	100.0
Argentina	0	0	0	0	22	100.0

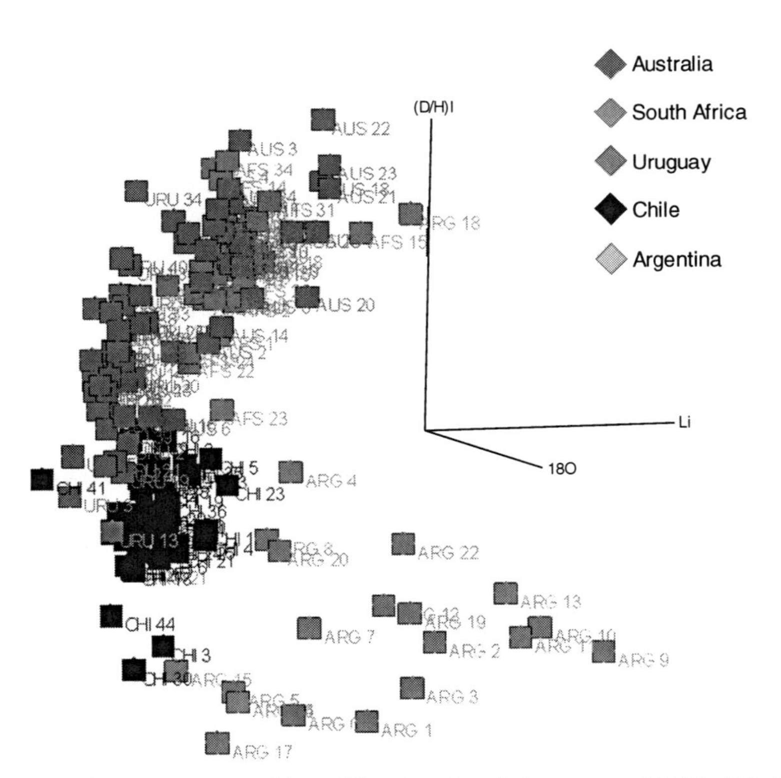

Fig. 8.11 3D pattern recognition of Southern hemisphere wines ($(D/H)_1$, ^{18}O, Li).

8.3.3 Carbon isotope ratios

Separating the different components of a wine and performing isotopic ratio determination of each of them elucidates the intimate biochemical processes that produced the wine throughout fermentation. GC/IRMS was the first methodology used and produced very interesting results. $^{13}C/^{12}C$ isotope ratios of the major higher alcohols, for example, 2-methylpropan-1-ol, 2- and 3-methylbutan-1-ol, butan-2,3-diol and 2-phenyl-1-ethanol were determined and set in relation to the ^{13}C values of the corresponding extracted genuine wine ethanol.[25]

High-performance liquid chromatography linked to isotope ratio mass spectrometry (HPLC-co-IRMS) via a Liquiface© interface was used to simultaneously determine ^{13}C isotope ratios of glucose, fructose, glycerol and ethanol in sweet wines.[26] Authentic wines from various French production areas and various vintages were analysed, and internal ratios of ^{13}C isotope ratios ($R^{13}C$) of the four compounds studied were found to be constant.

Thus, ratios of isotope ratios are found to be 1.00 ± 0.04 and 1.02 ± 0.08 for $R^{13}C$ glucose/fructose and $R^{13}C$ glycerol/ethanol, respectively. Moreover, $R^{13}C$ ethanol/sugar is found to be 1.15 ± 0.10 and 1.16 ± 0.08 for $R^{13}C$ glycerol/sugar. Additions of glucose, fructose and glycerol to a reference wine show a variation of the $R^{13}C$ addition as low as 2.5 g/L^{-1}, and concentrated analysis shows that ^{13}C isotope ratio can be used directly to determine their authenticity. The isotopic ratio of glucose, fructose, glycerol and ethanol determined by HPLC-co-IRMS is a promising tool toward the authentication of 'pure' wine.

8.4 Verifying compliance with specific regulations

Insuring wine quality requires high standards; the European regulation is the most stringent in the world; to insure fair competition sophisticated method for wine control are used.

8.4.1 Chaptalization

In 1991, chaptalization determination (the process of adding sugar to unfermented grape must in order to increase the alcohol content after fermentation) in wine became an official method in the European Union.[27] This determination permits to evaluate the percentage of ethanol coming naturally from the grape sugar and ethanol produced with added sugar of a different botanical origin (beet, cane, rice, orange, sorghum, etc.). Verification of the origin of alcohol in wine is based on comparing isotope values of ethanol samples to those of a set of reference samples. For the past 20 years at harvest time, samples of grapes from all European wine producing countries have been transferred to an official laboratory where they are turned into wine to build up an annual data bank.[28] This data bank, used for control practices, is also used for geographical localization and the determination of the year of production. Control of chaptalization is performed using carbon values and deuterium on specific sites of the ethanol. Performing these measurements involves instrumentation for techniques such as NMR and IRMS exclusively dedicated to these procedures. The result of an NMR measurement is given in Fig. 8.12.

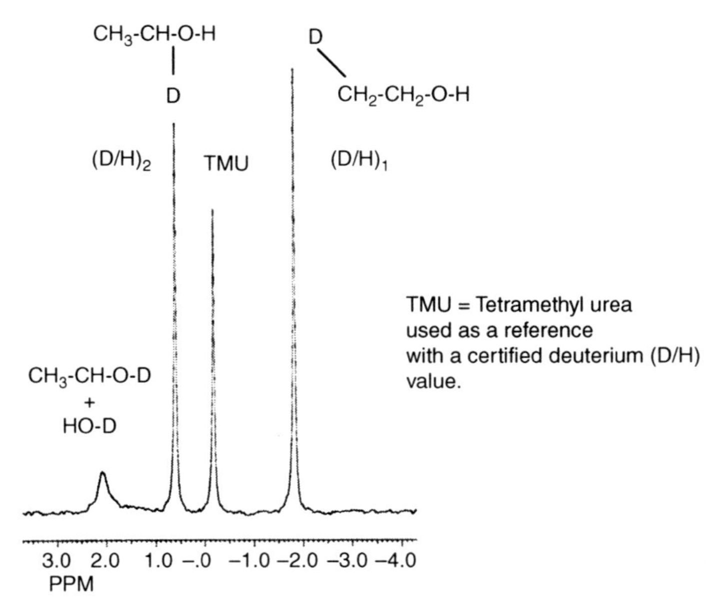

Fig. 8.12 Deuterium NMR spectrum of wine ethanol.

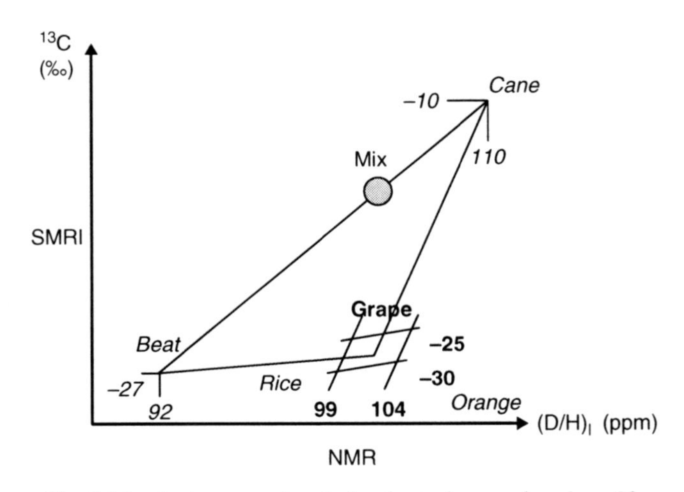

Fig. 8.13 Isotopes under study: deuterium and carbon-13.

8.4.2 Botanical origin of alcohol

The diagram in Fig. 8.13 identifies the origin of the alcohol in wine and allows a quantification of the contributions of various sugars to the final ethanol content. The analysis performed on spirits also show that these parameters provide valuable information on the origin of alcohol. Labelling can thus

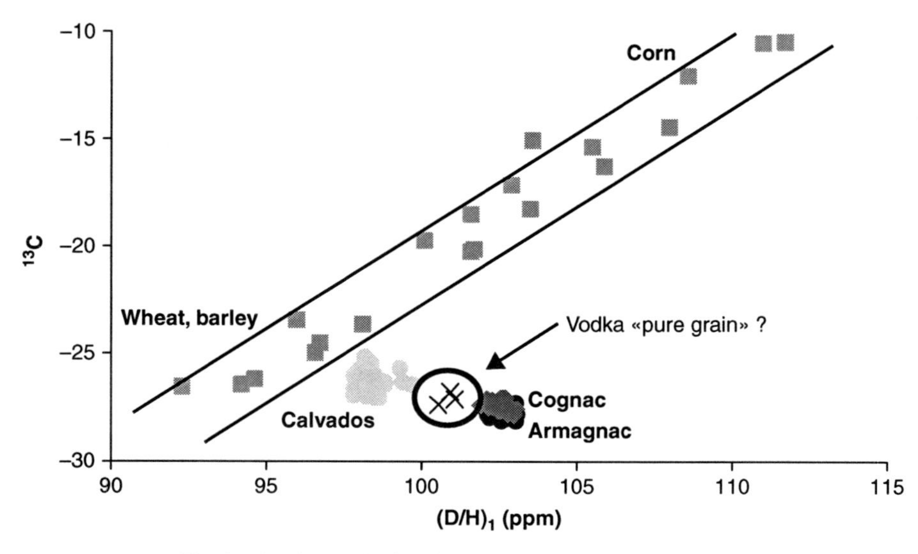

Fig. 8.14 Control of vodkas presented as ' pure grain'.

be controlled to reflect this additional information, to reflect the 'pure grain' content of Vodka, for example, as seen in Fig. 8.14. Tequila is another example where isotopic parameters are required, despite being insufficient on their own, requiring the introduction of other parameters such as FTIR spectrum[29] or ^{13}C specific sites of the ethanol measured with NMR to provide a definitive answer.[30]

8.4.3 Watering down

It is important to remember that water is the major component of wine. The oxygen-18 content of the water in the wine is used to control the 'watering down' of a wine.

8.4.4 Sulphur

Sulphur isotopes (of which 25 are known) can be tracers of both processes and pollution. Sulphur is an essential element in plant development and, as such, fertilizers contain significant amounts. It is used in the wine-making process as a preservative (usually in sulphur dioxide form) against spoilage and for its antioxidant properties. Owing to health concerns, its concentration has been lowered in wine and some labelling states, for example, 'no sulphur added' or even 'without sulphur'. Yeasts are known to produce some sulphur dioxide and use of HPLC-ICP/MS[31] to determine isotopic ratios of sulphur compounds will permit greater control, by

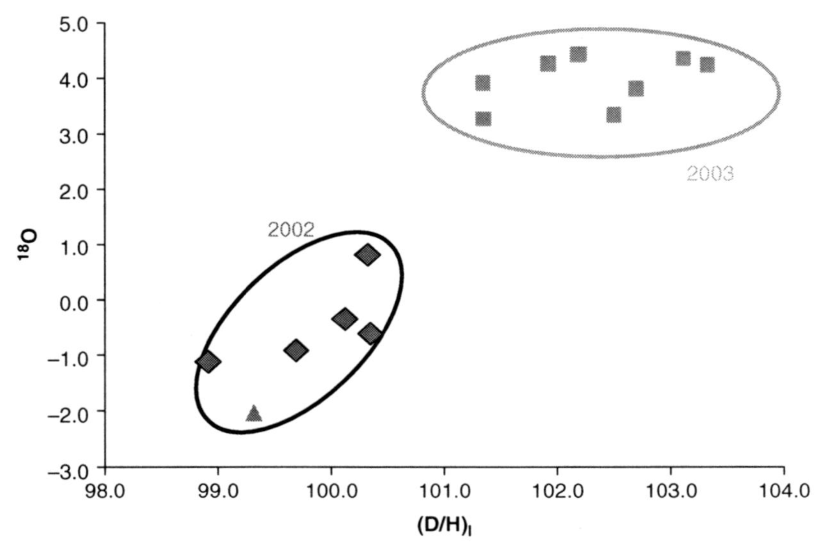

Fig. 8.15 Use of ^{18}O and $(D/H)_I$ data bank values, for assessing the year of production of a wine.

establishing which part is really natural and which has been intentionally added.

8.4.5 Using controls on production parameters to verify the year of production

The database referred to in Section 8.4.1, to which contributions are submitted annually, has allowed the development of other applications based on isotopic values and in some cases it is possible to control the vintage of a wine. In the following example (see Fig. 8.15), the label of a wine from a determined area (AOC: Appellation of Controlled Origin) indicated 2003 as the year of production, which was a good year according to the wine critics and the climatic parameters. The isotopic analysis conducted to check the production parameters (sugar or water added) pointed to a 'watering' of 50% according to the formula

$$\% = \frac{^{18}O_{ref} - {}^{18}O_{sample}}{^{18}O_{ref} - {}^{18}O_{water}} \times 100 \qquad [8.1]$$

where $^{18}O_{ref}$ is the average value of the water of the wine of the reference samples of the data bank for that year and specific area. When compared to routine analysis, this finding presents a chemical impossibility, leading to the only possible conclusion: the wine is mislabelled (excluding the possibility of another origin).

8.5 Contaminants

Mycotoxins occur in specific bio-environments, as is the case of ochratoxin A (OTA), produced by the mould *Aspergillus carbonarius*. This mycotoxin is present in wines at very low levels, but is only present in a 'Mediterranean' climate; it is unknown, for example, in an 'Atlantic' climate. The presence of this mycotoxin, even in very small amounts, is the mark of a specific biotope or a mixture. The International Organisation of Vine and Wine (OIV) has adopted a method for the determination of OTA in wines.[32]

Pesticides are used on grapes and as a consequence residues are found in wines. When used in a 'rational' way, according to good practice, those minute residues are the mark of the yearly climate linked to a location; in a hot year or climate, the crop is more prone to insect attacks, whilst in more humid years or locations it is more susceptible to fungal disease.

Leaving aside organic wines, not all wines have residues, even when researched with modern tools. However, if found, a pesticide residue profile is a mark of a specific region and a year of production. OIV has adopted (2012) a method based on the 'Quick Easy Cheap Effective Rugged and Safe' multi-residue method (QuEChERS).[33]

8.6 Wine dating

Wine labelling very often indicates the year of production. This mention is not mandatory regarding the EU legislation but provides a valuable information for the consumer and as such has to checked.

8.6.1 Radio isotopes

Ratios of $^{230}Th/^{234}U$ and $^{234}U/^{238}U$ are being used in archeology[34] for dating over the period 5000 – 400 000 years. The oldest trace of a wine from the Neolithic period is about 5000 years old[1] (proof that wine can be aged) so those ratios, while of some use, could rather seldom be applied in a lifetime.

8.6.2 Radioactivity

Early studies on the vintage of wines using radioactivity date from the 1950s,[35] and used tritium of the water in wine (beta radioactivity). These early results were profoundly modified by the nuclear fallout that followed (Fig. 8.16). More recent studies began in 1979 in Bordeaux[36] using ^{14}C and continued up to a recent period. This method was easier to use and gave more stable results that were influenced less by geographical location (Fig. 8.17).

Low background γ-ray spectrometry, based on the use of high performance (HP) Ge crystals, showed that besides the well-known isotope ^{40}K, wine also contains trace amounts of ^{137}Cs (less than 1 Bq/L) with an activity

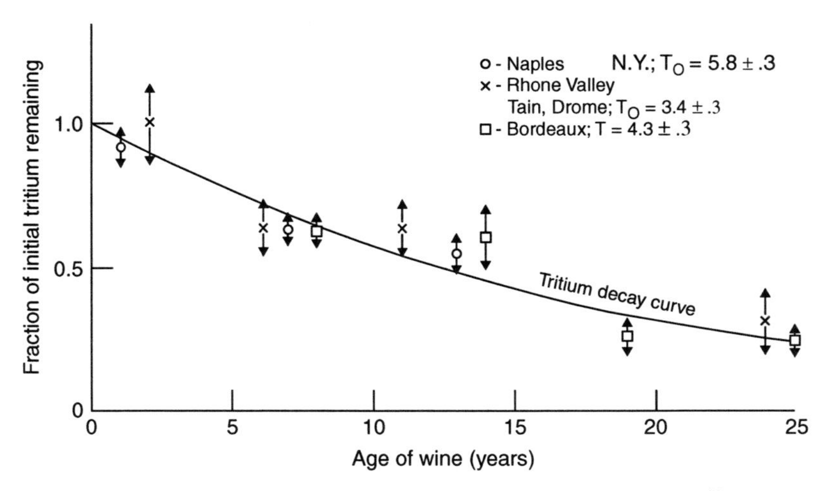

Fig. 8.16 Variation of water tritium in wine over 25 years.[35]

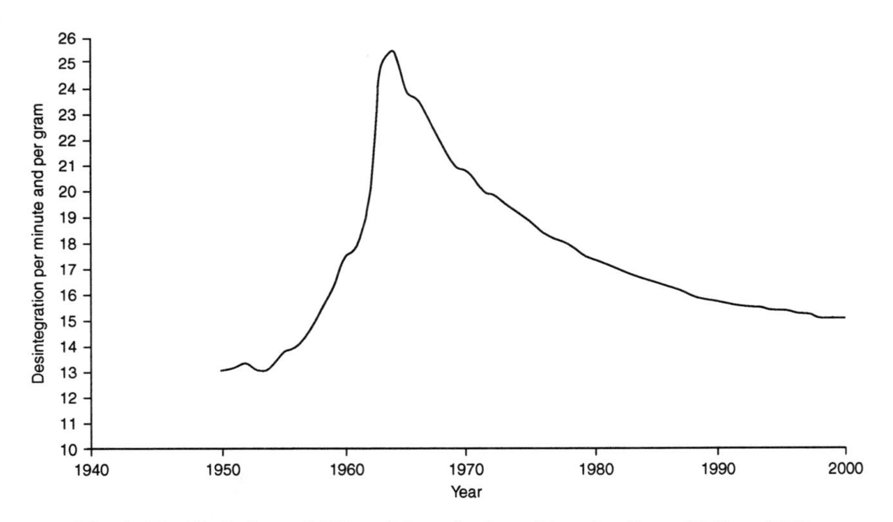

Fig. 8.17 Variation of ^{14}C activity of ethanol in wine from 1960 to 2000.

depending on the vintage.[37,38] Measures depending on the year give a characteristic of nuclear fallout, on which even the accident at Chernobyl in 1986 appears at an extremely low level. A curve was established for wine between 1952 and 2000 and this technique thus led to the possibility of verifying the year written on the label or cork (Fig. 8.18). Time consuming preparation, reducing one bottle of wine to a few grams of ashes, has given way to less labour intensive work and careful measurements that do not require opening

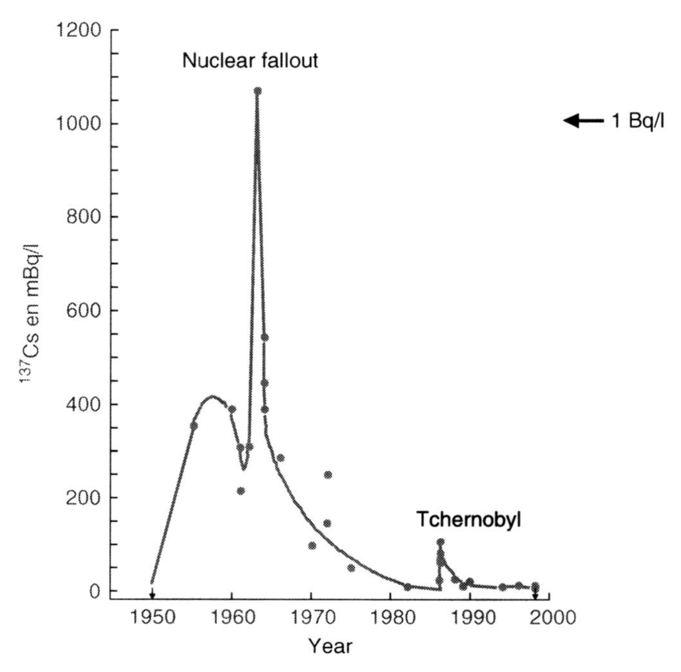

Fig. 8.18 Ultra-low ^{137}Cs measurements of wines for the period 1950–2000.

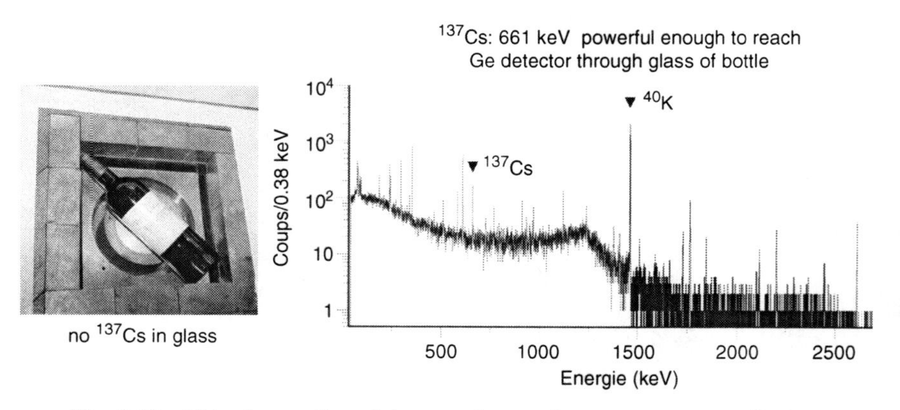

Fig. 8.19 Ultra-low radioactivity non-destructive measurements of wines.

the bottle[39] (Fig. 8.19). The technique has thus proved to be very useful for detecting counterfeit wines of the nineteenth century and first half of the twentieth century. Non-destructive ultra-low radioactivity (less than 1 Bq/L) was applied to wines of older vintages, such as prestigious wines of Bordeaux from 1900, and has also been used to try to solve historical puzzles on very old wines (1784–1787) from the time of the French Revolution which could have belonged to Thomas Jefferson, third President of the United States[40] These curves were obtained in Europe and it would be interesting to exploit them in other parts of the world, particularly in the Southern hemisphere. Further developments are underway[41] using other gamma radiation-emitting radioisotopes, at even lower levels, such as ^{226}Ra, ^{210}Pb and ^{228}Ra. These latter isotopes, having an adequate half-life, appear to be typical of the year of production. These non-destructive measurements could ultimately form the hallmark of an indelible and unforgeable wine.

8.6.3 Lead speciation

The form in which minerals are linked with organic compounds is often very complex. One of the first 'metallomics' studies, for example, showed that lead is engaged for the most part in a complex with the so called RGII (Rhamnogalacturonan II).[42] The biochemical lead properties are consequently greatly modified and probably render this heavy metal harmless, as the complex is probably indigestible.

Not all of the lead is engaged in this complex, and using techniques derived for analysing the ice of the Arctic Circle ice field, organo-lead compounds from petrol additives were found in wines. Ultra-traces of tetramethyl- and triethyl lead are detectable in wines of the 1950–1990 period.[43]

For a 'vertical' series of wines representing over 40 years in the same area (in which the study was specifically focused), extremely low levels were found in the range of 10–500 ng/L for trimethyl and 0–50 ng/L for triethyl. Between 1950 and 1991 there was a sharp increase in trimethyl after the introduction of the tetramethyl around 1960. This reached a maximum circa 1978 and has steadily decreased to the present day. The triethyl form was already present in 1960 but also decreased after 1976–1978. The use of these curves permits several deductions with respect to the year of production and the origin of the wine. A 1900 wine, for example, should not contain any organo-lead, but a 1964 wine has organo-leads and a very low trimethyl/triethyl ratio.

So far no systematic work has been done following this pioneering research (wine from the Rhône valley only), but this determination can be very useful in tandem with ultra-low radioactivity (see above). Unfortunately these determinations require techniques not easily found, even in a modern laboratory, including an ultra-clean room, GC/MIP or GC/ICP/MS and adequate know-how.

8.7 Analysis of packaging to identify counterfeit wines

Counterfeit wine scams appear from time to time in the press, usually in relation to valuable bottles considered as masterpieces, antiques or work of art. Some are considered as museum pieces (see 1787 Thomas Jefferson Bottles[44] or the billionaire vinegar[40]). We are leaving here the common ground of wine as an alimentary product and consequently the packaging has also to be looked into.

Multiple techniques can be applied here too, but few are cheap and robust enough to be used in a 'normal' laboratory. Analysis of packaging is an important avenue. All of the items that constitute the packaging are worth consideration, namely: glass of the bottle, paper of the label, capsule and cork.

8.7.1 Glass

Glass is known to have been used for wine storage since the Roman era and the production of glass bottles for wine has developed from the craftsmanship techniques of the eighteenth century to the computer controlled industrial processes of the twenty-first. One constant point of the technique is that bottles are 'blown'. Originally they were hand blown using manpower, making old bottles identifiable by their uneven shape, marks from mouth-blown spinning and bubbles. Modern bottles, in contrast, are produced using machinery and have the cast of the mould. Consequently the composition of the glass has also evolved significantly.

Container glass is made of sand (silicon), lime, soda, alumina and other fining agents melted together at fusion temperature around 1500°C. SiO_2, Na_2O, CaO, K_2O, MgO and Al_2O_3 account for more than 98% of the glass composition.[12] A high potassium content is found for old bottles in comparison to a much lower one nowadays.

Suspiciously-labelled wine was found with this technique, as seen in Fig. 8.20 (the potassium content shown by the triangles does not fit in with the general tendency).

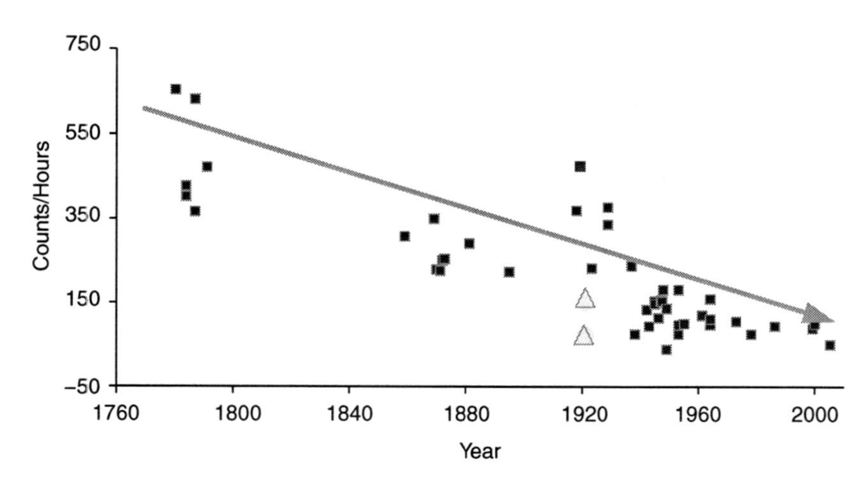

Fig. 8.20 Potassium content of wine glass bottles (^{40}K measured with a Ge detector).

Originally subject to poorly controlled colour (mainly due to its iron content), wine bottle glass has evolved toward a very well defined colour, as the precise knowledge of the oxidation state of the pigments has developed (chromium, selenium, manganese, etc.) and has gained great mechanical resistance (using titanium as a post-treatment).

To reduce the melting point temperature of the sand and calcium in the oven, recycled glass is increasingly being used, consequently leading to the addition of other traceable minerals (e.g., lead coming from old caps). Each glass wine bottle thus has a very distinguishable signature according to the factory it came from and can reveal a period of time when it was produced. Even historical events have influenced the composition, such as the switch from manganese to chromium following decolonization;[11] such events mark the composition of the glass with a landmark in time.

The mineral composition of the glass reflects a time scale, and a wide range of techniques are available to analyse its composition, including X-ray fluorescence analysis (with or without an electronic microscope) and PIXE (Proton Induced X-ray Emission), for example, many of which are non-destructive.

In cases where use of destructive techniques are possible, it is easy to drill a hole in the glass to perform a borax fusion or HF solubilization of the glass, followed by the use of AA, ICP or ICP/MS analysis. However, when dealing with potential highly valuable items ('Thomas Jefferson's' bottles were sold for around €100 000 each) the whole bottle must be treated with care.

Large instruments are required to generate protons and the technique, although very powerful, is restricted to specialized laboratories[45,46] (see Fig. 8.21). The spectra obtained permit the concentration of the major and minor mineral elements of the glass to be evaluated in a non-invasive way (Fig. 8.22). PIXE can be also applied to the label pigments with great precision (Fig. 8.23).

8.7.2 Raman spectroscopy: a multipurpose technique

Raman spectroscopy is a well-established analytical technique allowing chemical components of a material to be identified. Raman scattering results from the interaction of the electric field of an incident monochromatic and coherent excitation light (produced by a laser) with the electrons of atoms and molecules. The forced oscillations of the electrons produce a strong elastic scattering of photons (Rayleigh scattering) and a very weak inelastic scattering (Raman scattering) due to molecular and lattice vibrations.

In the 1990s, major improvements of this technology appeared with multichannel CCDs detectors and the use of photonic crystals or multilayer coatings for rejecting the Rayleigh elastic scattering. Thus the coupling of a Raman spectrometer with other technologies (an optical microscope in particular) became possible and portability increased thanks to constant progress in the development of sensitive detectors, with low cost lasers and optical fibres becoming more readily available. Today, low cost and miniaturized portable

Fig. 8.21 PIXE: Experimental design for wine glass analysis (courtesy of ARCANE).[46]

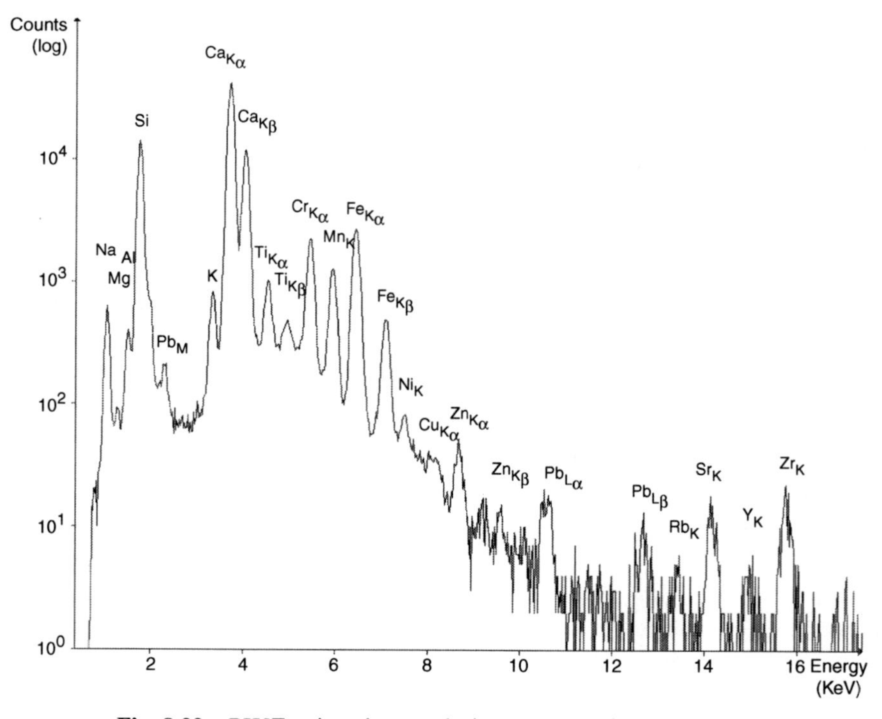

Fig. 8.22 PIXE: wine glass analysis (courtesy of ARCANE).[46]

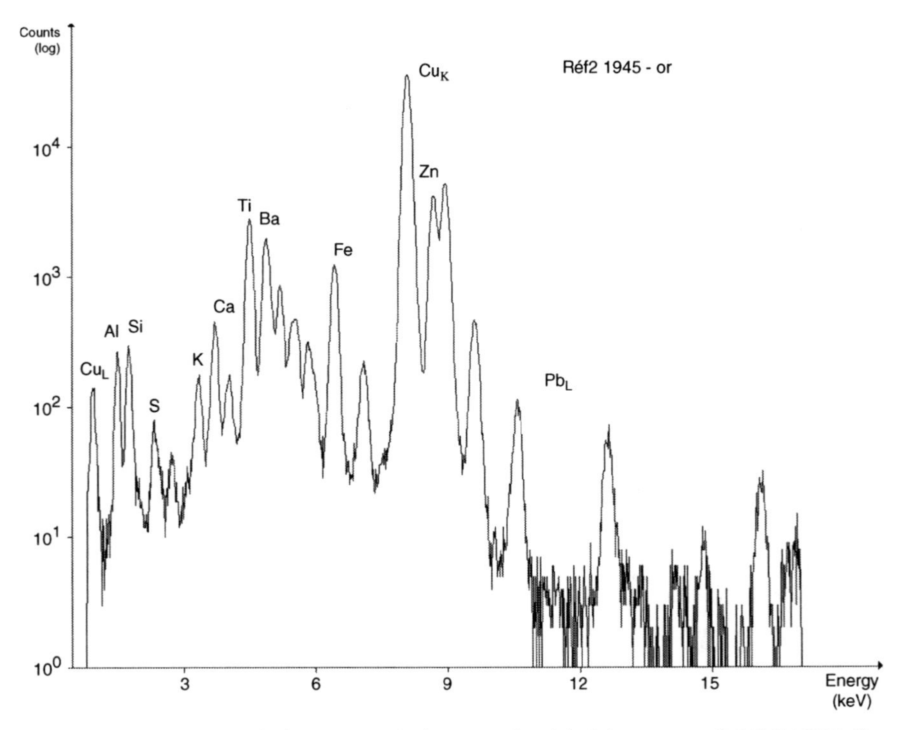

Fig. 8.23 PIXE: gold pigment analysis on a wine label (courtesy of ARCANE).[46]

Raman spectrometers are increasingly available, therefore opening a very wide range of new applications for this technique.

Raman spectroscopy requires no sample preparation, is non-invasive and non-destructive. The dimension of the sample may be very small as it depends on the size of the laser spot, which is usually approximately a few micrometres, depending on the optical components of the spectrometer and on the wavelength of the laser. This ability to obtain information on the molecular composition and structure of a microsample in a non-destructive way makes Raman microscopy a very powerful technique.

However, one of the main limitations of Raman spectroscopy is due to the sample itself, which in some cases may produce fluorescence, obscuring the Raman spectrum. Despite such a limitation, Raman spectroscopy has been quite widely used to identify the chemical structure of components in art works and cultural heritage materials,[47] to detect drugs dissolved in alcoholic beverages,[48] or to detect explosives.[49] In some cases Raman measurements can even be performed through the packaging. For example pharmaceutical products can be authenticated through non-invasive chemical analysis of products concealed within non-transparent plastic bottles, as well as other types of packaging.[50]

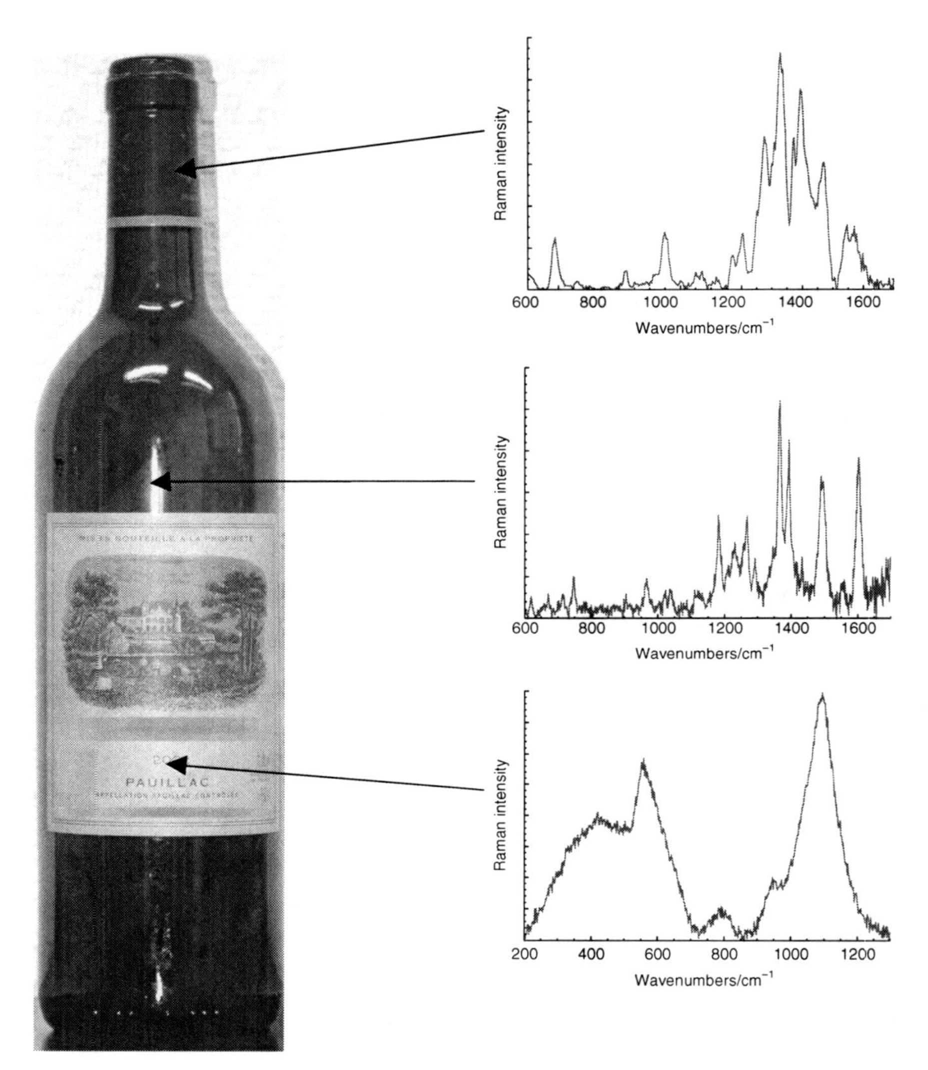

Fig. 8.24 Raman spectra recorded at various locations on a bottle of Bordeaux wine: Raman spectra of the pigment of the cap (top), the silica glass (middle) and the pigment of the red ink in the text of the label (bottom).

Raman spectroscopy can be used as an efficient, complementary tool with other techniques for detecting counterfeits. All components of a bottle of wine, for example, may be analysed by means of Raman microscopy. In Fig. 8.24, Raman spectra of the different pigments present in the composition of the labels and of the cap are shown.[46] The chemical composition of the bottle itself (glass) can be analysed qualitatively. These Raman spectra constitute fingerprints of the various components of the wine bottle and can obviously help in identifying an original product from counterfeits.

8.7.3 Label, cap and cork

The so called 'dry material' used around the wine, to make it presentable to the consumer can also be put to the test for traceability purposes.

Label

Paper manufacturing processes have changed enormously over the ages. If the label can be analysed and compared to an original, the determination of the different percentage of fibres (deciduous or coniferous) can be used as a comparison. Whether the paste was prepared chemically or mechanically, in conjunction with the determination of the paper coating composition (SIO_2 or $CaCO_3$), can be used as well.

Cap

'Old' capsules were made of lead. The migration of lead in the wine through old or defective corks,[51] in addition to ecological problems, potential lead pollution and health problems, led to the decision of the OIV to phase out the use of lead capsules in 1990. This decision was then followed worldwide. The use of aluminium is now widespread and tin has replaced lead for expensive wines. The composition of the alloy is specific and cannot easily be imitated.

Cork

Long used as a stopper for wine bottles, cork ages with the wine to the point that the pile of parallelepiped cells that constitute a cork slowly dissolves in the wine. Bottles are usually aged in a horizontal position to keep the cork moist and to guarantee 'wine-proofing'. As the bottles age, an old cork loses its mechanical ability; with time it is completely filled with wine and the capsule is the last barrier against definite leakage. The cork has to be replaced every 30 years or so, and the evolution of cork with time (slow dissolving) is impossible to imitate. The chemical composition of the printing on the cork can also be compared to authentic samples.

8.8 Future trends

8.8.1 Twenty-first century guarantee of authenticity

New innovative techniques are being developed for insuring the integrity of the packaging and providing means to authenticate the product for both the producer and the consumer.[52] These techniques even take advantage of 'smartphone' capabilities as a primary approach, whilst other techniques use bubble codes, laser inscription and foolproof labels. However, a bottle can still be uncorked or refilled through clever holes and ultimately the wine has to be analysed as final proof.[53]

8.2.8 Unforgeable chemical identity cards for wine: 1D and 2D NMR

NMR proton profiling was developed for pattern recognition of geographical origin, otherwise known as fingerprinting of wines. Amino acids, sugars and organic acids are easily identified[54] and constitute a solid base for establishing a chemical identity. The manipulation of radio frequency sequences permits a wide variety of experiments, and facilitates the study of such complex mixtures as wine; even the signal of water can be eliminated ('irradiation'). The power of NMR analysis has been demonstrated, for example, by its ability to differentiate between local wines produced in the aforementioned area of Drava,[55] successfully identifying the wines produced on the Slovenian side.

Several authors using a metabolomics approach combined with multivariate statistics used one-dimensional (1D) NMR in the regional assignment of both Muscat grapes and wines connected to local climate in Korea,[56] whilst also employing the technology for the characterization of wines by grapes,[57] as in, for example, the geographical classification of both Greek[58] and Chinese wines.[59] The identification of substances according to their chemical shifts was confirmed using COSY, TOCSY and JRES sequences.

The study of the compositional evolution of Port wine samples during aging was performed using DOSY,[60] and quantitative ^{1}H NMR was used during the wine fermentation process[61] to simultaneously follow the evolution of the main metabolites (ethanol and malic, lactic, succinic acids and amino acids). Care has to be taken for quantitative measurements when handling direct NMR spectrum data, due to the shift induced by pH variations between wines.[62]

It was shown that the mapping of 2D (^{13}C-^{1}H correlation spectra) NMR analysis of polyphenolic extracts of wine,[18] could unambiguously distinguish information such as local and continental geographic origin, along with types of grape. Figure 8.25 gives an idea of the complexity of the 2D data processing needed for such studies. Quantitative ^{13}C NMR was used to differentiate between alcohol of 'C4' origin (cane, corn, etc.) from those coming from pure tequila (*Agave tequilana*) or pineapple.

Recent work in our laboratory[63] demonstrated that the ^{13}C NMR spectrum from slightly concentrated wine with ethanol removed, permitted characterization of the origin of a pure wine varietal (Fig. 8.26). This technique has many advantages compared to the method described in Fig. 8.26b, including minimal preparation and ease of automation. However, it does still have a few drawbacks, such as the 6-hours acquisition time per sample.

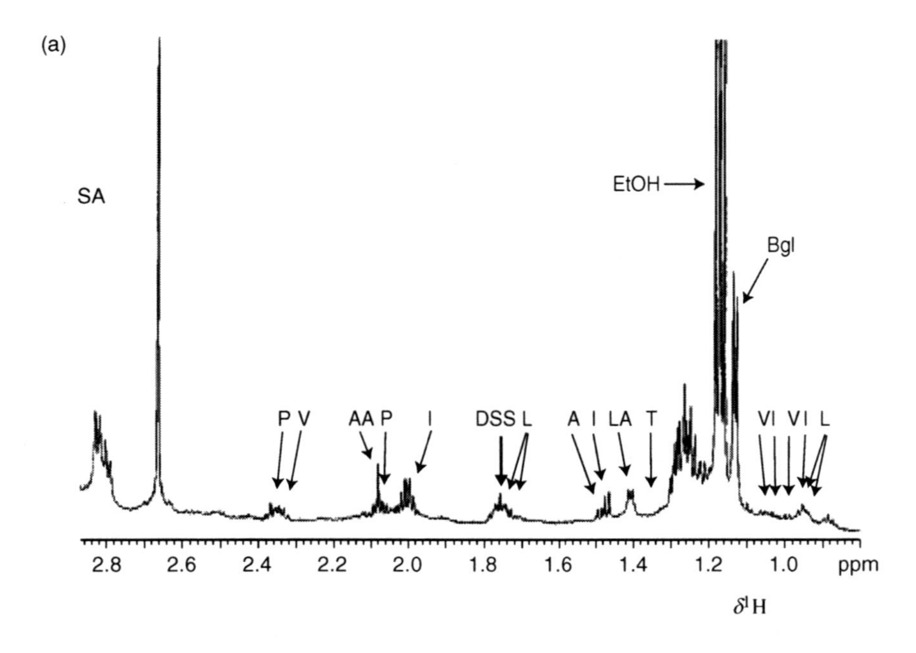

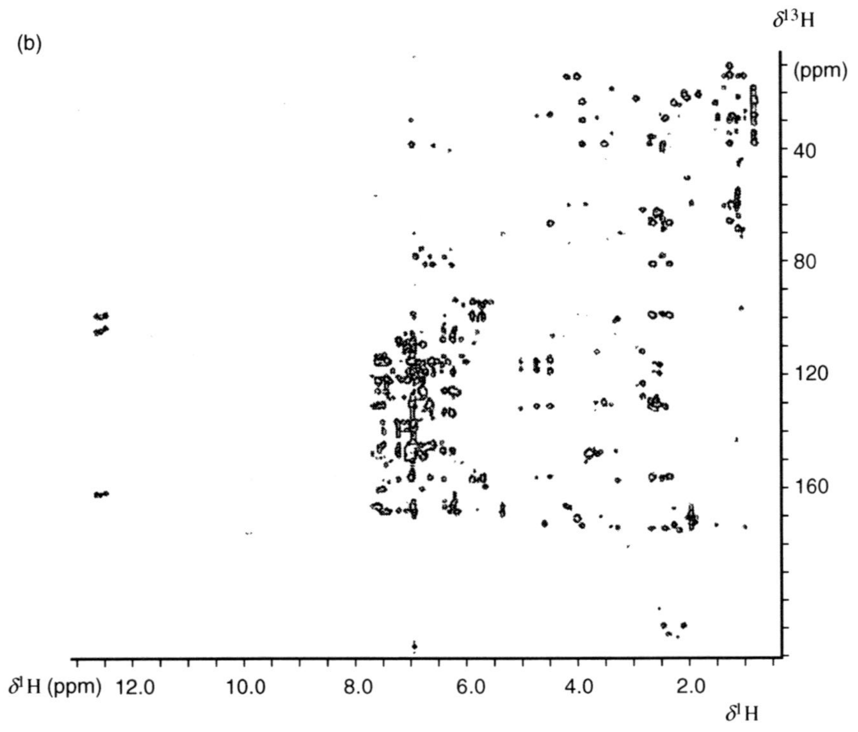

Fig. 8.25 1D and 2D NMR analysis of (a) a wine and (b) a polyphenolic wine extract.[18]

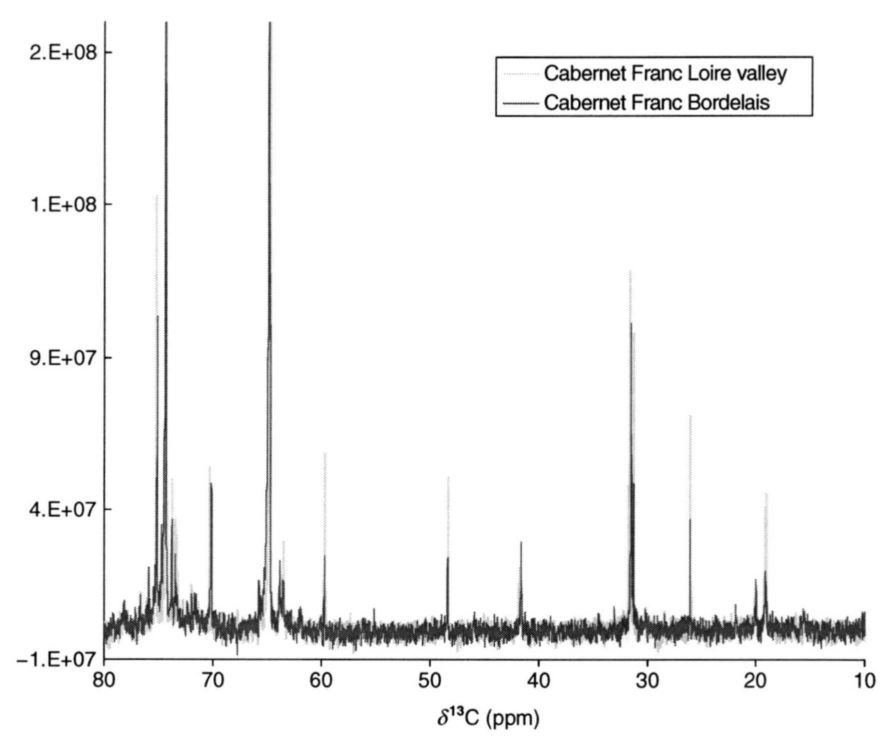

Fig. 8.26 ^{13}C NMR analysis of varietal wine from two different areas of France.

NMR analysis tests were even directly conducted non-destructively on bottles of wine.[64,65] The results of this study are limited because only the peak of acetic acid was observed in the given experimental conditions. Nevertheless, this non-invasive technique detects fraudulent products such as cocaine[66] dissolved in wine directly through the bottle using a standard clinical IRM scanner (Fig. 8.27).

This study allows us to imagine in the near future an NMR apparatus with acquisition parameters optimized for performing analysis directly on the bottles of wine. So far, no work has been finalized to ensure full traceability of wine from the NMR data alone. However, the information provided by the measures appears to be sufficient to characterize a unique wine, and the objective has been fully achieved for juices (orange juice in particular) using an HPLC/stop flow/J-resolve NMR sequence.[67]

The future of the NMR technique resides in the ability to refine it to a bench-top analytical tool, employing high-temperature magnets and 'cryo-coolers' that do not require huge quantities of expensive liquid gases like helium.

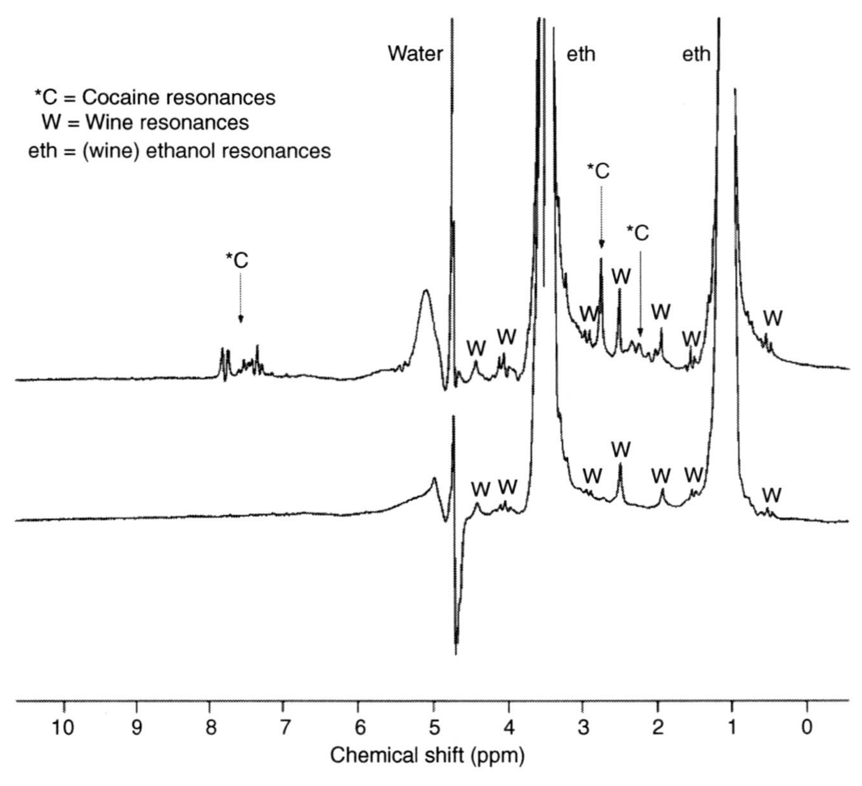

Fig. 8.27 Direct NMR analysis of a wine bottle.[66]

8.9 Conclusion

Tracing geographical origin and authenticity of food is a social demand and a scientific challenge. The employment of multiple techniques,[68] multi-element multi-isotopes[69] and multivariate mathematical tools[70] are all proposed as solutions for authenticating food and beverages, and wine is no exception.

As a very useful tool, each year, in response to the European directive,[28] the laboratories create and update a database whose isotopic values allow vintage and geographical location to be recognized. The influence of technological parameters on wine isotopes are studied carefully,[71] and the database is used as a baseline for grape varieties, ultra-low radioactivity research, NMR and isotope studies. Significant effort is needed to complete and maintain this database each year, but it is an invaluable tool.

The problems posed by the authenticity of food products and alcoholic beverages in particular are complex, but they can be largely resolved by implementing powerful, sophisticated analytical techniques. Backed by the combination of expert experience and advanced technology, it is unsurprising that wine is much more difficult to imitate than bank notes.

8.10 References

1. MCGOVERN, P.E., GLUSKER, D.L., EXNER, L.J. and VOIGT, M.M. Neolithic resinated wine, *Nature* (1996), **381**, 480–481.
2. SALAGOÏTY, M.H., LAGRÈZE, C., GAYE, J., DOMEC, A. and MÉDINA, B. Pratique de l'analyse sensorielle des vins sous assurance qualité (norme ISO 17025) Application a la reconnaissance de défauts, janvier 2012 Revue des oenologues N° 142.
3. SALAGOÏTY, M.H., GUYON, F., RENÉE, L., GAILLARD, L., LAGRÈZE, C., DOMEC, A., BAUDOUIN, M. and MÉDINA, B. Quantification method and organoleptic impact of added carboxymethyl cellulose to dry white wine, *Anal. Methods* (2011), **3**, 380–384.
4. CIVB, Conseil Interprofessionnel du Vin de Bordeaux, 1 cours du XXX juillet 33000 Bordeaux, A new look for Bordeaux Wine, Constellation Bordeaux, 2011 www.bordeaux.com.
5. SALAGOÏTY, M.H., GUYON, F., CHAUVET, S., VIATEAU, M., MORERE, M., CHARBREYRIE, D., JOLY, C. and MEDINA, B. Contribution à l'identification des cépages dans les vins blancs et les vins rouges. *Bull. OIV* (2006), **79**(904–906), 271–276.
6. CHABREYRIES, D., CHAUVET, S., GUYON, F., SALAGOÏTY, M.H., ANTINELLI, J.F. and MEDINA, B. Characterization and quantification of grape variety by means of shikimic acid concentration and protein fingerprint in still white wines, *J. Agric. Food Chem.* (2008), **56**, 6785–6790.
7. Pirouette, Infometrix, Inc.; 10634 E. Riverside Dr., Suite 250, Bothell, WA 98011 USA, http://www.infometrix.com/.
8. SIRET, R., L'ADN, un traceur des cépages dans les moûts et les vins, revue de 10 ans de recherche et d'expérimentations, Revue des œnologues N' 14017, juillet (2011).
9. FANZONE, M., PENA-NEIRA, A., JOFFRÉ, V., ASSOR, M. and ZAMORA, F. Phenolic characterization of Malbec wines trom Mendoza Province (Argentina), *J. Agric. Food Chem.* (2010), **58**, 2388–2397.
10. VON BAER, D., MARDONES, C.,GUTIÉRREZ, L., HOFMANN, G., HITSCHFELD, A. and VERGARA, C. Anthocyanin, flavonols and shikimic acid profiles as a tool to verify varietal authenticity in red wines produced in Chile; in Authentication of food and wine, (2007), 228–238, ACS symposium series 952 S.E Ebeler Editor.
11. MÉDINA, B., and SUDRAUD, P. Teneur des vins en chrome et nickel, causes d'enrichissement, *Connaissance de la Vigne et du Vin* (1980), **14**(2), 79–96.
12. SCHOLZE, H. LE VERRE, *Nature, Structure et Propriétés*, 1974, 2ème édition, Institut du Verre, Paris.
13. BERNARD, M. Wine Authentication, in P.R. ASHURST and M.D. DENNIS (Eds.), *Analytical Methods in Food Authentication*, Blackie Academic & Professional, London, (1998).
14. MARISA, C., ALMEIDA, R., TERESA, M., VASCONCELOS, M., BARBASTE, M. and MÉDINA, B. ICP-MS multi-element analysis of wine samples a comparative study of the methodologies used in two laboratories. *Anal. Bioanal. Chem.* (2002), **374**, 314–322.
15. CATARINO, S., TRANCOSO, I.M., MADEIRA, M., MONTEIRO, F., BRUNO DE SOUSA, R. and CURVELO-GARCIA, A.S. Rare earth data for geographical origin assignment of wine: a Portuguese case study. *BOIV* (2011), **84**(965–967), 233–246.
16. MÉDINA, B., AUGAGNEUR, S., BARBASTE, M., GROUSSET, F.E. and BUAT-MÉNARD, P. Atmospheric pollution influence on the lead content of the wines, *Food Add. Contam.* (2000), **17**, 435–445.

17. AUGAGNEUR, S. Etude de la composition isotopique du plomb dans une série séculaire de vins: mise en évidence de la pollution d'origine anthropique Thèse, Université de Bordeaux I (1996).
18. BARBASTE, M. thèse, Recherches sur l'origine géographique et le millésime des vins, Université de Pau et des pays de l'Adour (2001).
19. AUGAGNEUR, S., MEDINA, B. and GROUSSET, F. Measurement of lead isotope ratios in wine by ICP-MS and its applications to the determination of lead concentration by isotope dilution; Fresenius, *J. Anal. Chem.* (1996), **357**(8), 1149–1152.
20. HORN, P., HOLZL, S., TODT, W. and MATTHIES, D. Isotope abundance ratios of Sr in wine provenance determinations, in a tree-root activity study, and of Pb in a pollution study tree-rings. Isotopes, *Environ. Health. Stud.* (1998), **34**, 31–42.
21. BARBASTE, M., ROBINSON, K., GUILFOYE, S., MEDINA, B. and LOBINSKI, R.J. Precise determination of strontium isotope ratios in wine by inductively coupled plasma sector field multicollector mass spectrometry (ICP-SF-MC-MS), *Anal. At. Spectrom.*, (2002), **17**, 135–137.
22. MILLOT, R. and NÉGREL, P. *Wine* and *Champagne: Evidence from Lithium Isotopes; Forensic Isotope Ratio Mass Spectrometry, Fourth FIRMS Network Conference*, April 11–14 (2010), Washington DC, co-hosted by the Federal Bureau of Investigation.
23. DE MUYNCK, D., DEGRYSE, P. and VANHAECKE, F. The use of strontium, neodymium and lead isotope variations in archaeological applications, Federation of Analytical Chemistry and Spectroscopy Societies (FACSS); Symposium: Chemistry in Art and Archaeology,10/19/2009.
24. MANTON, W. Determination of the provenance of cocoa by soil protolith ages and assessment of anthropogenic lead contamination by Pb/Nd and lead isotope ratios, *J. Agric. Food. Chem.* (2010), **58**, 713–721.
25. SPITZKE, M.E. and FAUHL-HASSEK, C. Determination of the 13C/12C ratios of ethanol and higher alcohols in wine by GC-C-IRMS analysis, *Eur. Food Res. Technol.* (2010), **231**, 247–257.
26. GUYON, F., GAILLARD, L., SALAGOÏTY, M.H. and MÉDINA, B. Intrinsic ratios of glucose, fructose, glycerol and ethanol 13C/12C isotopic ratio determined by HPLC-co-IRMS: toward determining constants for wine authentication, *Anal. Bioanal. Chem.* (2011), **401**, 1551–1558.
27. COMMISSION REGULATION (EEC) No 2347/91 of 29 July 1991 on the collection of samples of wine products for the purposes of cooperation between Member States and for analysis by nuclear magnetic resonance, including analysis for the purposes of the Community databank.
28. COMMISSION REGULATION (EEC) No 2348/91 of 29 July 1991 establishing a databank for the results of analyses of wine products by nuclear magnetic resonance of deuterium.
29. LOPEZ, M.G. Authenticity: The case of Tequila.; in Authentication of food and wine (2007), 274–287, ACS symposium series 952 S.E. Ebeler Editor.
30. THOMAS, F., RANDET, C. GILBERT, A., SILVESTRE, V., JAMIN, E., AKOKA, S., REMAUD, G., SEGEBARTH, N. and GUILLOU, C. Improved characterization of the botanical origin of sugar by Carbon-13 SNIF-NMR applied to Ethanol, *J. Agric. Food. Chem.* (2010), **58**, 11580–11585.
31. GINER MARTÍNEZ-SIERRA, J., MORENO SANZ, F., HERRERO ESPÍLEZ, P., SANTAMARIA-FERNANDEZ, R., MARCHANTE GAYÓN, J. M. and GARCÍA ALONSO, J. I. Evaluation of different analytical strategies for the quantification of sulfur-containing biomolecules by HPLC-ICP-MS: Application to the characterisation of 34S-labelled yeast, *J. Anal. At. Spectrom.* (2010), **25**, 989–997.
32. OIV: Méthode de détermination de l'Ochratoxine A par colonne d'immuno-affinité. Résolution 16/2001.

33. OIV: dosage de residus de pesticides dans le vin (par gc/ms ou lc/ms-ms) après extraction par la methode QUECHERS. 2011 OENO-SCMA 10–436 Résolution Et5.
34. SCHWARTCZ, H.P. Absolute age determination of archeological sites by uranium series dating of travertines, *Archeometrix* (1980), **22**(1), 3–24.
35. KAUFMAN, S. and LIBBY, W.L. The natural distribution of tritium. *Phys. Rev.* (1954), **93**(6), 1337.
36. MARTINIÈRE, P., GAULTIER, J.-M., SUDRAUD, P., SEVERAC, J. ET CAUSSANEL, J.-L. Évolution de la radioactivité par le carbone 14 des vins de Gironde – Application à la recherche de millésimes. *Ann. Fals. Exp. Chim.* (1979), **72**, 263–274.
37. HUBERT, PH., HUBERT, F., OHSUMI, H., GAYE, J., MEDINA, B. and JOUANNEAU, J.M. Application de l'analyse par spectrométrie gamma bas bruit de fond à la datation des vins d'origine française, *Annal. Fals. Exp. Chim.* (2001), **957**, 357–368.
38. HUBERT, PH., PERROT, F., GAYE, J., MEDINA, B. and PRAVIKOFF, M.S. Radioactivity measurements applied to the dating and authentication of old wines, *C.R.A.S.* (2009), **10**(7), 587–700.
39. HUBERT, PH., GAYE, J. and MEDINA, B. Datation non destructive du vin et du verre des bouteilles. *Annales des Falsifications et de l'expertise Chimique et toxicologique* (2009), **971**, 54–56.
40. WALLACE, B. *The Billionaire's Vinegar: The Mystery of the World's Most Expensive Bottle of Wine,* Crown Publishers (2008), ISBN 0307338770, 9780307338778, 319.
41. HUBERT, PH., GAYE, J. and MEDINA, B. Unpublished results, 2012, CENBG/IN2P3 Université de Bordeaux, Centre d'Etudes Nucléaires de Bordeaux Gradignan. courriel: hubertp@cenbg.in2p3.fr.
42. SZPUNAR, S., PELLERIN, P., MAKAROS, A., DOCO, T., WILLIAMS, P., MÉDINA, B. and LOBINSKI, R. Speciation analysis for biomolecular complexes of lead in wine by size exclusion high-performance liquid chromatography-inductively coupled plasma mass spectrometry, *J. Anal. Atom. Spectrom.* (1998), **13**, 749–754.
43. LOBINSKI, R., WITTE, C., ADAMS, F.C., TEISSEDRE, P.L., CABANIS, J.C. and BOUTRON, C.F. Organolead in wine. *Nature* (1994), **370**, 6484.
44. HOLMBERG, L. Wine fraud, *Int. J. Wine Res.* (2010), **2**, 105–113, Dovepress journal. http://www.dovepress.com/wine-fraud-peer-reviewed-article-IJW
45. GUÉGAN, H., GUEGAN, H., RIDARD, B. and CHOPINET, M.H. Authentification de bouteilles de vin anciennes par faisceaux d'ions de haute énergie (2009), Les Techniques de l'Ingénieur, Ed. IN 109, 1–7.
46. ARCANE, http://www.cenbg.in2p3.fr/arcane/.
47. COLOMBAN, P. On-site Raman identification and dating of ancient glasses: a review of procedures and tools, *J. Cult. Herit.* (2008), **9**, e55.
48. ELIASSON, C., MADEOD, N., and MATOUSEK, P. Non-invasive detection of cocaine dissolved in beverages using displaced Raman spectroscopy, *Anal. Chim. Acta* (2008), **607**, 50.
49. MOORE, D.S. and SCHARFF, R.J. Portable Raman explosives detection, *Anal. Bioanal. Chem.* (2009), **393**, 1571–1578.
50. RICCI, C., ELIASSON, C., MACLEOD, N.A., NEWTON, P., MATOUSEK, P., KAZARIAN, S.G. Characterization of genuine and fake artesunate anti-malarial tablets using Fourier transform infrared imaging and spatially offset Raman spectroscopy through blister packs, *Anal. Bioanal. Chem.* (2007), **389**(5), 1525–1532.
51. MEDINA, B., GUIMBERTEAU, G. and SUDRAUD, P. Dosage du plomb dans les vins. Une cause d'enrichissement: les capsules de surbouchage, *Connaissance de la vigne et du vin* (1977), **11**(2), 183–193.
52. SECF – Société des Experts Chimistes de France, « Wine Track (2011 » Journée Scientifique et Professionnelle sur la Traçabilité des Vins et Spiritueux Sète, 13 octobre (2011); http://chimie-experts.org/.

53. SCHAMEL, G. Forensic economics: some evidence for new wine to be sold in old bottles. Conference abstract, AAWE (2009). http://www.wine-economics.org/meetings/Reims2009/programinfo/Abstracts/Schamel.pdf.
54. BUZAS, M.CH., CHIRA, N., DELEANU, C. and ROSCA, S. Identification and quantitative measurements by 1H-NMR spectroscopy of several compounds present in Romanian wines, *Rev. Chim. (Bucuresti)* (2003), **54**, 831–833.
55. KOSIR, I., KOCJANCIC, M. and KIDRIC, J. Wine analysis by 1D and 2D NMR spectroscopy. *Analusis* (1998), **26**, 97–101.
56. SON, H.S., HWANG, G.S., KIM, K.M., AHN, H.J., PARK, W.M., VAN DEN BERG, F., HONG, Y.S. and LEE, C.H. Metabolomic studies on geographical grapes and their wines using ^{1}HNMR analysis coupled with multivariate statistics, *J. Agric. Food Chem.* (2009), **57**, 1481–1490.
57. SON, H.S., KIM, K.M., VAN DEN BERG, F., HWANG, G.S., PARK, W.M., LEE, C.H. and HONG, Y.S. ^{1}H Nuclear magnetic resonance-based metabolomic characterization of wines by grape varieties and production areas, *J. Agric. Food Chem.* (2008), **56**(17), 8007–8016.
58. ANASTASIADI, M., ZIRA, A., MAGIATIS, P., HAROUTOUNIAN, S.A., SKALTSOUNIS, A.L. and MIKROS, E. NMR-based metabolomics for the classification of Greek wines according to variety, region, and vintage. Comparison with HPLC data, *J. Agric. Food Chem.* (2009), **57**, 11067–11074.
59. DU, Y.Y., ZHANG, X., BAI, G.Y. and LIU, M.L. Classification of wines based on combination of ^{1}H NMR spectroscopy and principal component analysis, *Chinese J. Chem.* (2007), **25**, 930–936.
60. NILSSON, M., DUARTE, I.F., ALMEIDA, C., DELGADILLO, I., GOODFELLOW, B.J., GIL, A.M. and MORRIS, G.A. High-resolution NMR and diffusion-ordered spectroscopy of Port Wine. *J. Agric. Food Chem.*(2004), **52**, 3736–3743.
61. LOPEZ-RITUERTO, E., CABREDO, S., LOPEZ, M., AVENOZA, A., BUSTO, J.H. and PEREGRINA, J.M. A thorough study on the use of quantitative ^{1}H NMR in Rioja Red Wine Fermentation Processes, *J. Agric. Food Chem.* (2009), **57**, 2112–2118.
62. LARSEN, F.H., FRANS VAN DEN BERG, F. and ENGELSEN, S.B. An exploratory chemometric study of 1H NMR spectra of table wines. *J. Chemometrics* (2006), **20**, 198–208.
63. GUYON, F., SABATHIÉ, N., GAILLARD, L. and MEDINA. Unpublished results, (2012) SCL: Laboratoire de Bordeaux, Ministère de l'économie des finances et de l'industrie, ministère du Budget des Comptes Publics et de la Réforme de l'Etat; courriel: labo33@scl.finances.gouv.fr.
64. WEEKLEY, A.J., BRUINS, P. and AUGUSTINE, M.P. Nondestructive method of determining acetic acid spoilage in an unopened bottle of wine, *Am. J. Enol. Vitic.* (2002), **53**, 318–321.
65. WEEKLEY, A.J., BRUINS, P., SISTO, M. and AUGUSTINE, M.P. Using NMR to study full intact wine bottles, *J. Magn. Reson.* (2003), **161**, 91–98.
66. GAMBAROTA, G., PERAZZOLO, C., LEIMGRUBER, A., RETO MEULI, R., MANGIN, P., MARC AUGSBURGER, M. and GRABHERR, S. Non-invasive detection of cocaine dissolved in wine bottles by ^{1}H magnetic resonance spectroscopy. (2010), Drug Test. Analysis, http://www.drugtestinganalysis.com/.
67. SPRAUL, M., SCHÜLTZ, B., RINKE, P., KOSWIG, S., HUMPFER, E., SCHÄFER, H., MÖRTTER, M., FANG, F., MARX, U.C. and MINOJA, A. NMR-based multi parametric quality control of fruit juices: SGF profiling. *Nutrients* (2009), **1**, 148–155.
68. WEN, H., LEE, M.Y., SONG, Y., MOON, S. and PARK, S. Combined genomic-metabolomic approach for the differentiation of geographical origins of natural products: deer antlers as an example, *J. Agric. Food Chem.* (2011), **59**, 6339–6345.
69. KELLY, S., HEATON, K. and HOOGEWERFF, J. Tracing the geographical origin of food: the application of multi-element and multi-isotope analysis, *Trends Food Technol.* (2005), **16**, 555–567.

70. PREYS, S., VIGNEAU, E., MAZEROLLES, G., CHEYNIER, V. and BERTRAND, D. Multivariate prototype approach for authentication of food products, *Chemometr. Intell. Lab.*(2007), **87**, 200–207.
71. GUYON, F., DOUET, CH., COLAS, S., SALAGOITY, M.H. and MEDINA, B. Effects of must concentration techniques on wine isotopic parameters, *J. Agric. Food. Chem.* (2006), **54**, 9918–9923.
72. GUILLAUME, F.; unpublished results (2012) ISM: Institut des Sciences Moléculaires, Université de Bordeaux, 351 cours de la Libération 33405 Talence, France, f.guillaume@ism.u-bordeaux1.fr.

9

Using new analytical approaches to verify the origin of fish

J. Martinsohn, European Commission Joint Research Centre, Italy

DOI: 10.1533/9780857097590.3.189

Abstract: Globally, fish and fish products are a major food commodity, and the fish processing industry is an economically important sector. Persistent overexploitation of fish stocks, and illegal fishing, threaten to a great extent the future sufficient availability of the important natural resource fish. Additionally fraud along the supply chain, false labelling of fish and fish products, mislead the consumer and has a negative impact on consumer trust and the industry. Also, in many areas of the world, aquaculture production is on a steep rise which might lead to environmental challenges, for example, when fish escape from farms and breed with their wild counterparts which is why it is important to be able to distinguish between wild and farmed fish of the same species. These examples show that the fisheries and aquaculture sector relies on efficient traceability schemes allowing to determine the origin of fish and fish products. This chapter discusses measures taken to ensure fish and fish product traceability and presents modern analytical technologies and applications, which can considerably strengthen traceability 'Ocean to Fork'. These analytical technologies have to answer three key questions, ideally also on processed products: 'What species is it?'; 'Where does it come from?' and 'Is it of farmed or wild origin?'

Key words: mislabelling, DNA technology, microchemistry, forensics, species identification, origin assignment, capacity building.

9.1 Introduction: the commercial importance of fish and fish product origin

It is generally accepted that the fisheries sector depends on the development of efficient traceability schemes that ensure that the source of a product can be determined at any step of the supply chain ('Ocean to Fork') in order to better support sustainable fisheries, fight illegal activities and to ensure consumer protection and information. Such traceability is currently almost exclusively

Fig. 9.1 The three key questions new analytical approaches based on genetics and chemistry can help to answer when verifying the origin of fish and fish products. (Fish symbols courtesy of the Integration and Application Network, University of Maryland Centre for Environmental Science. Map: European Union, 2010.)

based on labelling requirements and certification. It follows that the capacity to independently determine the origin of fish and fish products constitutes a valuable asset for a traceability scheme.

Putting emphasis on marine fisheries and aquaculture, this chapter briefly discusses current measures taken to ensure traceability with a focus on the international level, the USA and the European Union (EU). It follows a presentation of modern analytical technologies and applications, which can considerably strengthen 'Ocean to Fork' traceability. These analytical technologies have to answer three key questions, ideally also for processed products: 'What species is it?'; 'Where does it come from?' and 'Is it of farmed or wild origin?' (Fig. 9.1). All three questions are important when determining the true origin of fish and fish products. Species identification, the easiest of the three questions to answer with the technologies described in this chapter (see below), can be used when a fish product is sold under a label indicating a certain species and coming from a distinct region, but the true species used does not exist in the indicated region. Relevance and added value of the presented technologies for the fisheries and aquaculture sector are considered and needs for sufficient capacity building identified. Finally what future trends in this field will most probably arise is reflected upon.

9.2 Legal standards for fish origin and problems of counterfeiting

The need for efficient traceability and traceability schemes which are powerful enough to ensure compliance with rules and to fight fraud in the fisheries sector is generally and globally acknowledged. However, as stated above, to be fully efficient, such traceability schemes must be legally binding and be supported by independent control measures. Historically the agricultural sector has been a forerunner in the development of traceability schemes (Blancou, 2001). Systems have been developed to keep track of data on livestock and products along the supply chain ('Farm to Fork') for consumer protection and in support of the industry. In the EU the development and implementation of traceability schemes has been boosted by the free trade within the community market as well as by a number of food crises caused by epidemics, in particular Bovine Spongiform Encephalopathy (BSE) (European Commission, 2004). In the context of this chapter it is worth looking more closely at the implementation of traceability for bovine animals in the EU, which is binding to member states through Regulation (EC) 1760/2000 (Council, 2000). Each animal has to be individually tagged and is thereby identified by a unique number and barcode. Every trade related transfer is recorded and the complete information stored in national databases, which are compulsory for EU member states. At the end of the product chain the tracing consists of labels on the final meat product indicating the place of birth, fattening and slaughter. To protect consumers against incorrectly labelled food, the 'Farm to Fork' principle has been implemented in the EU agricultural sector since 2002 through Regulation (EC) 178/2002. This principle implies the tracking and identification of food products during their transit through the supply chain 'through all stages of production, processing and distribution' (European Parliament and Council, 2002). While not explicitly referring to fish and fish products, these are covered by Regulation (EC) 178/2002 as the rules refer to food and feed in general.

In the EU traceability legislation is also implemented specifically covering fish and fish products and complementing Regulation (EC) 178/2002: Regulation (EC) 104/2000 establishes that commercial name, production method and catch area or country of production are required for labelling on each product (European Council, 2000). Regulation 104/2000 was subsequently extended by Regulation (EC) 2065/2001, adding more specific details to the final information given to the customer in terms of production method and geographic origin. The labelling information has to be given at every step of the production and retailing chain (European Commission, 2001).

The comprehensive legal framework established in the EU enabling traceability to ensure food safety and consumer protection but also to deter fraud is paralleled by similar legislation in other countries, as recently reviewed by Smith *et al.* (2008). In the USA, the so-called 'Lacey Act' establishes rules for the labelling of fish and wildlife products (US Department of Justice, 2009). This law was originally introduced to fight illegal hunting at the beginning of the last century, but its scope has since been broadened considerably through numerous amendments. Nowadays it prohibits the selling of unlabelled fish and wildlife products and penalises mislabelling. The Lacey Act is a very powerful law since under its remit any US citizen is liable if breaking an underlying foreign fisheries or wildlife law and subsequently imports, exports, transports, sells or receives that product in the USA. Any misdoing is regarded as a felony provided that the matter under investigation amounts in value to more than $350 and that the investigating authorities can prove that the defendants had knowledge of their wrongdoings, that is, acted intentionally (if no knowledge can be proven the wrongdoing is regarded as a misdemeanour).

Traceability of fish and fish products has also been on the international agenda for quite a while through the Food and Agriculture Organization (FAO) of the United Nations (Committee of Fisheries, 2003), and a recent FAO report emphasises the need for and value of traceability schemes in the fisheries and aquaculture sector (Washington and Ababouch, 2011).

The FAO also takes a strong stance against illegal, unreported and unregulated (IUU) fishing with the International Plan of Action to Prevent, Deter and Eliminate Illegal, Unreported and Unregulated Fishing (IPOA IUU) (FAO, 2001). This voluntary plan, which is endorsed by numerous nations and also the EU, has been elaborated within the framework of the FAO Code of Conduct for Responsible Fisheries, and foresees a variety of measures to deter IUU fishing.

The EU has recently undertaken significant steps to improve fisheries control and to step up the fight against IUU fishing. The EU's control and enforcement policy is built on three complementing pillars: a new control regulation (European Council, 2009); the IUU Regulation to combat illegal, unreported and unregulated fishing (European Council, 2008), which applies to both EU and non-EU vessels; and the Regulation on Fisheries Authorisations (European Council, 2008a) which deals with the control of EU vessels fishing outside EU waters and of third country vessels fishing in EU waters.

All three regulations are meanwhile fully implemented and the control regulation and the IUU regulation are of particular interest in the context of this chapter. The control regulation lays down that fisheries and aquaculture products on the market in the Community shall be adequately labelled to ensure traceability (European Council, 2009) in line with regulation (EC) 178/2002 (see above). Minimum labelling requirements include the FAO alpha-3 code of each species and information for consumers provided for in regulation (EC) No. 2065/2001 (see above) such as commercial designation, scientific name, relevant geographical area and production method. Additionally, according to the IUU regulation, fishery products can only be imported into the EU when accompanied by a catch certificate, affirming that fish derive from legal catches. The certification scheme also aims to ensure product traceability at all stages of production, from catch to processing, and transport to marketing, and to be a tool to enhance compliance with conservation and management rules. Interestingly, in article 13 'New technologies', the control regulation refers explicitly to genetics in the context of traceability tools and in paragraph 16 of the introductory part it states: '*It should be for the Council to decide on the future use of electronic monitoring devices and* ***traceability tools such as genetic analysis*** *and other fisheries control technologies if these technologies lead to an improved compliance with rules of the common fisheries policy in a cost-effective way*' (Highlighting by the author).

In summary it is apparent that the need for efficient traceability in the fisheries sector is acknowledged by the industry, but also globally at the regulator and policy level. As stated above however, fish product traceability is currently almost entirely based on the 'written word' that is, certificates and labelling. Undoubtedly the ability to verify such information with independent analytical methods would be a great asset in support of control, enforcement and also consumer information and protection.

In the following section a concise overview is provided on new analytical approaches with a high potential to support fish product traceability in a cost-efficient way.

9.3 Applications of new analytical approaches to verify fish origin

The scope of this book does not permit extensive coverage of the high number of analytical procedures developed and applied to analyse the authenticity of fish products. Here only a selection of new and potentially highly powerful and cost-effective methods for origin assignment of fish and fish products will be discussed. The interested reader will find a summary in recently published literature (Kochzius, 2008; Martinsohn, 2011) and be provided with sources of further information at the end of this chapter.

To be valuable as independent authentication tools as part of traceability schemes, analytical approaches have to be reliable and cost-effective applications which are beyond or very advanced in the research and development phase. Such tools must enable control authorities to conclude whether labels on fish and fish products identify the correct species, correct origin and whether fish are derived from aquaculture or the wild (Fig. 9.1). Moreover, ideally such technologies can be applied on processed products, and allow a short response time, that is, quickly lead to results. In the last few decades many different methods have been applied, mostly for fish species identification but also in support of origin assignment. For example, species identification has been based on the description of morphological characteristics and parasite load, that is, the description of parasitic species and their abundance on fish. As these parameters are also influenced by environmental conditions surrounding the fish, their analysis has been employed for origin assignment (MacKenzie and Abaunza, 1998)).

A drawback is that most morphological markers are influenced by environmental factors (e.g., temperature), which renders their interpretation complex, in contrast to analysing molecular genetic markers (see below) which are transmitted across generations independently of environmental factors.

The general progress in molecular biology and genetics led to an increased usage of molecular and genetic markers for species identification and origin assignment. However this is only possible if robust reference datasets are available as baselines: if the analyst wishes to know whether a can of fish does indeed contain the species indicated on the label, her or his data derived from the analysis of the content of the can must be comparable to a set of

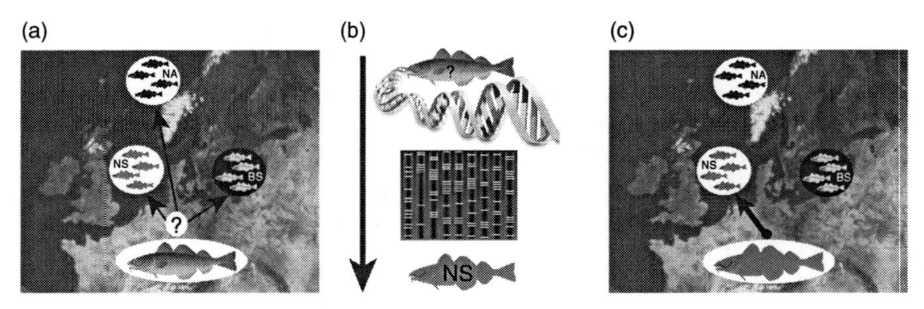

Fig. 9.2(a–c) The stages involved in using the principle of individual origin assignment (IA), to trace individuals or fish products back to the most likely source. See text for explanation.(Graphical elements courtesy of the Integration and Application Network, University of Maryland Centre for Environmental Science. Map: European Union, 2010.)

previously validated data for species identification. The same is true for origin assignment: if a fish of dubious origin is being analysed the resulting data must be matched to data from previously analysed fish of diverse geographical regions to which the fish under investigation might belong (Fig. 9.2).

Individual fish of unknown origin, which might be sampled in a harbour, supermarket or at a fishmonger, are assigned to a set of baseline stocks or populations (Fig. 9.2a). In this scenario the three baseline populations North Atlantic (NA), North Sea (NS) and Baltic Sea (BS) are used as hypothetical regions of origin. The fish in question is analysed to determine its genotype (Fig. 9.2b). Based on its genotype, the individual is assigned probabilistically to the baseline populations (Fig. 9.2c). Here the sample can be assigned with certainty as originating from the North Sea.

In the following sections some advanced analytical technologies with potential to support fish product traceability are outlined.

9.3.1 Fatty acid analysis

In all living organisms fatty acids constitute a major source of metabolic energy and are essential materials for the formation of structural elements at the cellular level such as cell and tissue membranes. In marine fish about 20 different fatty acids are present at relative amounts of greater than 1%, and the amount of each fatty acid in a tissue can vary substantially at the species and population level. This variation can be measured by spectrometric analysis to establish a distinct fatty acid profile, which in turn can be used as a biomarker (Guil-Guerrero *et al.*, 2005).

While the fatty acid composition of fish tissues is influenced by biological (e.g., age, maturity) and environmental features (e.g., diet, water temperature; reviewed in Grahl-Nielsen (2005)), it is also to a large extent genetically determined (Kwetegyeka *et al.*, 2008). As populations or stocks of a fish species inhabiting different geographical areas are in general characterised by distinct environmental conditions and nutriments, they might be distinguishable by qualitative and quantitative analysis of the fatty acid composition. Such an approach has been followed for stock identification in a presumed fisheries fraud case investigating alleged false reporting of herring landed by a Norwegian purse seiner (Grahl-Nielsen, 2005). The use of fatty acid profiling to distinguish wild from farmed fish was proposed some years ago (Moretti *et al.*, 2003), and meanwhile nuclear magnetic resonance (NMR) spectroscopy of lipids has been successfully applied to distinguish farmed from wild Atlantic salmon, to discriminate between different geographical origins and to verify the origin of market samples (Aursand *et al.*, 2009). Fatty acid profiling can be performed on fresh or frozen fish tissues, but its application on processed fish products, such as smoked or canned fish, is not established and needs further assessment.

9.3.2 Microchemistry and stable isotope analysis

Isotopic and elemental markers have recently gained importance in solving questions of natal origin, spawning site fidelity, connectivity between populations and also for traceability (Campana, 2005). Because chemical elements are taken up from the surrounding waters, the chemical composition of hard tissues such as otoliths (the fish ear-stones) and scales reflect chemical properties of the environment to which the fish have been exposed, which can be used to establish 'elemental fingerprints'. For this purpose the elemental concentrations from isotopes of various elements such as Sr, Ba, Mb, Fe and Pb are determined (Campana, 2004).

Otoliths are highly suitable for the identification and analysis of geochemical signatures (Fig. 9.3). Figure 9.3a depicts otoliths of common sole (*Solea solea* L.), Atlantic herring (*Clupea harengus* L.), Atlantic cod (*Gadus morhua* L.) and European hake (*Merluccius merluccius* L.). As can be seen, otoliths tend to be species specific in size and shape. Figure 9.3b depicts a common sole otolith of which a section is magnified, clearly showing the growth increments. Figure 9.3c depicts how otolith microchemistry can be used to determine where fish come from. In this example herring, sampled from different populations and regions of the Baltic Sea, have been analysed. The symbols (triangles, squares and dots) represent individual fish. Fish with similar chemical otolith composition tend to group together in distinct geographical areas. Horizontal axis: otolith sodium (Na) and barium (Ba) concentration. Vertical axis: otolith strontium (Sr) and lead (Pb) composition.

Otoliths grow throughout the life of the fish and the material deposited in annual growth increments is neither reabsorbed nor altered. This is ideal to establish elemental fingerprints and the relation of those to migration and stock structure of fish (Campana, 2004). A wide variety of elemental analysis techniques are available, grouping into two main categories: bulk analysis, in which the otoliths are dissolved and assayed using chemical approaches; and probe analysis, where specific zones in the otolith may be targeted for analysis of specific life stages. Both approaches have been employed to

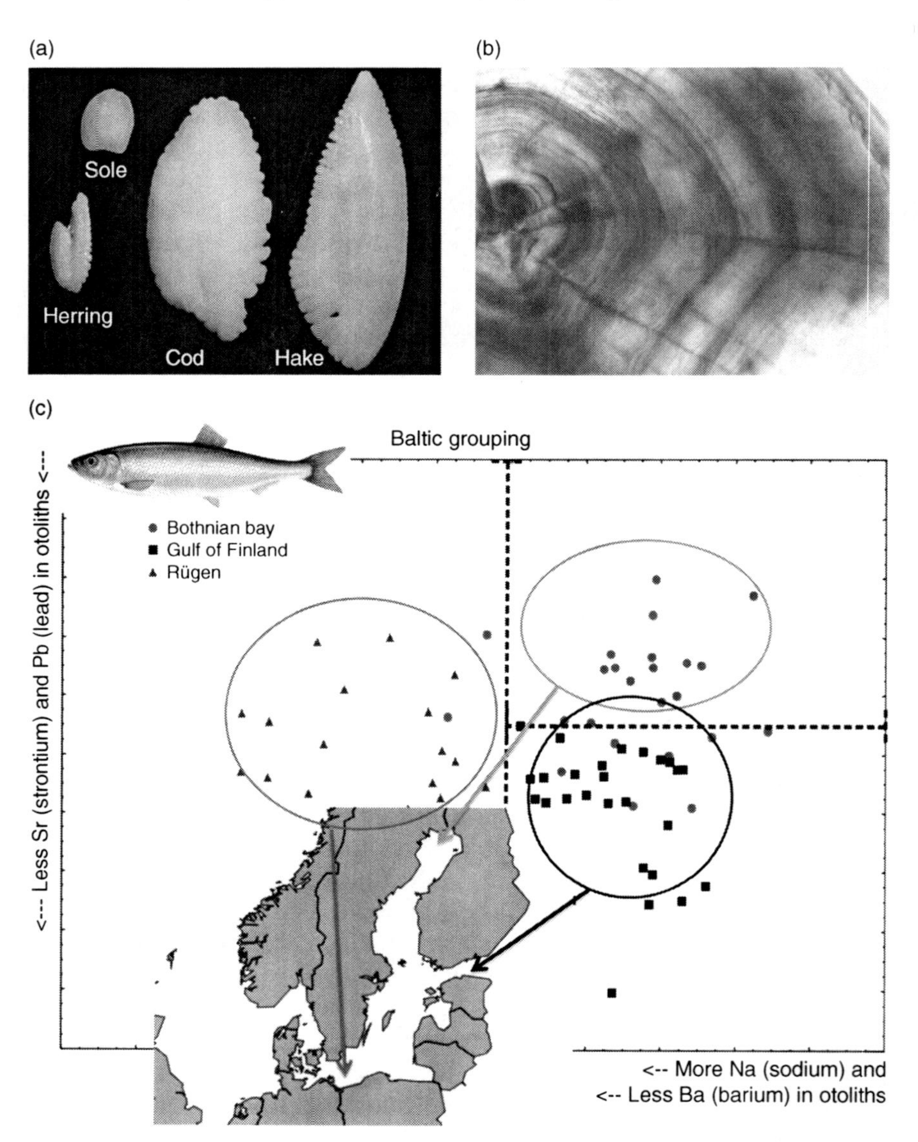

Fig. 9.3 (a–c) Otoliths: Fish ear-stones as precious gems for origin assignment.See text for explanation. (Pictures and graph courtesy of A. Geffen, University of Bergen.)

examine population structures and stock tracing of various marine fish species (Campana *et al.*, 2000; Geffen *et al.*, 2003; Swan *et al.*, 2006). The discrimination power of otolith microchemistry, even over relatively short geographical distances, has been demonstrated in a variety of studies. It could be shown that individual juvenile common soles (*Solea solea*) originating from two estuaries along the French coast, separated by a distance of about 200 km, could be discriminated by elemental otolith fingerprints (De Pontual *et al.*, 2000). Also a recent study on Atlantic cod (*Gadus morhua* L.) of the north-east Atlantic used otolith microchemistry analysis, in combination with otolith shape analysis, body morphometry, bacterial assemblages, parasite load and DNA markers, to distinguish wild from farmed fish and to determine the precise harvest origin of individuals (Higgins *et al.*, 2010). Otolith trace element analysis for traceability is suitable for all stages along the supply chain as long as the heads remain with the fish. This is a disadvantage compared to the use of DNA analysis which is also possible on processed products (see below). An asset of otoliths is that the elemental composition is resilient, and does not degrade or change over time after death (Thresher, 1999).

Other hard tissues, such as scales or spines, can also be used to establish elemental fingerprints (Gillanders, 2001). For example, scale microchemistry has been employed as a tool to investigate the origin of wild and farmed Atlantic salmon (*Salmo salar*) (Adey *et al.*, 2009). For inspection and enforcement this constitutes an interesting alternative to otoliths, as the isolation of scales for analytical purposes is less damaging to the individual fish than the extraction of otoliths. Recently, in studies performed in the frame of the EU funded project FishPopTrace (FishPopTrace, 2008–2011) on the highly commercial fish species cod, hake and herring, it was shown that otolith microchemistry can reveal population structure of marine fish even on small geographical scales (Fig. 9.3). Therefore this kind of analysis can be applied for origin assignment and is valuable for the support of traceability schemes.

9.3.3 Genetic marker analysis

The analysis of genetic markers in fisheries science started in the 1950s, and addressed a wide range of research topics and scientific hypotheses. Many of the questions scientifically addressed with genetic marker analysis are of rather academic nature and discussed in detail elsewhere (Kochzius, 2008). The analysis of genetic markers can be separated into DNA analysis, revealing DNA sequence variation, and protein analysis, that reveals differences in amino acid composition. For analysis samples of fish (or products) are collected, and DNA or protein is extracted from tissue, which nowadays is a routine process facilitated by the availability of numerous commercial kits. Protein analysis, such as isoelectric focusing (IEF) and related techniques (Rehbein, 2003), or assays based on antibodies binding to proteins with high specificity (Taylor *et al.*, 1994), can provide high species resolution. While these techniques are still applied for fish food authenticity control, their use is progressively being replaced by DNA-based technologies (see below) as DNA-marker analysis has increasingly become the standard approach in fisheries genetics and fish food research in general (Hauser and Carvalho, 2008; Rasmussen and Morrissey, 2008).

For fish species identification as well as population structure studies, which are required to enable origin assignment, the genome of mitochondria (sub-cellular organelles creating energy for cellular activity) and the nuclear genome have been valuable (Kochzius, 2008). Particularly 'DNA barcoding', which is based on sequence analysis of the mitochondrial *cytochrome oxidase I* (COI) gene for species identification is highly amenable for routine fish species identification purposes (Ward *et al.*, 2009). Interestingly this approach has recently been officially approved by the US Food and Drug Administration (FDA, 2011). For the analysis of species identifying DNA markers, a panoply of different technologies has been employed, which is reviewed in detail elsewhere (Teletchea, 2009).

This section focuses on DNA markers most suitable for traceability applications. Microsatellites, also called short tandem repeats (STRs) in forensics, are tandem sequence repeats of one to six nucleotides (e.g., 'cgtacgtacgtacgtacgta') in the genome. Their extensive polymorphism is characterised by a high variability in the number of repeats (between 5 and 100 repeats) even between individuals. This extreme variability has rendered microsatellites the standard marker for human identity testing (genetic fingerprinting) through DNA profiling, and is also used in forensic genetic crime scene investigation (Butler, 2005). Microsatellites have also been used in numerous fish population studies, and their potential value as markers for origin assignment in traceability schemes is high. However, despite this common application of microsatellites, there are considerable disadvantages, particularly when the intention is to use microsatellites as standard markers in monitoring and control schemes on a national or supra-national level. The interpretation of microsatellite data is susceptible to mistakes, often caused by mistyping and scoring errors, and also there is a prevailing lack of comparability among laboratories (Dewoody *et al.*, 2006). While both phenomena pose a major impediment to routine applications, numerous examples exist where microsatellites have been used for fish population/stock analysis, management and also origin assignment (Hauser and Carvalho, 2008; Manel *et al.*, 2005). These examples encompass highly commercial species such as Atlantic salmon (Primmer *et al.*, 2000) and Pacific salmon (DFO, 2011).

More recently single nucleotide polymorphisms (SNPs), genome sites where more than one nucleotide (A, C, G or T) is present in a species, have increasingly been used as genetic markers in fisheries genetics (Hudson, 2008). SNPs are in fact the most abundant polymorphism in the genome (Brumfield *et al.*, 2003), but normally only two alleles exist (bi-allelic markers) per locus, thus they are less variable than microsatellites, where often numerous alleles exist (see above). However, the lack of information per SNP marker on genetic differentiation, for example, between individuals of two different populations, is by far outweighed by their high abundance. A significant asset of SNPs is that they, unlike other genetic markers where routine genotyping and transfer of protocols between laboratories proves difficult, provide categorical information. This allows data to be standardised across laboratories, for use in (supra-)national control schemes and also for forensic applications (Sobrino *et al.*, 2005). However, since the use of SNPs as genetic markers is relatively recent, considerable further research effort targeting all commercial marine fish species will be necessary before they can be employed routinely for origin assignment in traceability schemes. Studies on marine fish using SNPs show that investing in such research is worthwhile: SNPs as markers to distinguish stocks of Atlantic cod (*Gadus morhua*) provided a high resolution power for stock identification, comparable to the resolution obtained when using microsatellite loci as genetic markers (Wirgin *et al.*, 2007). An impressive example for the increasing use of SNPs is the North Pacific Anadromous Fish Commission (NPAFC) that is developing SNP arrays for Pacific salmon (http://www.npafc.org). The current rapid progress of DNA analysis technologies will have significant effects on the development of traceability tools. High-throughput sequencing has declined dramatically in cost, while speed and quality of analysis has increased by orders of magnitude. This is well illustrated by human genome sequencing, which can currently be accomplished in a few weeks at costs below US$ 5000, compared to more than a year and more than US$ 100 million in 2007, and this trend is continuing with DNA sequencing costs dropping by half every 5 months (Metzker, 2010; Stein, 2010). New high-throughput DNA sequencing technologies provide great opportunities for genetic fish population analysis and consequently also for traceability in the fisheries sector as shown by FishPopTrace (FishPopTrace, 2008–2011) where in a few weeks over 100 million bases of sequence data have been generated for the three highly commercial fish species European hake (*Merluccius merluccius*), Atlantic herring (*Clupea harengus*) and common sole (*Solea solea*).

For species identification, DNA-microarrays ('DNA-Chips') potentially have great value. Microarrays consist of a surface with thousands of covalently attached DNA oligonucleotides. This allows monitoring of thousands of different species-identifying DNA sequences simultaneously with one array (size about 1 × 1 cm). Theoretically, just one chip would enable the screening for all major economic fish species simultaneously (Kochzius *et al.*, 2010). The development of DNA-microarrays is laborious but the running costs are moderate. Other high-throughput and parallel processing methodologies for fish species identification have also been developed (Dooley *et al.*, 2005).

Current technology development and engineering might lead to the development of hand-held DNA-analytical devices, enabling field use and analysis on the spot. This would be a major advantage with respect to the response time, the lapse of time between the onset of an investigation and the reception of analytical results, since inspectors in the fisheries sector carry considerable responsibility: if they decide to put landings 'on hold' because of suspect content, and since this can have austere consequences for fishers and stakeholders, a reduction in response time is indeed crucial. Engineering of hand-held DNA analysis machines is currently being carried out in the context of forensic genetic analysis at crime scenes (Liu *et al.*, 2008), and recent publications show that progress has been made in this area (Arnaud, 2008). However, at present no cost-effective hand-held analytical device supporting fisheries control and enforcement or traceability is available.

Numerous examples of successful applications clearly show that DNA analysis, chemistry and forensics have enormous potential for the support of fisheries control and enforcement. This is particularly true for species identification, even on processed products, which is fully established. Also origin assignment will increasingly be applicable thanks to scientific endeavours which allow progressive refinement of the level of marine fish population structure description. This is manifested by a recent collaboration between the Danish Fisheries Inspectorate and the Danish Technical University (DTU) where, based on DNA evidence admitted in a court case, a fisherman, who had claimed a false claim origin for his catch, was convicted (see further below). Remarkably, regular cooperation between the Danish Fisheries Inspectorate and the DTU has been established in the meatntime: the inspectors are equipped with tissue sampling kits and can send samples for genetic analysis to a laboratory of the DTU if needed (Lars Bonde Erikson; Officer of the Danish Fisheries Inspectorate – Personal communication). Further applications of genetic marker analysis to determine the origin of fish and fish products are described in the following section.

9.3.4 The application of genetic marker analysis to determine the origin of fish and fish products

In many cases false origin declaration is linked to false species labelling, and can be addressed by DNA-based species identification. This has recently been demonstrated by DNA analysis on hake products commercialised in Spanish and Greek, in a study revealing more than 30% mislabelling, on the basis of species substitution (Garcia-Vazquez *et al.*, 2011). African species were substituted for products labelled as American and European species. The authors suspected that the products were deliberately mislabelled for profit reasons because real market prices of European and American hake products are higher than those of their African counterparts in sold in Spain. Similar examples exist in other parts of the world (von der Heyden *et al.*, 2010).

Inferring the origin of fish products directly from DNA analysis, through individual origin assignment, is more challenging. Origin assignment is based on the probabilistic matching of individual fish to groups of fish with similar characteristics, caused by common environmental factors, common genetic makeup or both. Such groups are generally referred to as populations or, in a fisheries management context, as stocks. Importantly the terms 'stock' and 'population' are not necessarily interchangeable, which in the context of traceability is an important consideration: 'stock' is a technical term describing a group of individuals under management considerations for exploitation, while a population is determined by genetic and/or demographic characteristics among individuals. This in turn means that a commercially exploited stock can be composed of two or more populations (Carvalho and Hauser, 1994). This means that stocks might not be genetically distinct groups of fish, but rather reflect differences in phenotypic life history parameters in response to environmental variation. Normally, stocks occupy well-defined spatial fishing areas. The term 'population' takes into account the reproductive relatedness of organisms. Simplified, it describes individuals of a particular species, which can be defined as a local interbreeding group (Waples and Gaggiotti, 2006). Such a grouping has reduced genetic exchange (gene flow) with other groups of the same species, which leads to isolation and genetic differentiation from other groups of the same species (Waples *et al.*, 2008). Importantly this means that groupings based on demographic relationships (e.g., microchemistry – see above) might not necessarily reveal identical patterns to those dependent on interbreeding and resultant population structure (genetic markers). Such potentially discordant patterns revealed by different tools necessarily demand sufficient data on salient features of the population and demographic structure of fish for interpretation, and also stress the value of integrating data of different sources in a holistic approach.

Assigning individuals to their origin is based on the probability estimation of encountering a combination of characteristic features (e.g., genotypes or chemical composition), established for these individuals, in a number of potential source populations (Fig. 9.2). This can only be accomplished if a baseline exists, that is, the composition of the potential source populations has previously been determined. Genetic markers combine various assets with regards to creating such baselines. They are heritable and the genetic structure of populations typically remains stable over generations. Also, genetic loci are particularly accessible by statistical analysis and a large body of underlying population genetic theory supporting the statistical analysis is readily available (Waples, 1990). Moreover, a variety of software applications using genotype data to infer population structure and assign individuals to populations, have been developed (Banks and Eichert, 2000; Cornuet *et al.*, 1999; Piry *et al.*, 2004; Pritchard *et al.*, 2000, 2009), and recently approaches have been devised to analyse population structure in the context of the seascape and environmental variables (Hansen and Hemmer-Hansen, 2007). These developments will greatly support traceability approaches.

Using genetic markers such as SNPs and new statistical methods for their analysis, it has been shown that for commercial marine fish species, including herring and cod, population structuring can be evident even at small geographic scales (Hauser and Carvalho, 2008; Nielsen *et al.*, 2009), which is favourable for traceability. Its feasibility has been proven, for example, by scientists of the DTU who were able to assign Atlantic cod (*Gadus morhua*) individuals to a North Sea, Baltic Sea or north-east Arctic Ocean population with almost 100% certainty, using microsatellite markers (Nielsen *et al.*, 2001). The same principle has been applied to generate evidence in court during a case where a fisherman claimed the wrong origin for his catch. The evidence provided the basis for a conviction (Personal Communication: Lars B. Erikson, Danish Directorate of Fisheries; Inspectorate of Fisheries; Einar E. Nielsen, DTU). Determining the geographical origin of fish and fish products throughout the supply chain also addresses the question of whether a fish originates from the wild or from aquaculture. For example in Norway genetic assignment was being used to identify the farm of origin for escaped Atlantic salmon (*Salmon salar*) (Glover *et al.*, 2008). Genetic species identification is well established and can be routinely used for traceability purposes. This is different for genetic origin assignment and the distinction between wild and farmed animals, where the feasibility has been demonstrated, through cutting-edge research, but which is not yet fully established. To support an enhanced integration of genetic origin assignment and the distinction between wild and farmed animals, using genetics, more research and technology development will be needed, but also, to enable further technology transfer and capacity building, a close and informed dialogue between researchers and stakeholders is required (see also below). To enable further technology transfer and capacity building a close and informed dialogue between researchers and stakeholders is required (see also below).

9.3.5 Gene expression and proteomics

Gene expression, the process by which information encoded by a gene is used in the synthesis of a gene product (DNA to mRNA, or other RNA products, by transcription, and mRNA to proteins by translation), can differ between species (interspecies level). Gene expression also changes in response to the environment, and is subject to adaptive evolution on an intraspecies level. For the past 10 years microarrays have been used to analyse gene expression in fish to investigate questions related to ecology, evolution and environment. Gene expression variation has also been assessed in natural populations of marine fish (Larsen *et al.*, 2007), to study speciation, and to examine host-pathogen interactions (Goetz and MacKenzie, 2008). Microarrays used in the study of population differences in gene expression target thousands of genes simultaneously ('transcriptomics' – reviewed in (Nielsen *et al.*, 2009)). Such microarrays have now been developed for a large number of marine fish, for example, gilthead sea bream (*Sparus aurata*) (Ferraresso *et al.*, 2008), Atlantic halibut (*Hippoglossus hippoglossus*) (Douglas *et al.*, 2008) and Senegalese sole (*Solea senegalensis*) (Cerda *et al.*, 2008). They are presently used particularly for aquaculture species as they help to elucidate transcriptional changes under specific farming conditions or during infections. This can improve knowledge about reproduction, development, nutrition and immunity thereby supporting the optimisation of production under culture conditions.

Proteomics, investigation into the sets of proteins expressed by the genome of an organism under given environmental conditions, has recently been employed to understand protein diversity across and within human populations (Biron *et al.*, 2006).

Analysis of gene expression, both at the RNA (transcriptomics) and protein (proteomics) level, is in principle applicable to develop suitable markers for traceability, be it for origin assignment (if different expression patterns between populations of fish occupying different regions/environments can be established, this can be used for traceability) or to distinguish between wild and farmed fish. This has recently been demonstrated in a study for European hake (*Merluccius merluccius*), where proteomics have been used to establish differential protein expression patterns in hake from the Mediterranean Sea, the Cantabrian Sea and the Atlantic Ocean (Gonzalez *et al.*, 2010). However, while assessing the potential of such novel tools is currently subject to explorative research, none is yet at an applicable stage for control and enforcement.

9.3.6 The application of forensics for food authentication

In the scientific literature it has recently become popular to discuss forensics as a powerful factor in ensuring food authentication (e.g., Primrose *et al.*, 2010; Teletchea *et al.*, 2005; Woolfe and Primrose, 2004), and also as valuable in fisheries control and enforcement, including the supply chain (e.g., Glover, 2010; Martinsohn and Ogden, 2008; Ogden, 2008; Stokstad, 2010; Supernault *et al.*, 2010; Triantafyllidis *et al.*, 2010).

The application of forensics – the provision of information to criminal investigations, resulting from the application of scientific methods, while applying strict standards – can certainly constitute a great asset to investigations in the fisheries sector and along the fish product supply chain. However in utilising forensics, the question of costs and benefits must also be considered and in many cases modern analytical technologies do powerfully support control and monitoring as illustrated in Fig. 9.4.

The power of forensic evidence for investigations of fish product fraud has been demonstrated recently in a spectacular case of illegal import into the USA and sale of over 1000 tons of catfish, worth US$ 15.5 million, which was labelled as high-value species such as grouper. The investigation of this large-scale conspiracy involved diverse US federal agencies and species identification was undertaken with forensic DNA testing carried out by scientific laboratories. The test results provided evidence during the court trials and led to convictions (further documented in Martinsohn, 2011).

To ensure that criminal courts routinely accept DNA testing or chemical analysis in cases involving animals, rigorous validation of the techniques, sampling and evidence handling following strict guidelines, and statistical evaluation of analytical results is needed. As all elements contributing to evidence, such as sample collection and transfer, data compilation and analysis, as well as reference data sets will be scrutinised for flaws during court trials they must be validated (Ogden, 2008). In fact, the handling of physical evidence is one of the most important factors in an investigation. Samples must be collected, inventoried, preserved, transported and submitted for testing without compromising the evidential chain of custody. It might be appropriate to transfer

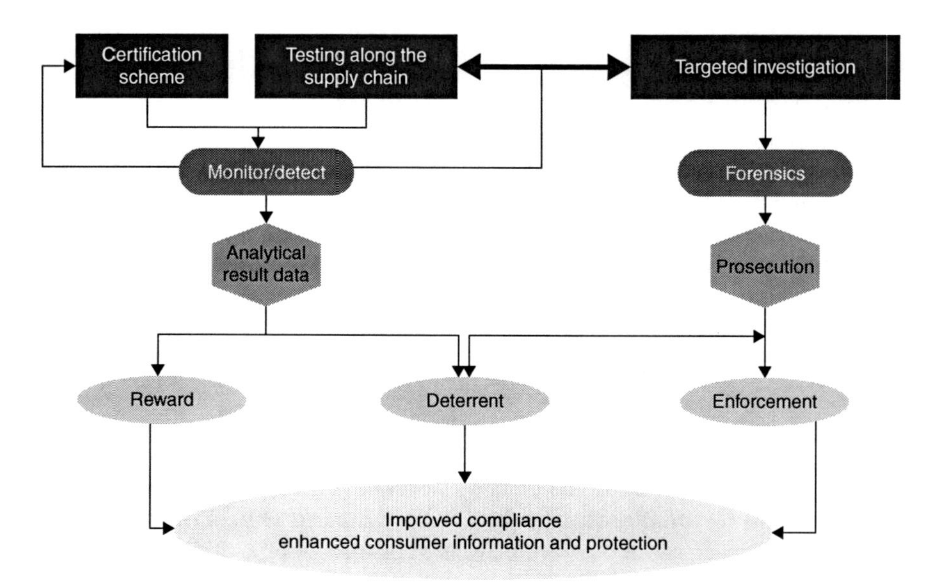

Fig. 9.4 Flow diagram depicting the role of genetic and chemical analysis and the application of forensics in fish authentication and traceability schemes (Courtesy of R. Ogden, TRACE Network).

guidelines established for processing human crime scenes, such as approved by the Technical Working Group on Crime Scene Investigation (Reno *et al.*, 2000), into enforcement for the fisheries and fish processing sector. This approach would help to avoid errors that result in inadmissibility of evidence. Also to this end the production of Standard Operating Procedures (SOPs), documents containing instructions on how to perform procedures that are routine, standardised and for which no *ad-hoc* modification is acceptable, is necessary. SOPs help to ensure quality and integrity of data, provide for uniformity and accountability and facilitate the transfer of analytical procedures between laboratories (Bowker *et al.*, 2006). Apart from guaranteeing acceptance of evidence in courts, applying forensic standards to traceability technologies could also greatly facilitate international collaboration, technology transfer and capacity building (see below) also in developing countries.

9.4 The need for capacity building

As outlined above, a variety of modern and advanced analytical technologies are available to support traceability in the fisheries sector but none of them is yet a standard application. Reasons for this lack of uptake will be considered and followed by a discussion of what is needed to build the necessary capacity and infrastructure as to render the implementation of these technologies possible.

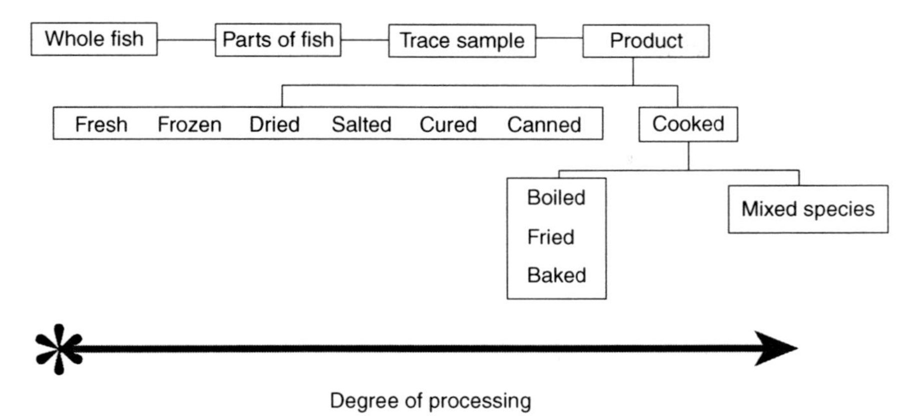

Fig. 9.5 Degree of processing of fish products.

Many examples, compiled in a recent report on the subject (Martinsohn, 2011), show that modern applications based on DNA technology, chemistry and forensics are not only theoretically available, but have already found their way into the field of fisheries control, enforcement and traceability, even when addressing the rather challenging question of geographical origin. Moreover, such technologies are producing evidence that is admitted and used in court cases of fisheries and fish product fraud. In particular DNA-based analytical technologies bear many obvious advantages when used to determine the origin of fish and fish products. To start with the genome of animals shows high variability at the level of individuals, which is taken advantage of in human forensics for DNA-fingerprinting (Butler, 2005), but also between populations (Allendorf and Luikart, 2006) and species (Waugh, 2007). Therefore the genome of living organisms, including fish, provides an endogenous natural tag. Furthermore, a great asset of DNA-based analytical approaches for fish and fish product authentication is that DNA can be extracted for genetic analysis from basically any tissue and a wide range of processed products (Fig. 9.5), and the costs of DNA sequencing are currently plummeting (Anonymous, 2010).

Yet describing these technologies as routine applications would be a gross overstatement. Rather, considering their great potential value as independent control methods in traceability schemes, they tend to be underutilised, and so we must ask 'Why?'

This lack of uptake into traceability schemes of the fisheries sector is even more surprising, since the agricultural sector has clearly shown that modern technologies allowing to trace animals and products back to their origin can be and are readily implemented. For example, radio-frequency identification (RFID) is being progressively introduced to ensure traceability of livestock in various countries (Ribeiro *et al.*, 2010), and has been compulsory for newborn sheep and goats in the EU since 2010 (Council, 2004). Interestingly an RFID-enabled traceability system has also been proposed for the supply chain of live fish (Hsu *et al.*, 2008). This example shows clearly that a robust legal framework making compulsory the use of a certain technology has a major boosting effect for their uptake. In this context it is interesting to note that the recently implemented new Common Fisheries Policy control regulation explicitly highlights the potential value of genetics for traceability schemes in the fisheries sector. The text however stops short of any obligation to use gene-based technology and merely proposes that EU member states should look into the feasibility of their application (European Council, 2009).

At the root a major impediment to the efficient uptake of DNA analysis, molecular biology and chemistry, is a prevailing gap between the academic research realm and stakeholders, caused by insufficient communication and mutual information transfer. This is true for fisheries management in general (Waples *et al.*, 2008), as well as for control and enforcement along the supply chain.

As explained above and illustrated in Fig. 9.2, baseline reference data has to be available to enable origin assignment be it genetic or based on chemical data. Another considerable obstacle to technology transfer is the current lack of a central 'data-hub', where DNA and chemical data sets establishing such baselines and of relevance to control and enforcement are centrally stored, professionally managed and easily accessible. Such a data-hub could be similar to the European Bioinformatics Institutes (EBI) database (www.ebi.ac.uk), and its two counterparts, the US National Center for Biotechnology Information GenBank (www.ncbi.nlm.nih.gov/Genbank) and the DNA Database of Japan (DDBJ) (www.ddbj.nig.ac.jp). The EBI database is produced in an international collaboration with GenBank and DDBJ. Each of the three entities collects a portion of the total sequence data reported worldwide, and all new and updated database entries are shared regularly as part of the International Nucleotide Sequence Database Collaboration (INSDC).

Obviously a similar infrastructure, compiling genetic and other data and tailored to the needs of fisheries managers, control authorities and the industry, would constitute an invaluable asset. While this is being tackled for species identification with the DNA-barcoding approach (http://www.fishbol.org/), an overarching strategy including population baseline data for origin assignment, is currently unavailable (WGAGFM, 2007).

Additionally building a network of test laboratories, certified to carry out analysis for control and enforcement purposes, and sharing information, validated protocols, as well as expertise should ideally be set up. This does not imply that such laboratories have to be newly established: in many countries facilities with the necessary capacity are available. However they work on a rather isolated and *ad hoc* basis at the moment, as part of, or when contacted by, authorities.

Another crucial component of capacity building is training targeted at control personnel working on the spot who will have to be familiarised with how to take tissue samples for analyses, and also for laboratory personal and enforcement officers.

The current steep fall in technology costs, especially for DNA analysis, indicate that the methods discussed here are cost-effective. Moreover these technologies have an added value in that they are beneficial to fisheries management beyond control, enforcement and traceability (Waples *et al.*, 2008). A rough estimate for species identification on 100 samples based on DNA extracted from fish fillets would cost about 35 Euros per sample for monitoring purposes and 100 Euros for enforcement purposes, if forensics are applied (Rob Ogden, TRACE Wildlife Forensics, personnel communication). However costs and benefits do greatly depend on the actual fraud scenario, the value of fraud, the number of samples analysed, to what extent the applied methods are routine applications etc. This is why in order to boost the uptake of the technologies discussed in this chapter, a careful assessment of costs and benefits (Pearce *et al.*, 2006) would be valuable, as it would provide a valuable reference and tool for stakeholders and regulators and rationalising and facilitating the decision finding process on their implementation.

9.5 Conclusion and future trends

Major efforts, on national and global level, will be needed to attain sustainable exploitation levels of the natural fish resource, and to build a healthy and thriving fisheries sector. This is generally acknowledged, as also testified by the recent United Nations Conference on Sustainable Development 'Rio + 20', where the sustainable exploitation of oceans was one of the prominent topics (United Nations, 2012). Aquaculture activity is currently experiencing a major boost, and might help to satisfy the ever growing demand for fish as food and contribute to food security (Bostock *et al.*, 2011; The Government Office for Science, 2011; Godfray *et al.*, 2010; Hishamunda and Ridler, 2006). However the growth of the aquaculture industry implies environmental risks which will have to be addressed. Also, particularly in the light of the complexity of global trade patterns, consumer information and protection will face major challenges.

Enabling traceability is and will be crucial to addressing those issues and the verification of fish and fish product origin is an indispensible element of any powerful traceability scheme. This includes all stages along the fish food supply chain, starting from control and enforcement of the fishing activity, right to the plate of the consumer. It is important to emphasise that the independent verification of origin does not only have a deterrent or enforcement aspect but can also provide a rewarding asset, in that it allows the fishing industry and producers to proof compliance with rules or certificates such as eco labels (Fig. 9.5). Indeed eco-labelling and certification schemes in the fisheries sector become increasingly popular (Brécard *et al.*, 2009): regular, independent control of these labels, including the verification of the true origin of the fish (product) sold under the label, will reassure the consumer and constitute an incentive to purchase the product.

All three questions 'Which species?, 'Where from?', 'Wild or farmed?' are fundamental with respect to the verification of fish origin and the technologies presented can provide powerful verification tools to control authorities and the industry.

This is particularly true for DNA-based technologies, as this field is advancing at a staggering pace (Metzker, 2010), and the fisheries industry, including control authorities, can and should profit from this phenomenon. Of course this rapid advancement also has risks: there is an immense data generation and accumulation task to deal with (Creighton, 2010; Zhang *et al.*, 2011), and the availability of many different genetic marker types and development of new analytical approaches can easily be confusing and a strain for stakeholders rather than being supportive. In fact most routine control applications do not need ever increasing levels of sophistication but would be much better relying on solutions tailored to the needs of end-users, and standardisation across national and, ideally, international laboratories. To ensure an effective technology transfer from research and technology development to end-users, an informed dialogue between all stakeholders is needed and decision making should be rationalised, by also taking into account costs and benefits.

To maximise the power of fish product origin verification a holistic approach taking into account different and complementary technologies is advisable. This has gone on in recent high-level workshops held by the FAO[1] and the European Commission[2], involving regulators, policy makers, control authority personnel and scientists. This chapter has outlined a selection of different analytical approaches and in addition it should be discussed whether, for traceability along the fish product supply chain, these can be combined with other technologies such as RFID tags for example. For the fisheries sector this extends to modern technologies used to monitor and control fishing activity, particularly those based on remote sensing such as vessel monitoring systems (VMS) (Kourti *et al.*, 2005; Saitoh *et al.*, 2011). A platform for an interdisciplinary discussion in the area of fisheries monitoring, control and surveillance (MCS) is provided at the global level by the International MCS Network (iMCS) through its Global Fisheries Enforcement Training Workshop (GFETW) series (iMCS Network, 2011). Similar activities, focussing more on the supply chain and with stronger industry involvement would be highly beneficial to the fisheries sector as a whole.

This chapter has shown that state-of-the-art technologies based on genetics and chemistry and supported by forensics, can provide powerful tools for the independent verification of fish and fish products. This provides a great asset for

[1]Informal Workshop on the Use of Forensic Technologies in Fisheries Monitoring, Control and Surveillance; FAO, Rome, Italy, 9–10 December 2009.

[2]Interdisciplinary Workshop on the Potential and Applicability of Advanced Technologies based on Genetics, Chemistry and Forensics for Traceability and Control in the Fisheries Sector. European Commission DG MARE, DG JRC, Brussels, Belgium, 17 June 2008.

traceability schemes and efforts should be put into boosting their uptake by the fisheries sector and industry. Properly implemented such technologies will help to ensure that a rich and beautiful display of fish and fish products on markets will not one day be restricted to descriptions in the literature (e.g., Zola, 1873), and that fish will remain one of the most precious food resources for human beings.

9.6 Acknowledgements

I would like to thank the members of the FishPopTrace consortium for their support and valuable discussions throughout the writing of this chapter.

9.7 Sources of further information and advice

- Books and Articles

  Washington, S. and Ababouch, L. (2011) Private standards and certification in fisheries and aquaculture: current practice and emerging issues. Fisheries and aquaculture Technical Paper. FAO Fisheries and Aquaculture Department. Rome. ISBN 978–92–5-106730–7.

  Martinsohn, J. T. (2011) Deterring Illegal Activities in the Fisheries Sector: genetics, genomics, chemistry and forensics to fight IUU fishing and in support of fish product traceability. JRC Reference Reports. European Commission. Luxemburg, ISBN 978–92–79–15905–3.
- Websites

  FishPopTrace: The Structure of Fish Populations and Traceability of Fish and Fish Products. URL: http://fishpoptrace.jrc.ec.europa.eu

  FISH-BOL: Fish Barcode of Life Initiative. URL: www.fishbol.org

  Illegal-Fishing Info. URL: http://www.illegal-fishing.info

  Fisheries and Oceans Canada. URL: http://www.dfo-mpo.gc.ca
- Agencies

  Community Fisheries Control Agency (CFCA).

  European Food Safety Authority (EFSA).

  US Food and Drug Administration (FDA).

9.8 References

ADEY, E. A., BLACK, K. D., SAWYER, T., SHIMMIELD, T. M. and TRUEMAN, C. N. (2009) Scale microchemistry as a tool to investigate the origin of wild and farmed Salmo salar. *Marine Ecology Progress Series*, **390**, 225.

ALLENDORF, F. W. and LUIKART, G. (2006) *Conservation and the Genetics of Populations*, Malden, MA, Blackwell Pub.

Anonymous (2010) Human genome at ten: the sequence explosion. *Nature*, **464**, 670–671.

ARNAUD, C. H. (2008) Moving from bench to bedside. *Chemical and Engineering News*, **86**, 56.

AURSAND, M., STANDAL, I. B., Praël, A., MCEVOY, L., IRVINE, J. and AXELSON, D. E. (2009) 13C NMR pattern recognition techniques for the classification of Atlantic salmon (*Salmo salar* L.) according to their wild, farmed, and geographical origin. *Journal of Agricultural and Food Chemistry*, **57**, 3444–3451.

BANKS, M. A. and EICHERT, W. (2000) WHICHRUN (version 3.2): a computer program for population assignment of individuals based on multilocus genotype data. *Journal of Heredity*, **91**, 87.

BIRON, D. G., LOXDALE, H. D., PONTON, F., MOURA, H., MARCHÉ, L., BRUGIDOU, C. and THOMAS, F. (2006) Population proteomics: an emerging discipline to study metapopulation ecology. *Proteomics*, **6**, 1712–1715.

BLANCOU, J. (2001) A history of the traceability of animals and animal products. *OIE Revue Scientifique et Technique*, **20**, 420–425.

BOSTOCK, J., MCANDREW, B., RICHARDS, R., JAUNCEY, K., TELFER, T., LORENZEN, K., LITTLE, D., ROSS, L., HANDISYDE, N., GATWARD, I. and CORNER, R. (2011) Aquaculture: global status and trends. *Philosophical Transactions of the Royal Society B: Biological Sciences*, **365**, 2897–2912.

BOWKER, J., FISH, U. S. and Wildlife Services – Aquatic Animal DRUG, A. (2006) *Definition and Use of the Standard Operating Procedure (SOP)*, U.S. Fish & Wildlife Services – Aquatic Animal Drug Approval.

BRÉCARD, D., HLAIMI, B., LUCAS, S., PERRAUDEAU, Y. and SALLADARRÉ, F. (2009) Determinants of demand for green products: an application to eco-label demand for fish in Europe. *Ecological Economics*, **69**, 115–125.

BRUMFIELD, R. T., BEERLI, P., NICKERSON, D. A. and EDWARDS, S. V. (2003) The utility of single nucleotide polymorphisms in inferences of population history. *Trends in Ecology and Evolution*, **18**, 249.

BUTLER, J. M. (2005) *Forensic DNA Typing: Biology, Technology, and genetics of STR markers*, Amsterdam, Boston, Elsevier Academic Press.

CAMPANA, S. E. (2004) Otolith elemental composition as a natural marker of fish stocks. In CADRIN, S. X., FRIEDLAND, K. D. and WALDMAN, J. R. (Eds.) *Stock Identification Methods.* San Diego, Elsevier.

CAMPANA, S. E. (2005) Otolith science entering the 21st century. *Marine and Freshwater Research*, **56**, 485.

CAMPANA, S. E., CHOUINARD, G. A., HANSON, J. M., FRÉCHET, A. and BRATTEY, J. (2000) Otolith elemental fingerprints as biological tracers of fish stocks. *Fisheries Research*, **46**, 343.

CARVALHO, G. R. and HAUSER, L. (1994) Molecular genetics and the stock concept in fisheries. *Reviews in Fish Biology and Fisheries*, **4**, 326–350.

CERDA, J., MERCADE, J., LOZANO, J., MANCHADO, M., TINGAUD-SEQUEIRA, A., ASTOLA, A., INFANTE, C., HALM, S., VINAS, J., CASTELLANA, B., ASENSIO, E., CANAVATE, P., MARTINEZ-RODRIGUEZ, G., PIFERRER, F., PLANAS, J., PRAT, F., YUFERA, M., DURANY, O., SUBIRADA, F., ROSELL, E. and MAES, T. (2008) Genomic resources for a commercial flatfish, the Senegalese sole (*Solea senegalensis*): EST sequencing, oligo microarray design, and development of the Soleamold bioinformatic platform. *BMC Genomics*, **9**, 508.

Committee of Fisheries (2003) Report of the twenty-fifth session of the Committee of Fisheries. In *FAO Fisheries Report.* Rome, Food and Agriculture Organization of the United Nations.

CORNUET, J. M., PIRY, S., LUIKART, G., ESTOUP, A. and SOLIGNAC, M. (1999) New methods employing multilocus genotypes to select or exclude populations as origins of individuals. *Genetics*, **153**, 1989.

COUNCIL, E. (2004) Council Regulation (EC) No 21/2004 of 17 December 2003 establishing a system for the identification and registration of ovine and caprine animals and amending Regulation (EC) No 1782/2003 and Directives 92/102/EEC and 64/432/EEC. IN L5, O. J. O. T. E. U. (Ed.), Official Journal of the EU L5.

COUNCIL, E. P. A. E. (2000) Regulation (EC) No 1760/2000 of the European parliament and of the Council of 17 July 2000 establishing a system for the identification and registration of bovine animals and regarding the labelling of beef and beef products and repealing Council Regulation (EC) No 820/97. In COUNCIL, E. P. A. E. (Ed.), Official Journal of the European Communities L 204/1.

CREIGHTON, C. J. (2010) Inaugural Genomics Automation Congress and the coming deluge of sequencing data. *Expert review of molecular diagnostics*, **10**, 849–851.

DE PONTUAL, H., LAGARDÉRE, F., TROADEC, H., BATEL, A., DÉSAUNAY, Y. and KOUTSIKOPOULOS, C. (2000) Otoliths imprinting of sole (*Solea solea*) from the Bay of Biscay: A tool to discriminate individuals from nursery origins? *Oceanologica Acta*, **23**, 497–513.

DEWOODY, J., NASON, J. D. and HIPKINS, V. D. (2006) Mitigating scoring errors in microsatellite data from wild populations. *Molecular Ecology Notes*, **6**, 951.

DOOLEY, J. J., SAGE, H. D., CLARKE, M. - A. L., BROWN, H. M. and GARRETT, S. D. (2005) Fish Species Identification Using PCR-RFLP Analysis and Lab-on-a-Chip Capillary Electrophoresis: Application to Detect White Fish Species in Food Products and an Interlaboratory Study. *Journal of Agricultural and Food Chemistry*, **53**, 3348–3357.

DOUGLAS, S. E., KNICKLE, L. C., WILLIAMS, J., FLIGHT, R. M. and REITH, M. E. (2008) A first generation Atlantic halibut *Hippoglossus hippoglossus* (L.) microarray: application to developmental studies. *Journal of Fish Biology*, **72**, 2391–2406.

European Commission (2001) Commission Regulation (EC) No 2065/2001 of 22 October 2001 laying down detailed rules for the application of Council Regulation (EC) No 104/2000 as regards informing consumers about fishery and aquaculture products. *Official Journal of the European Communities L*, **278**, 6.

European Commission (2004) Report from the Commission to the Council and the European Parliament on the implementation of Title II of the Regulation (EC) No 1760/2000 of the European Parliament and of the Council establishing a system for the identification and registration of bovine animals and regarding the labelling of beef and beef products. In COMMISSION, E. (Ed.) Brussels, European Commission.

European Council (2000) Council Regulation (EC) No 104/2000 of 17 December 1999 on the common organisation of the markets in fishery and aquaculture products. *Official Journal of the European Communities L*, **17**, 22.

European Council (2008) Council Regulation (EC) No 1005/2008 of 29 September 2008 establishing a Community system to prevent, deter and eliminate illegal, unreported and unregulated fishing, amending Regulations (EEC) No 2847/93, (EC) No 1936/2001 and (EC) No 601/2004 and repealing Regulations (EC) No 1093/94 and (EC) No 1447/1999. *Official Journal of the EU L*, **286**(1), 32.

European Council (2008a) Council Regulation (EC) No 1006/2008 of 29 September 2008 concerning authorisations for fishing activities of Community fishing vessels outside Community waters and the access of third country vessels to Community waters, amending Regulations (EEC) No 2847/93 and (EC) No 1627/94 and repealing Regulation (EC) No 3317/94. *Official Journal of the EU L*, **286**(33), 44.

European Council (2009) Council Regulation (EC) No 1224/2009 of 20 November 2009 establishing a Community control system for ensuring compliance with the rules of the common fisheries policy, amending Regulations (EC) No 847/96, (EC) No 2371/2002, (EC) No 811/2004, (EC) No 768/2005, (EC) No 2115/2005, (EC) No 2166/2005, (EC) No 388/2006, (EC) No 509/2007, (EC) No 676/2007, (EC)

No 1098/2007, (EC) No 1300/2008, (EC) No 1342/2008 and repealing Regulations (EEC) No 2847/93, (EC) No 1627/94 and (EC) No 1966/2006. *Official Journal of the European Union L*, L **343**(1), 50.

European Parliament and European Council (2002) REGULATION (EC) No 178/2002 of the European Parliament and of the Council of 28 January 2002, laying down the general principles and requirements of food law, establishing the European Food Safety Authority and laying down procedures in matters of food safety. *Official Journal of the European Communities L*, **31**(1), 1.

FAO (2001) International Plan of Action to Prevent, Deter and Eliminate Illegal, Unreported and Unregulated Fishing (IPOA IUU). In FAO (Ed.) Rome, Food and Agriculture Organization of the United Nations (FAO).

FDA – U.S. Food and Drug Administration (2011) Retrieved 09 January 2012, from http://www.fda.gov/Food/FoodSafety/Product-SpecificInformation/Seafood/DNAspeciation/default.htm.

FERRARESSO, S., VITULO, N., MININNI, A. N., ROMUALDI, C., CARDAZZO, B., NEGRISOLO, E., REINHARDT, R., CANARIO, A. V. M., PATARNELLO, T. and BARGELLONI, L. (2008) Development and validation of a gene expression oligo microarray for the gilthead sea bream (*Sparus aurata*). *BMC Genomics*, **9**.

FishPopTrace (2008–2011) The Structure of Fish Populations and Traceability of Fish and Fish Products. At http://fishpoptrace.jrc.ec.europa.eu. Accessed 31 January 2012. *FishPopTrace was funded by the European Community's Seventh Framework Programme under grant agreement no KBBE-212399 in the area of "Food, Agriculture and Fisheries, and Biotechnology"*. The FishPopTrace Consortium.

GARCIA-VAZQUEZ, E., PEREZ, J., MARTINEZ, J. L., PARDIÑAS, A. F., LOPEZ, B., KARAISKOU, N., CASA, M. F., MACHADO-SCHIAFFINO, G. and TRIANTAFYLLIDIS, A. (2011) High level of mislabeling in Spanish and Greek hake markets suggests the fraudulent introduction of African species. *Journal of Agricultural and Food Chemistry*, **59**, 475–480.

GEFFEN, A. J., JARVIS, K., THORPE, J. P., LEAH, R. T. and NASH, R. D. M. (2003) Spatial differences in the trace element concentrations of Irish Sea plaice Pleuronectes platessa and whiting Merlangius merlangus otoliths. *Journal of Sea Research*, **50**, 245.

GILLANDERS, B. M. (2001) Trace metals in four structures of fish and their use for estimates of stock structure. *Fishery Bulletin*, **99**, 410.

GLOVER, K. A. (2010) Forensic identification of fish farm escapees: the Norwegian experience. *Aquaculture Environment Interactions*, **1**, 1–10.

GLOVER, K. A., SKILBREI, O. T. and SKAALA, Ø. (2008) Genetic assignment identifies farm of origin for Atlantic salmon Salmo salar escapees in a Norwegian fjord. *ICES Journal of Marine Science*, **65**, 912.

GODFRAY, H. C. J., CRUTE, I. R., HADDAD, L., MUIR, J. F., NISBETT, N., LAWRENCE, D., PRETTY, J., ROBINSON, S., TOULMIN, C. and WHITELEY, R. (2010) The future of the global food system. *Philosophical Transactions of the Royal Society B: Biological Sciences*, **365**, 2769–2777.

GOETZ, F. W. and MACKENZIE, S. (2008) Functional genomics with microarrays in fish biology and fisheries. *Fish and Fisheries*, **9**, 378.

GONZALEZ, E., KREY, G., ESPIÑEIRA, M., DIEZ, A., PUYET, A. and BAUTISTA, J. (2010) Population proteomics of the European hake (*Merluccius merluccius*). *Journal of Proteome Research*, **9**(12), 6392–6404.

GRAHL-NIELSEN, O. (2005) Fatty acid profiles as natural marks for stock identification. In CADRIN, S. X., FRIEDLAND, K. D. and WALDMAN, J. R. (Eds.) *Stock Identification Methods: Applications in Fishery Science*. London, Elsevier Academic Press.

GUIL-GUERRERO, J. L., VENEGAS-VENEGAS, E., RINCÓN-CERVERA, M. T. and SUÁREZ, M. D. (2005) Fatty acid profiles of livers from selected marine fish species. *Journal of Food Composition and Analysis*, **24**, 217–222.

HANSEN, M. M. and HEMMER-HANSEN, J. (2007) Landscape genetics goes to sea. *Journal of Biology*, **6**, 6.

HAUSER, L. and CARVALHO, G. R. (2008) Paradigm shifts in marine fisheries genetics: ugly hypotheses slain by beautiful facts. *Fish and Fisheries*, **9**, 333.

HIGGINS, R. M., DANILOWICZ, B. S., BALBUENA, J. A., DANIELSDOTTIR, A. K., GEFFEN, A. J., MEIJER, W. G., MODIN, J., MONTERO, F. E., PAMPOULIE, C., PERDIGUERO-ALONSO, D., SCHREIBER, A., STEFANSSON, M. and WILSON, B. (2010) Multi-disciplinary fingerprints reveal the harvest location of cod Gadus morhua in the northeast Atlantic. *Marine Ecology Progress Series*, **404**, 197–206.

HISHAMUNDA, N. and RIDLER, N. B. (2006) Farming fish for profits: a small step towards food security in sub-Saharan Africa. *Food Policy*, **31**, 401–414.

HSU, Y. C., CHEN, A. P. and WANG, C. H. (2008) A RFID-enabled traceability system for the supply chain of live fish. *Proceedings of the IEEE International Conference on Automation and Logistics, ICAL* 2008. Qingdao.

HUDSON, M. E. (2008) Sequencing breakthroughs for genomic ecology and evolutionary biology. *Molecular Ecology Resources*, **8**, 3–17.

ICES (2007) Report of the Working Group on the Application of Genetics in Fisheries and Mariculture (WGAGFM), 19–23 March 2007, Ispra, Italy. ICES cM 2007/MCC:03, 70 pp.

IMCS NETWORK, (2011) International Monitoring, Control, and Surveillance Network for Fisheries-Related Activities. At http://www.imcsnet.org. Accessed 31 January 2012.

KOCHZIUS, M. (2008) Trends in fishery genetics. In BEAMISH, R. J. and ROTHSCHILD, B. J. (Eds.) *The Future of Fisheries Science in North America, Springer*, 451–491. 1 ed., Springer.

KOCHZIUS, M., SEIDEL, C., ANTONIOU, A., BOTLA, S. K., CAMPO, D., CARIANI, A., VAZQUEZ, E. G., HAUSCHILD, J., HERVET, C., HJÖ RLEIFSDOTTIR, S., HREGGVIDSSON, G., KAPPEL, K., LANDI, M., MAGOULAS, A., MARTEINSSON, V., NÖ LTE, M., PLANES, S., TINTI, F., TURAN, C., VENUGOPAL, M. N., WEBER, H. and BLOHM, D. (2010) Identifying fishes through DNA barcodes and microarrays. *PLoS ONE*, **5**, e12620.

KOURTI, N., SHEPHERD, I., GREIDANUS, H., ALVAREZ, M., ARESU, E., BAUNA, T., CHESWORTH, J., LEMOINE, G. and SCHWARTZ, G. (2005) Integrating remote sensing in fisheries control. *Fisheries Management and Ecology*, **12**, 295–307.

KWETEGYEKA, J., MPANGO, G. and GRAHL-NIELSEN, O. (2008) Variation in fatty acid composition in muscle and heart tissues among species and populations of tropical fish in lakes Victoria and Kyoga. *Lipids*, **43**, 1017–1029.

LARSEN, P. F., NIELSEN, E. E., WILLIAMS, T. D., HEMMER-HANSEN, J., CHIPMAN, J. K., KRUHØFFER, M., GRØNKJAER, P., GEORGE, S. G., DYRSKJØT, L. and LOESCHCKE, V. (2007) Adaptive differences in gene expression in European flounder (*Platichthys flesus*). *Molecular Ecology*, **16**, 4674–4683.

LIU, P., YEUNG, S. H. I., CRENSHAW, K. A., CROUSE, C. A., SCHERER, J. R. and MATHIES, R. A. (2008) Real-time forensic DNA analysis at a crime scene using a portable microchip analyzer. *Forensic Science International Genetics*, **2**, 301.

MANEL, S., GAGGIOTTI, O. E. and WAPLES, R. S. (2005) Assignment methods: matching biological questions with appropriate techniques. *Trends in Ecology and Evolution*, **20**, 136.

MARTINSOHN, J. T. (2011) Deterring Illegal Activities in the Fisheries Sector: genetics, genomics, chemistry and forensics to fight IUU fishing and in support of fish product traceability. In CENTRE, E. C. J. R. (Ed.) *JRC Reference Reports: Deterring Illegal Activities in the Fisheries Sector: genetics, genomics, chemistry and forensics to fight IUU fishing and in support of fish product traceability*, Luxemburg, European Commission.

MARTINSOHN, J. T. and OGDEN, R. (2008) A forensic genetic approach to European fisheries enforcement. *Forensic Science International: Genetics Supplement Series*, **1**, 610.

METZKER, M. L. (2010) Sequencing technologies – the next generation. *Nature Reviews. Genetics*, **11**, 31.

MORETTI, V. M., TURCHINI, G. M., BELLAGAMBA, F. and CAPRINO, F. (2003) Traceability issues in fishery and aquaculture products. *Veterinary Research Communications*, **27**, 497.

NIELSEN, E. E., HANSEN, M. M., SCHMIDT, C., MELDRUP, D. and GRØNKJÆR, P. (2001) Population of origin of Atlantic cod. *Nature*, **413**, 272.

NIELSEN, E. E., HEMMER-HANSEN, J., LARSEN, P. F. and BEKKEVOLD, D. (2009) Population genomics of marine fishes: Identifying adaptive variation in space and time. *Molecular Ecology*, **18**, 3128.

OGDEN, R. (2008) Fisheries forensics: the use of DNA tools for improving compliance, traceability and enforcement in the fishing industry. *Fish and Fisheries*, **9**, 462.

PEARCE, D., ATKINSON, G. and MOURATO, S. (Eds.) (2006) *Cost-Benefit Analysis and the Environment: Recent Developments*, OECD.

PIRY, S., ALAPETITE, A., CORNUET, J. M., PAETKAU, D., BAUDOUIN, L. and ESTOUP, A. (2004) GENECLASS2: A software for genetic assignment and first-generation migrant detection. *Journal of Heredity*, **95**, 536–539.

PRIMMER, C. R., KOSKINEN, M. T. and PIIRONEN, J. (2000) The one that did not get away: Individual assignment using microsatellite data detects a case of fishing competition fraud. *Proceedings of the Royal Society – Biological Sciences (Series B)*, **267**, 1699.

PRIMROSE, S., WOOLFE, M. and ROLLINSON, S. (2010) Food forensics: Methods for determining the authenticity of foodstuffs. *Trends in Food Science and Technology*, **21**, 582–590.

PRITCHARD, J. K., STEPHENS, M. and DONNELLY, P. (2000) Inference of population structure using multilocus genotype data. *Genetics*, **155**, 945.

PRITCHARD, J. K., XIAOQUAN, W. and FALUSH, D. (2009) *Documentation for Structure Software: Version 2.3*, Oxford, University of Oxford.

RASMUSSEN, R. S. and MORRISSEY, M. T. (2008) DNA-based methods for the identification of commercial fish and seafood species. *Comprehensive Reviews in Food Science and Food Safety*, **7**, 280.

REHBEIN, H. (2003) Identification of fish species by protein- and DNA-analysis. In PEREZ-MARTIN, R. and SOLETO, C. G. (Eds.) *Authenticity of species in meat and seafood products*. Association 'International Congress on Authenticity of Species in Meat and Seafood Products'.

RENO, J., MARCUS, D., ROBINSON, L., BRENNAN, N. and TRAVIS, J. (2000) Crime Scene Investigation: A Guide for Law Enforcement. In INVESTIGATION, U. S. D. O. J. T. W. G. O. C. S. (Ed.) Washington, U.S. Department of Justice; National Institute of Justice.

RIBEIRO, P. C. C., SCAVARDA, A. J. and BATALHA, M. O. (2010) RFID in the international cattle supply chain: Context, consumer privacy and legislation. *International Journal of Services and Operations Management*, **6**, 149–164.

SAITOH, S. I., MUGO, R., RADIARTA, I. N., ASAGA, S., TAKAHASHI, F., HIRAWAKE, T., ISHIKAWA, Y., AWAJI, T., IN, T. and SHIMA, S. (2011) Some operational uses of satellite remote sensing and marine GIS for sustainable fisheries and aquaculture. *ICES Journal of Marine Science*, **68**, 687–695.

SMITH, G. C., PENDELL, D. L., TATUM, J. D., BELK, K. E. and SOFOS, J. N. (2008) Post-slaughter traceability. *Meat Science*, **80**, 66–74.

SOBRINO, B., BRIÓN, M. and CARRACEDO, A. (2005) SNPs in forensic genetics: a review on SNP typing methodologies. *Forensic Science International*, **154**, 181.

STEIN, L. D. (2010) The case for cloud computing in genome informatics. *Genome Biology*, **11**, 207.

STOKSTAD, E. (2010) To fight illegal fishing, forensic DNA gets local. *Science*, **330**, 1468–1469.

SUPERNAULT, K., DEMSKY, A., CAMPBELL, A., MING, T., MILLER, K. and WITHLER, R. (2010) Forensic genetic identification of abalone (*Haliotis* spp.) of the northeastern Pacific Ocean. *Conservation Genetics*, **11**, 10.

SWAN, S. C., GEFFEN, A. J., MORALES-NIN, B., GORDON, J. D. M., SHIMMIELD, T., SAWYER, T. and MASSUTI, E. (2006) Otolith chemistry: an aid to stock separation of *Helicolenus dactylopterus* (bluemouth) and *Merluccius merluccius* (European hake) in the Northeast Atlantic and Mediterranean. *ICES Journal of Marine Science*, **63**, 504.

TAYLOR, W. J., PATEL, N. P. and JONES, J. L. (1994) Antibody-based methods for assessing seafood authenticity. *Food and Agricultural Immunology*, **6**, 305.

TELETCHEA, F. (2009) Molecular identification methods of fish species: reassessment and possible applications. *Reviews in Fish Biology and Fisheries*, **19**, 265.

TELETCHEA, F., MAUDET, C. and HÄ NNI, C. (2005) Food and forensic molecular identification: update and challenges. *Trends in Biotechnology*, **23**, 359.

The Government Office for Science. (2011) *Foresight. The Future of Food and Farming. Challenges and Choices for Global Sustainability*. Final Project Report. London, The Government Office for Science.

THRESHER, R. E. (1999) Elemental composition of otoliths as a stock delineator in fishes. *Fisheries Research*, **43**, 165–204.

TRIANTAFYLLIDIS, A., KARAISKOU, N., PEREZ, J., MARTINEZ, J. L., ROCA, A., LOPEZ, B. and GARCIA-VAZQUEZ, E. (2010) Fish allergy risk derived from ambiguous vernacular fish names: Forensic DNA-based detection in Greek markets. *Food Research International*, **43**, 2214–2216.

UNITED NATIONS, (2012) *RIO+20: United Nations Conference on Sustainable Development*, New York, United Nations. http://www.uncsd2012.org/. Accessed 15 March 2013.

U.S. Department of Justice: Environment and Natural Resources Division (2009) Lacey Act Amendments of 1980 ('Lacey Act'), 16 U.S.C. 3371 et seq. 18 U.S.C. 42. *Selected federal Wildlife Statues.* Washington, U.S. Department of Justice.

von der HEYDEN, S., BARENDSE, J., SEEBREGTS, A. J. and MATTHEE, C. A. (2010) Misleading the masses: detection of mislabelled and substituted frozen fish products in South Africa. *ICES Journal of Marine Science*, **67**, 176–185.

WAPLES, R. S. (1990) Conservation Genetics of Pacific Salmon: I. Temporal Changes in Allele Frequency. *Conservation Biology*, **4**, 144.

WAPLES, R. S. and GAGGIOTTI, O. E. (2006) What is a population? An empirical evaluation of some genetic methods for identifying the number of gene pools and their degree of connectivity. *Molecular Ecology*, **15**, 1419.

WAPLES, R. S., PUNT, A. E. and COPE, J. M. (2008) Integrating genetic data into management of marine resources: how can we do it better? *Fish and Fisheries*, **9**, 423.

WARD, R. D., HANNER, R. and HEBERT, P. D. N. (2009) The campaign to DNA barcode all fishes, FISH-BOL. *Journal of Fish Biology*, **74**, 329.

WASHINGTON, S. and ABABOUCH, L. (2011) Private standards and certification in fisheries and aquaculture: Current practice and emerging issues. FAO Fisheries and Aquaculture Department. (Ed.) *FAO Fisheries and Aquaculture Technical Paper.* Rome, Food and Agriculture Organization of the United Nations.

WAUGH, J. (2007) DNA barcoding in animal species: Progress, potential and pitfalls. *BioEssays*, **29**, 188.

WIRGIN, I., KOVACH, A. I., MACEDA, L., ROY, N. K., WALDMAN, J. R. and BERLINSKY, D. L. (2007) Stock identification of Atlantic cod in U.S. waters using microsatellite and single nucleotide polymorphism DNA analyses. *Transactions of the American Fisheries Society*, **136**, 375.

WOOLFE, M. and PRIMROSE, S. (2004) Food forensics: using DNA technology to combat misdescription and fraud. *Trends in Biotechnology*, **22**, 222.

ZHANG, J., CHIODINI, R., BADR, A. and ZHANG, G. (2011) The impact of next-generation sequencing on genomics. *Journal of Genetics and Genomics*, **38**, 95–109.

ZOLA, E. (1873) Le Ventre de Paris. Chapitre III pp. 697–998. In Les Rougon-Macquart. Histoire naturelle et sociale d'une famille sous le Second Empire. (1960). Édition d'Armand Lanoux, Henri Mitterand. Collection Bibliothèque de la Pléiade (No 146), Gallimard -rom. ISBN 9782070105892.

10

Using new analytical approaches to verify the origin of honey

K. McComb, University of Otago, New Zealand and R. Frew, University of Otago, New Zealand and FAO/IAEA Division of Nuclear Techniques in Food and Agriculture, Austria

DOI: 10.1533/9780857097590.3.216

Abstract: Honey is a valuable food commodity that is utilised worldwide, as a direct food source or as an ingredient in a large number of manufactured foods. Honey is also of interest due to the antibacterial properties that it exhibits and many honeys are known for their antimicrobial action, some more than others. The honey trade has long been a target for fraudulent practices including the adulteration of honey with cheaper sugars or the passing off of honey from one source (regional or botanical) as that from a more valued source. This chapter reviews the literature on the use of chemical measures to differentiate honey to determine its botanical source and geographical origin. The chemical techniques applied include trace element analysis, stable isotopes, infrared and NMR spectroscopy.

Key words: honey, manuka, trace elements, stable isotopes, provenance, adulteration, botanical origin.

10.1 Introduction

Honey is produced by honey bees (*Apis mellifera*) from the carbohydrate exudates of plants, namely, nectar and honeydew. Nectar originates from the lymph of plants, whereas honeydew is a secretion produced from plant lymph by parasitic insects such as aphids (Pisani *et al.*, 2008). Honey is a valuable food commodity that is utilised worldwide, as a direct food source or as an ingredient in a large number of manufactured foods (Rashed and Soltan, 2004). Honey is also of interest due to the antibacterial properties that it exhibits; many honeys are known for their antimicrobial action, some more than others (Lusby *et al.*, 2005).

Honey is a complex material and is made up of various chemical components, the main constituents being various carbohydrates, including glucose, sucrose, fructose and maltose, which make up approximately 77% of the average honey. Honey also has a very low moisture content with the majority of honeys containing between 13.4% and 22.9% water (Doner, 1977). Honey also contains small amounts of various minerals (0.04–0.2%), proteins (>0.5%) (Anklam, 1998) and other molecules that are of interest, including hydrogen peroxide and methylglyoxal which are key to the bactericidal activity of some honeys (Lusby *et al.*, 2005; Stephens *et al.*, 2010). Mineral content also affects the colour of the honey, with higher mineral content honeys such as honeydew honey being darker than honeys with a lower mineral content (Gonzalez-Miret *et al.*, 2005). The exact composition of honey varies depending on the source: honeys of different botanical origins have been seen to differ in the ratios of sugars present (Doner, 1977), free amino acids (Rebane and Herodes, 2008) and mineral content (Necemer *et al.*, 2009). The variability and complexity of honey composition is dependent upon the nectar-providing florae, the bee species, geographic area, season and storage (Kaskoniene and Venskutonis, 2010). This complexity has allowed for a number of different strategies and techniques to be employed in efforts to characterise and authenticate this valuable food commodity (Anklam, 1998; Kaskoniene and Venskutonis, 2010).

10.1.1 Honey fraud

Honey is a food product with a long history of fraudulent activity, mainly by adulteration with cheaper sugars and other substances to bulk it out. As the global demand for honey has increased it has become an important commodity to many nations, including New Zealand. At the same time as global market prices have increased for honey so has the incidence of fraudulent practices as some producers try to make quick profits. This can, however, impact negatively upon a country's honey industry. In 2001 the United States introduced heavy import levies on Chinese honey, over US$ 2.00 per pound of honey (Leeder, 2011). This was due to large amounts of cheap, poor quality Chinese honey being dumped into the US market, so much that US producers could not compete. China is the world's biggest honey producer but their industry had a serious setback in 1997 with an outbreak of a bacterial disease, foulbrood. This required the use of antibiotics to control the banned antibiotic, chloramphenicol, was one used by Chinese beekeepers (Dharmananda, 2003). Subsequently, traces of chloramphenicol have been found in shipments of honey from China leading to the North American and European markets placing further restrictions on Chinese honey imports (e.g., Health Canada, 2004).

To get around the ban in the EU and the crippling US import levies some honey producers transhipped Chinese honey through other countries, including Singapore, India, Malaysia, Indonesia, the Philippines and Australia, where it was relabelled a product of said nations and exported to the US and EU. The honey laundering scheme was first detected by Australian investigators in 2002, when a large shipment of honey from Singapore that was bound for the US was intercepted. At the time Singapore had no honey producing capability and the honey consignment was ultimately traced back to China (Leeder, 2011). Importation alerts for a number of countries are still in effect for honey moving into the US, including Thailand, Vietnam, Mexico, Brazil, Peru and especially China (USFDA, 2010). Table 10.1 shows the value of honey imports to the US from several countries for the period 2000–2010, the decreasing trend in imports from China should be noted, as should the large increases in exports from several countries for the year 2001 compared to the previous year. Imports of honey from China decreased dramatically for the year 2002 compared to 2001, while imports from Thailand, Australia and Malaysia increased dramatically. Similar fraud has been detected in Europe with illegal imports of tonnes of South American and Chinese honey being discovered in the UK (Sapsted, 2005) and Italy (Mieliditalia, 2011).

10.1.2 Definition of honey

The definition of what can be classed as honey is found in the United Nations Food and Agriculture Organisation (FAO) commission Codex Alimentarius. The Codex standard for honey can be found at: http://www.codexalimentarius.net/web/standard_list.jsp.

The Codex states that:

> Honey sold as such shall not have added to it any food ingredient, including food additives, nor shall any other additions be made other than honey. Honey shall not have any objectionable matter, flavour, aroma, or taint absorbed from foreign matter during its processing and storage. The honey shall not have begun to ferment or effervesce. No pollen or constituent particular to honey may be removed except where this is unavoidable in the removal of foreign inorganic or organic matter.

The key phrases in the Codex standard for defining honeys by their floral or nectar source are:

> 6.1.6 Honey may be designated according to floral or plant source if it comes wholly or mainly from that particular source and has the organoleptic, physicochemical and microscopic properties corresponding with that origin.

and:

> 6.1.7 Where honey has been designated according to floral or plant source (6.1.6) then the common name or the botanical name of the floral source shall be in close proximity to the word "honey".

The floral designation of a honey is primarily based upon the measurement of the pollens contained within that honey. There are cases however where other plants may contribute significantly to the nectar of a honey without contributing significant pollen.

Table 10.1 Value (thousands of $US) of honey imports into the US from selected countries for the ten-year period from 2000 to 2010

Country	2000	2001	2002	2003	2004	2005	2006	2007	2008	2009	2010
China	22 378	15 585	7942	35 387	30 335	21 500	24 893	9598	6336	153	2443
Australia	175	564	2939	362	1866	553	942	1063	799	752	483
Indonesia	0	0	0	610	2072	1508	1247	642	2946	8709	13 653
Thailand	124	1431	5288	1863	1485	578	2003	927	1034	2668	3866
Malaysia	0	0	1217	5829	684	489	236	2678	6073	14019	24915
Taiwan	105	90	114	181	988	2596	510	1156	6672	9788	3290
New Zealand	254	306	197	756	504	1114	654	1101	1872	2804	4177

Source: Data from the Department of Commerce, US Census Bureau, Foreign Trade Statistics.

Honey may additionally be defined by its region of production;

> 6.1.5 Honey may be designated by the name of the geographical or topographical region if the honey was produced exclusively within the area referred to in the designation.

The EU standards for honey are discussed in Chapter 2 of this book. The pertinent details being that, in the EU, honey is one of the food products where there are specific rules requiring that consumers receive information about the country where the honey was harvested. More specific geographical or regional origins can be given, provided the honey is all from that source. In addition, the floral origin of the honey (except for filtered and baker's honey) can be given if the honey is derived wholly or mainly from the indicated source, but this is optional.

10.1.3 Traceability

The export of honey from New Zealand is regulated by the New Zealand Food Safety Association. Information including production region, honey type and the results of analytical tests are accumulated in an electronic certificate. This E-cert system allows online verification of certificates and the efficient passing of information along the supply line. Internationally the recent scandals in the honey industry demonstrate that the official traceability systems are being skirted.

10.1.4 New Zealand Manuka honey

While all honey has been shown to be susceptible to fraud, the New Zealand manuka honey is a special case. Manuka honey is gathered in New Zealand from two closely-related plants, both of which are often referred to as manuka. The most common of these is *Leptospermum scoparium,* the other is *Kunzea ericoides* and is commonly known as kanuka. Both plants grow uncultivated throughout the country. Honey produced from either of these plants is usually designated manuka as the pollens of the two plants are so similar they are almost impossible to tell apart.

Manuka honey has become a special case since the discovery of high levels of non-peroxide activity (NPA) against bacteria (Molan and Russell, 1988). It was further found that not all manuka honey exhibited high levels of NPA (Allen *et al.*, 1991). It was from this work that the term Active Manuka Honey came into use. Thereafter manuka honey was classified into two categories. To distinguish between the two types of manuka honey the term UMF (Unique Manuka Factor) was coined to express the measurement of the additional antibacterial property.

UMF is measured by use of the assay technique developed by Allen *et al.* (1991) where the antibacterial effectiveness of manuka honey against *Staphylococcus aureus* is compared to that of solutions of differing concentrations of phenol.

The manuka honey is tested after treatment with the enzyme catalase to breakdown all the hydrogen peroxide so only the NPA is being measured. A honey rated as UMF 10 has an antibacterial action against *Staph. aureus* that is equivalent to a 10% solution of phenol. The exact mechanism of this extra antimicrobial activity is still yet to be understood, though studies have identified methylglyoxal as the principally responsible component (Adams *et al.*, 2008). The presence of methylglyoxal in manuka honey has been attributed to the non-enzymatic conversion of dihydroxyacetone originating from the *L. scoparium* nectar (Adams *et al.*, 2009). UMF varies for manuka honey produced in different regions and different seasons, the reason for which is not well known. Owing to the extra antibacterial properties exhibited, high UMF manuka honey demands higher prices in the global honey market and has become a lucrative industry for the New Zealand economy, grossing upwards of NZ$ 100 million per annum (Infoshare, 2011). With the propensity for increased gain from this market, the motivation exists for some of the industry to establish fraudulent practices for monetary gain. While producers may get NZ$ 5 per kilogram for premium table honey, they can get prices in excess of ten times this for active manuka. This is good for the manuka honey producers but wherever there is a price differential there is the motivation for fraud.

10.2 Chemical authentication of honey

The authentication of honey has become an issue of particular interest over the last decade. Several studies have been undertaken to confirm the origins of various botanical honey types produced in differing geographical locations based on a range of chemical parameters (Anklam, 1998; Kaskoniene and Venskutonis, 2010). Amino acid composition (Cometto *et al.*, 2003; Cotte *et al.*, 2004; Rebane and Herodes, 2008), protein composition (Wang *et al.*, 2009), volatile aroma compounds (Allisandrakis *et al.*, 2007; Jerkovic *et al.*, 2010), carbohydrate ratios (Nozal *et al.*, 2005; Ruiz-Matute *et al.*, 2010), flavonoids and phenolic compounds (Hadjmohammadi *et al.*, 2009; Yao *et al.*, 2004) as well as mineral (Gonzalez-Paramas *et al.*, 2000; Madejczyk and Baralkiewicz, 2008; Terrab *et al.*, 2004; Tuzen *et al.*, 2007; Yarsan, 2007) and stable isotope composition (Schellenberg *et al.*, 2010) have been utilised in efforts to characterise and confirm botanical and geographic origin of honeys. Particular emphasis can be laid upon the mineral or elemental content and isotopic composition of honey for authentication purposes, as these parameters can easily be related back to the geographic environment.

10.2.1 Mineral content

The mineral content of honey has been investigated extensively for a number of reasons: the mineral content of honeys can be an indicator of environmental pollution (DeVillers *et al.*, 2002), increased levels of toxic minerals

Table 10.2 Average concentrations of mineral elements determined in 91 honeys from Galicia, NW Spain

Element	Average concentration (mg kg^{-1})
K	1572
Na	138
Ca	102
Mg	106
Cu	1.11
Fe	5.12
Mn	4.02

Source: Rodriguez-Otero *et al.*, 1994.

Table 10.3 Average concentrations in ng g^{-1} of elements in sunflower honey at different stages of the production process

Type of honey	As	Cd	Cr	Cu	Fe	Ni	Pb	Sn	V
Freshly collected	<0.5	<0.5	–	172.5	535	25	<4	<2	<2
Sealed	<0.5	<0.5	–	155.5	226.5	(20	<4	<2	<2
Extracted	0.6	0.61	1.21	187.5	621	51	3.59	19	1.41
Ripened	0.5	0.72	3.71	184.5	548.5	42	146.5	25	1.58
Commercial	3.89	0.7	3.03	121	2320	6.77	23.9	13.1	3.83

Source: Caroli *et al.*, 1999.

are an inherent risk to human health (Caroli *et al.*, 1999) and more recently characterisation of the mineral fraction of honey has been shown to be useful for chemical authentication (Anklam, 1998). An earlier characterisation by Rodriguez-Otero *et al.* (1994) of the minerals found in 91 honeys from Galicia, north western Spain was indicative of the concentrations of the more abundant mineral elements present in honey, including Na, K, Ca, Mg, Cu, Fe and Mn. K was observed to be the most abundant element in the honeys, as shown in Table 10.2 followed by Na, Mg and Ca. This is seen to be the case for the majority of honey. Fe, Mn and Cu were observed at much lower average concentrations, 1–5 mg kg^{-1}. Of note is that, when compared with data obtained for honey from other regions, including other areas of Europe and North America, differences in these element concentrations from region to region are reported (Rodriguez-Otero *et al.*, 1994).

The mineral content of honey was expanded upon by the investigation of Caroli *et al.* (1999) which analysed the concentrations of As, Cd, Cr, Cu, Fe, Mn, Ni, Pb, Pt, Sn and V in sunflower honey originating from the Sienna region of Italy. The honey was sampled and analysed at four different stages of the production process: freshly collected by the bees, sealed in the honeycomb, after the honey is extracted by centrifugation and filtration, and after the honey has been stored and allowed to ripen. The results obtained (Table 10.3) using

Table 10.4 Average concentrations (ppm unless otherwise stated) of elements analysed in honey studied by Lattore *et al.* (1999)

Element	Galician honey	Non-Galician honey
Li (ppb)	9.2	3.2
Na	115	54.2
K	1345	618
Rb	1.5	0.3
Mg	77	105
Zn	2	1.5
Fe	3.7	3.8
Mn	5.2	1.1
Cu	0.89	0.14
Ni	<0.02	<0.02
Co	<0.05	<0.05

inductively-coupled plasma techniques were compared to a commercial honey sample and give an indication of how the mineral element profile of honey evolves during the production process.

Caroli *et al.* (1999) concluded that leaching of some elements, in particular Cr and Pb, could occur during the extraction and ripening steps of production, which were performed in stainless steel containers and facilitated by the acidic nature of the honey itself.

The application of statistical pattern recognition procedures to elemental data was undertaken by Latorre *et al.* (1999) to discriminate honey from Galicia, in north-west Spain, from non-Galician honey. The study was one of the earliest to utilise the characterisation of the mineral element profile of honey as a tool for classification and possible authentication. The concentrations of Li, Rb, Na, K, Mg, Zn, Cu, Fe, Mn, Ni and Co were determined by either atomic absorption or atomic emission spectroscopy for 22 honeys sourced from the Galician region and 20 sourced from outside of Galicia. It should be noted that the honeys analysed were of a number of botanical origins. The resulting data (summarised in Table 10.4) was analysed using several statistical techniques, including principal components analysis (PCA), linear discriminant analysis (LDA), k nearest neighbours (kNN) and soft independent modelling of class analogy (SIMCA).

Upon visualisation using PCA to look for groupings, a natural separation between the Galician honey and non-Galician honey was observed (Fig. 10.1). Within the PCA space formed by the Galician honeys only two non-Galician honeys were present.

Of the statistical analyses applied by Latorre *et al.* (1999), LDA gave the most promising result with 100% discrimination of the training set into groups of Galician and non-Galician honey, however the other two classification procedures, KNN and SIMCA, also gave a high percentage of correct classification. Lattore *et al.* (1999) were able to demonstrate that useful

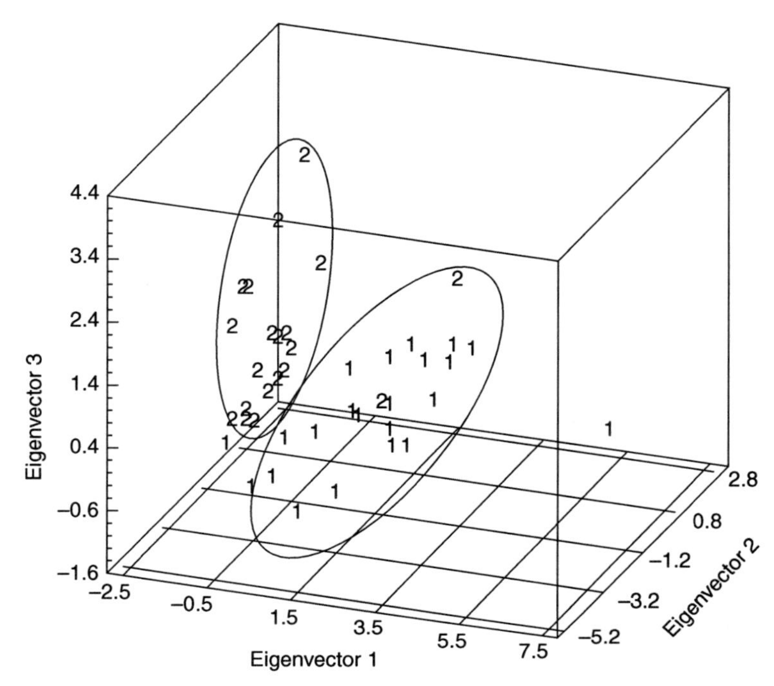

Fig. 10.1 Visualisation using PCA to determine groupings based upon the mineral element compositions of Galician honey (1) and non-Galician honey (2) by Latorre *et al.* (1999).

information could be extracted from mineral data for the possible purpose of authentication of honey from a particular geographic locale.

Fernandez-Torres *et al.* (2005) investigated the relationship between mineral composition of honey and botanical origin and, with application of statistical pattern recognition procedures, probed the efficacy of mineral elements in differentiating the botanical origin of honey. Forty honey samples were analysed by inductively-coupled plasma atomic emission spectroscopy, ten from each of four differing botanical origins, eucalyptus, orange blossom, rosemary and heather. The honey samples were obtained from various regions of Spain, and samples of similar and different geographical origins were included in each botanical class in an effort to guarantee that classification was due only to botanical origin. The concentrations of eleven elements were measured, including Zn, P, B, Mn, Mg, Cu, Ca, Sr, Ba, Na and K, of which the average values for each botanical origin are shown in Table 10.5.

The data obtained by Fernandez-Torres *et al.* (2005) was analysed using cluster analysis, PCA and LDA to visualise groupings and classify the data according to botanical origin. Cluster analysis was utilised as an unsupervised method of visualising groups within the data based on the similarity of the honeys in terms of the elements measured. The cluster analysis resulted in a

Table 10.5 Average concentrations in mg kg^{-1} of elements analysed in honey of differing botanical origin by Fernandez-Torres *et al.* (2005)

Botanical origin	Zn	P	B	Mn	Mg	Cu	Ca	Sr	Ba	Na	K
Eucalyptus	6.1	115	4.1	7.5	40.5	0.6	(204	0.78	0.46	168	1328
Orange blossom	3.8	63	6.5	0.71	21.3	<0.5	64	0.85	0.32	40	638
Rosemary	3.3	87	6.1	0.85	28.3	0.6	148	0.42	0.13	26	687
Heather	2.7	141	5.0	3.48	65.8	1.2	257	0.36	0.22	68	1844

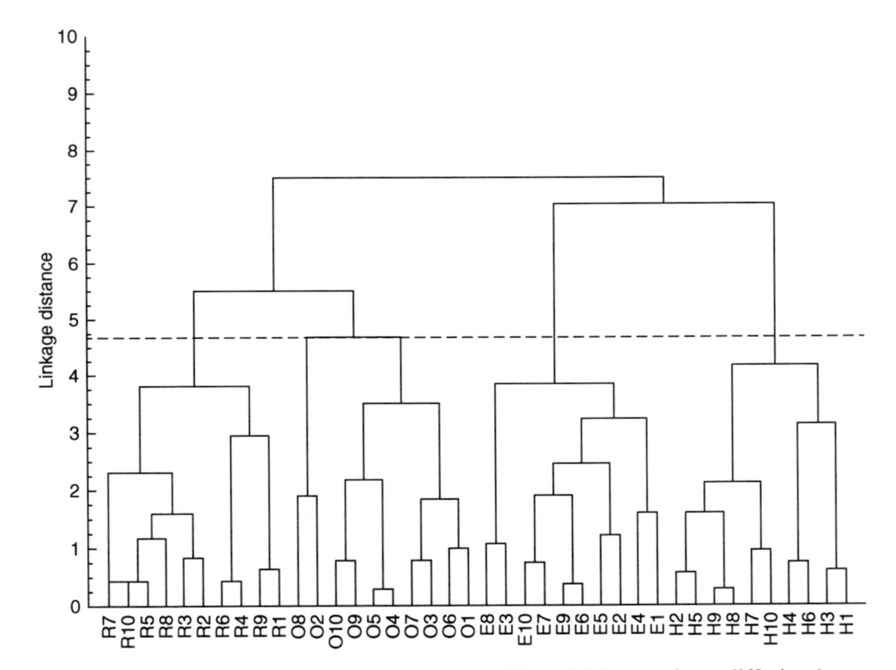

Fig. 10.2 Dendrogram of the cluster analysis of Spanish honey from differing botanical origins undertaken by Fernandez-Torres *et al.* (2005).

dendrogram in which four clusters with linkage distance level of 4.6 (Fig. 10.2) were observed, the first cluster from the left consisted of rosemary honey (R), the next orange blossom (O), then eucalyptus (E), with heather on the right (H). The linkage distance was a relative measure of similarity between the honeys.

Visualisation of the data using PCA by Fernandez-Torres *et al.* (2005) also resulted in four distinct groupings that were determined to result from variation in the elements measured due to different botanical origin (Fig. 10.3). The elements P, Mg, Ca, K and Zn were considered to be the dominant variables in the determination of this result; Ba, Sr and Cu were considered of less importance. It was concluded from the visualisation of the data by cluster

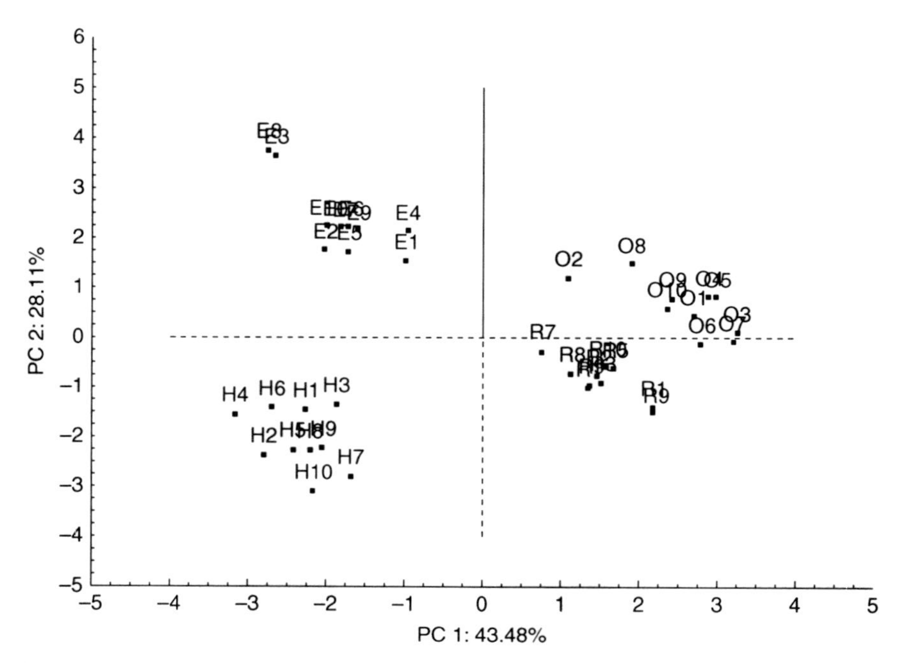

Fig. 10.3 Visualisation using PCA to determine groupings based on mineral element composition of eucalyptus (E), orange blossom (O), rosemary (R) and heather (H) honeys investigated by Fernandez-Torres *et al.* (2005).

analysis and PCA that enough information was contained within the mineral element composition to develop a classification method to determine botanical origin.

Based upon the visualisation results, Fernandez-Torres *et al.* (2005) applied the supervised classification technique LDA to develop and test functions for the discrimination of botanical origin for the honeys investigated. The LDA resulted in three discriminant functions of which the first two were plotted (Fig. 10.4) in which the four botanical groups appeared to be completely separated. Validation of the LDA model was undertaken by cross validation using 25% of the honeys as the evaluation set and was repeated several times, the result of which was evaluation of the recognition and prediction abilities of the derived model. The recognition ability, the proportion of members of the training set correctly classified, was 100%; the prediction ability, the percentage of members of the evaluation set correctly classified, was 97%.

Fernandez-Torres *et al.* (2005) concluded that the mineral element composition of honeys in conjunction with multivariate statistical pattern recognition procedures were a useful tool for the authentication and discrimination of the botanical origin of honeys.

An investigation of minerals in honey pertaining to the effects of environmental, geographical and botanical factors by Bogdanov *et al.* (2007) is

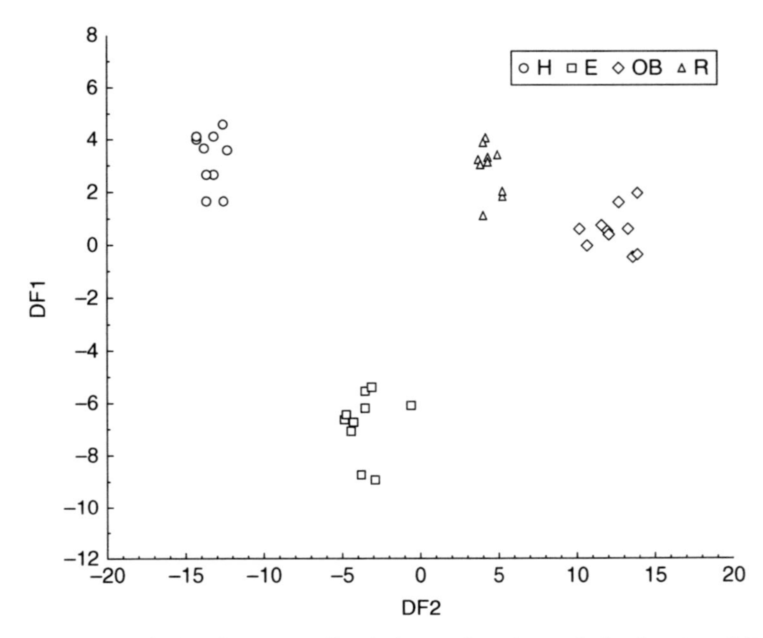

Fig. 10.4 Plot of the first two discriminant functions derived using LDA by Fernandez-Torres *et al.* (2005).

of particular importance. The analysis of 95 honey samples of known geographical and botanical origins by inductively-coupled plasma sector field mass spectrometry for the elements Cd, Pb, Cr, Mn, Fe, Ni, Cu and Zn was undertaken. The electrical conductivity (EC) of the honeys were also measured. The honey samples were all sourced from known Swiss production areas and were divided into four groups shown in Table 10.6 with respect to climatic and anthropogenic factors (Bogdanov *et al.*, 2007).

The data acquired by Bogdanov *et al.* (2007) was investigated further by use of several descriptive statistics based on parametric analysis of variance (ANOVA) as well as non-parametric methods. A tentative classification of botanical origin was also undertaken utilising PCA and LDA. The results were shown to correlate significantly with EC which is a widely used parameter for the determination of botanical origin (Bogdanov, 2004). The use of EC in conjunction with the trace elements Ni, Fe, Mn and Cd were significant with respect to the classification by botanical origin of the monofloral honeys studied, classification of polyfloral honey was considered of greater difficulty. Bogdanov and co-workers (2007) indicate that one of the problems of routine honey control is the differentiation between mono- and polyfloral honey and the use of trace elements for this purpose is promising. Of greatest importance from this study by Bogdanov *et al.* (2007) is that the trace elements measured could be from natural sources (soil, plants) and anthropogenic sources, however the botanical factors were observed to have the

Table 10.6 Groupings of honey investigated by Bogdanov *et al.* (2007)

Group (definition)	Number of samples	Honey types present
City (honey from the cities of Berne and Basel, >100 000 inhabitants)	13	10 mixed blossom, 2 mixed honeydew, 1 blend blossom/honeydew
Village (honey from villages with <5000 inhabitants)	18	10 mixed blossom, 1 linden, 1 dandelion, 3 mixed honeydew, 3 blend blossom/honeydew
Rural (>3 km from the nearest village, <700 m above sea-level)	51	7 acacia, 7 chestnut, 7 dandelion, 7 linden, 6 rape, 7 fir, 5 mixed blossom, 5 mixed honeydew
Mountain (>3 km from the nearest village, >700 m above sea-level)	13	7 rhododendron, 6 mixed mountain blossom

greatest influence upon the mineral composition of honeys and, because of this, botanical origin should be taken into consideration while undertaking other analyses.

The trace elemental composition can be successfully used to determine the provenance of honey provided a suitable reference database is available for building the statistical models. At this stage this limits the application of this tool to inclusion/exclusion tests for samples from regions that have been sufficiently sampled to define the trace elemental specifications for the honey types produced there. For this tool to be used in a predictive fashion will required a large increase in the availability of trace element data from authentic (botanical and geographical origin) honey samples and definition of the relationship with the composition of the underlying geology.

10.2.2 Stable isotope analysis

Another chemical measure of honey that has been successfully applied is the stable isotope composition. Recently Schellenberg *et al.* (2010) investigated the stable isotope ratios of honey from different regions of Europe as part of the EU's food traceability project, 'TRACE'. The stable isotope ratios of H, C, N and S in honey protein were analysed by isotope ratio mass spectrometry (IRMS) to determine if the acquired data could be utilised to differentiate honeys produced in regions with different climatic and geological characteristics. The results obtained by Schellenberg *et al.* (2010) from analysis of 617 samples are summarised in Table 10.7.

It was observed that the $\delta^{13}C$ values were higher (~ −24‰) in honeys from regions with a higher average temperature and decreased average humidity such as the Mediterranean sampling regions, whereas regions of

Table 10.7 Average δ values of C, H, N and S for honey protein from different regions of Europe as determined by Schellenberg *et al.* (2010)

Region of origin	$\delta^{13}C$	$\delta^{2}H$	$\delta^{15}N$	$\delta^{34}S$
Algarve (Portugal)	−25.4	−73	2.0	5.9
Allgau (Germany)	−26.1	−121	1.8	2.0
Barcelona (Spain)	−25.2	−105	0	4.7
Carpentras (France)	−24.6	−106	1.9	4.1
Chalkidiki (Greece)	−25.1	−111	1.3	5.6
Cornwall (UK)	−25.8	−86	5.4	9.5
Franconia (Germany)	−26.2	−118	2.6	3.6
Gauboden (Germany)	−26.3	−118	3.3	4.8
Iceland	−26.8	−107	0.5	8.0
Ireland	−25.7	−99	4.2	9.3
Jylland (Denmark)	−26.3	−112	4.5	4.2
Lakonia (Greece)	−25.8	−96	2.0	9.2
Limousin (France)	−25.4	−106	2.0	5.2
Marchfeld (Austria)	−25.9	−108	4.3	4.0
Muhlviertel (Austria)	−26.7	−114	3.2	4.5
Orkneys (UK)	−26.5	−88	0.9	11.1
Poland	−26.2	−101	3.9	3.5
Sicily (Italy)	−24.2	−99	3.9	2.0
Trentino (Italy)	−24.8	−110	0.8	5.2
Tuscany (Italy	−24.4	−99	1.7	3.7

lower average temperature and increased average humidity such as Orkney, Muhlviertel, Gauboden and Jylland exhibited lower $\delta^{13}C$ values (~ −26‰). The more temperate regions were observed to have more central $\delta^{13}C$ values (~ −25‰). The carbon isotopes in plants and resulting products are known to be influenced by plant metabolism (carbon fixation) (Smith and Epstein, 1971); it has also been determined that the $\delta^{13}C$ values in honey protein increases with increasing number of sunshine days/sunshine duration (Jonasson *et al.*, 1997). Schellenberg *et al.* (2010) determined that the European climate was distinct enough on the continental scale to influence $\delta^{13}C$ in honey and make it an applicable parameter for discriminating geographic origin, despite larger variation observed in $\delta^{13}C$ values on a smaller regional scale.

The $\delta^{2}H$ values measured for the honeys were also seen to vary systematically with climate, with honeys from Western Europe, close to the Atlantic Ocean, having higher values (−86‰, Cornwall). This was compared to inland and alpine regions which were observed to have lower $\delta^{2}H$ values (−121‰, Allgau; −118‰, Franconia). The $\delta^{2}H$ values of meteoric water are known to decrease with increasing distance to the sea, increasing altitude and decreasing average temperature for a specific region (Clark, 1997; Criss, 1999). These phenomena are due to the concentration of the lighter isotopologues with each evaporation and precipitation event relative to the heavy ones which evaporate more slowly and precipitate out to a greater extent (Clark, 1997; Criss, 1999). Schellenberg *et al.* (2010) observe that the systematic variation of δ^2H in the hydrosphere is conserved in the distribution of δ^2H in their analysed honey.

Nitrogen for plant material as amino acids and proteins is derived from soluble nitrogen-containing substances in the topsoil in which the plant is growing (Schellenberg *et al.*, 2010). Schellenberg *et al.* (2010) expected δ^{15}N values to not only reflect the soil conditions of the area in which the honey was collected but also the botanical origins as nitrogen is an important nutrition parameter for plants and many plants have adapted to nitrogen availability. Other factors such as water stress, altitude, distance from the ocean and fertilizer use were also expected to affect measured δ^{15}N values (Camin *et al.*, 2004). Schellenberg *et al.* (2010) were unable to statistically evaluate by botanical origin due to a lack of monofloral honeys present in the sample honey population. Statistical evaluation by geographical origin exhibited low δ^{15}N values for samples from Barcelona, 0.0‰, and Trentino, 0.8‰ compared to high values for Cornwall, 5.4‰. However no discernible trends were identified and this was attributed to the extensive variation exhibited by honey samples in this parameter for each region. More detailed interpretation of δ^{15}N was expected once analysis of soil samples had been undertaken for comparison.

Schellenberg *et al.* (2010) also measured the δ^{34}S values of honey and observed separation of honey from varying regions into possible groupings. One group consisted of regions that exhibited low δ^{34}S values (Sicily, Allgau, 2.0‰), an intermediate group that consisted of 60% of the sampling areas, a group consisting of the samples from Iceland (8.0‰), as well as a group consisting of regions exhibiting high δ^{34}S values (Ireland, 9.3‰; Cornwall, 9.5‰; Lakonia, 9.2‰ and Orkneys, 11.1‰). The δ^{34}S values of plants and plant materials are influenced strongly by soil sulphur composition which is determined by soil geology, therefore the sulphur isotope composition of honey can reflect the geology of the plant growing region (Forstel, 2007). Additionally, coastal areas can be influenced by 'sea spray effect' (Thode, 1991), where aerosolised seawater, containing sulphates, is deposited onto the terrestrial environment and can overlay the geological source of sulphur. The δ^{34}S value of seawater is close to 22‰. Schellenburg *et al.* (2010) attributed the δ^{34}S values of honeys from the high δ^{34}S group to the sea spray effect, as the regions in this group were coastal or island regions.

LDA was utilised by Schellenberg *et al.* (2010) as a classification tool for geographic origin based upon the acquired isotope data. Of the 617 honey samples analysed 516 were used for the classification procedure which resulted in a recognition ability of 60.2% and prediction ability after leave-one-out or 'jackknife' cross validation of 58.3%. The separation of geographic regions in the discriminant space was observed by plotting the first two discriminant functions (Fig. 10.5); function one, representing 70% of the variation between groups was highly correlated with δ^{34}S values and δ^2H values, while function two, representing 13% of the variation between groups was highly correlated

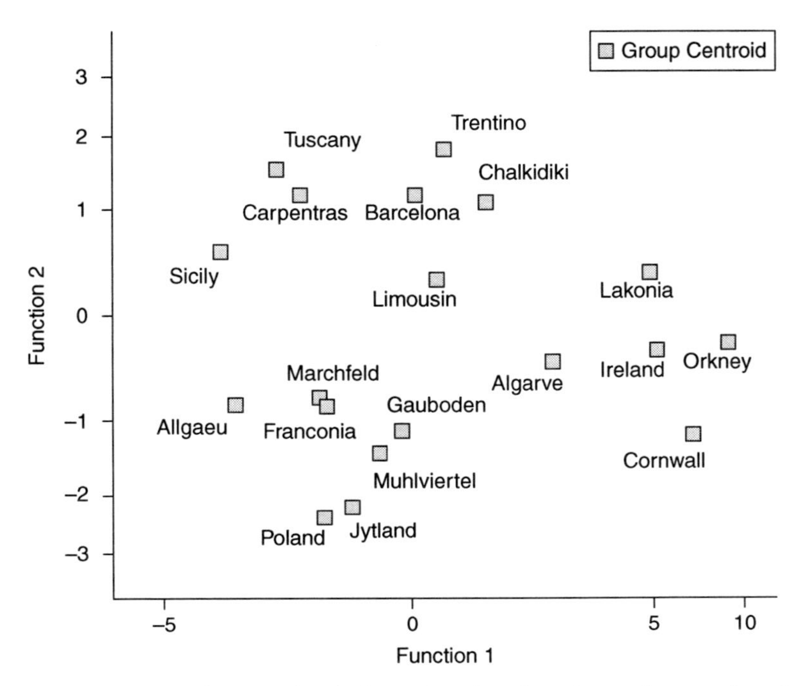

Fig. 10.5 Plot of the first two discriminant functions based upon isotope data determined by Schellenberg *et al.* (2010). Centroids for each group of honey samples are shown for the purpose of clarity.

with $\delta^{34}S$ values and $\delta^{13}C$ values. It was concluded that the isotopes in honey reflect the influences of precipitation, climate and soil geology and accordingly may be useful as parameters for geographical discrimination, although more discriminating parameters would be needed for regions of similar climate and/or geology.

Kropf *et al.* (2010) used common physico-chemical properties in conjunction with mineral element composition and stable isotope measurements to authenticate the geographical origin of three types of monofloral honey produced in Slovenia. For the study, 122 honeys of confirmed botanical origin (55 of black locust, 30 of lime and 37 of chestnut) were obtained from different natural geographic regions of Slovenia. Elemental analysis of the honeys was undertaken using total-reflection X-ray fluorescence spectroscopy; C and N isotopes were determined by IRMS. The statistical analyses undertaken by Kropf *et al.* (2010) on the obtained data included ANOVA, cluster analysis, PCA and LDA. With reference to the conclusions of Bogdanov *et al.* (2007), each monofloral honey type was considered separately from the others. For the purpose of geographical authentication Kropf *et al.* (2010) divided the honeys into groups that represented four geographic macroregions of Slovenia, designated as Alpine, Dinaric, Mediterranean and Pannonian. The data obtained is summarised in Table 10.8.

Table 10.8 Average values of parameters measured by Kropf (2010) for the authentication of three monofloral honeys produced in Slovenia

Honey type	Parameter measured	Geographic macroregion			
		Alpine	Dinaric	Mediterranean	Pannonnian
Black locust	Number of samples	0	3	30	32
	Electrical conductivity (mS cm^{-1})	–	0.193	0.178	0.162
	Total ash (g 100 g^{-1})	–	0.06	0.07	0.06
	pH	–	4.09	4.04	4.16
	Total acids (meq kg^{-1})	–	16.98	16.68	15.67
	Lactones (meq kg^{-1})	–	2.99	3.01	5.20
	Free acids (meq kg^{-1})	–	13.99	13.99	10.48
	Proline (mg kg^{-1})	–	322	291	222
	Protein (g 100g^{-1})	–	0.156	0.164	0.144
	S (mg kg^{-1})	–	48.6	46.0	46.7
	Cl (mg kg^{-1})	–	55.4	103	89.2
	K (mg kg^{-1})	–	313	286	261
	Ca (mg kg^{-1})	–	19.1	18.5	15.4
	Mn (mg kg^{-1})	–	3.09	15.4	1.65
	Rb (mg kg^{-1})	–	0.84	0.68	0.75
	$\delta^{13}C_{honey}$ (‰)	–	−24.2	−24.8	−24.8
	$\delta^{13}C_{protein}$ (‰)	–	−23.5	−24.0	−23.9
	$\delta^{15}N$ (‰)	–	2.2	3.1	2.7
Lime	Number of samples	11	16	3	0
	Electrical conductivity (mS cm^{-1})	0.727	0.836	0.823	–
	Total ash (g 100g^{-1})	0.23	0.30	0.25	–
	pH	4.80	4.99	4.59	–
	Total acids (meq kg^{-1})	15.43	14.57	18.73	–
	Lactones (meq kg^{-1})	1.79	2.87	1.62	–
	Free acids (meq kg^{-1})	13.64	12.78	17.11	–
	Proline (mg kg^{-1})	341	321	258	–
	Protein (g 100g^{-1})	0.196	0.516	0.154	—

Table 10.8 Continued

Honey type	Parameter measured	Geographic macroregion			
		Alpine	Dinaric	Mediterranean	Pannonnian
	S (mg kg^{-1})	27.1	41.8	58.5	–
	Cl (mg kg^{-1})	339	443	272	–
	K (mg kg^{-1})	1670	1940	1610	–
	Ca (mg kg^{-1})	63.6	72.0	60.3	–
	Mn (mg kg^{-1})	2.86	3.70	4.38	–
	Rb (mg kg^{-1})	4.04	6.53	5.14	–
	$\delta^{13}C_{honey}$ (‰)	−25.6	−25.4	−26.2	–
	$\delta^{13}C_{protein}$ (‰)	−24.8	−25.0	−25.2	–
	$\delta^{15}N$ (‰)	1.9	1.9	1.8	–
Chestnut	Number of samples	17	12	0	8
	Electrical conductivity (mS cm^{-1})	1.348	1.549	–	1.619
	Total ash (g 100g^{-1})	0.66	0.66	–	0.72
	pH	5.28	5.61	–	5.39
	Total acids (meq kg^{-1})	15.11	14.55	–	15.14
	Lactones (meq kg^{-1})	2.13	2.88	–	2.24
	Free acids (meq kg^{-1})	12.98	11.67	–	13.03
	Proline (mg kg^{-1})	599	486	–	563
	Protein (g 100g^{-1})	0.352	0.344	–	0.338
	S (mg kg^{-1})	52.6	33.2	–	38.4
	Cl (mg kg^{-1})	226	25.2	–	263
	K (mg kg^{-1})	3290	3940	–	3790
	Ca (mg kg^{-1})	143	164	–	132
	Mn (mg kg^{-1})	(20.6	25.4	–	22.7
	Rb (mg kg^{-1})	14.8	15.2	–	23.9
	$\delta^{13}C_{honey}$ (‰)	−25.8	−25.8	–	−25.8
	$\delta^{13}C_{protein}$ (‰)	−25.3	−25.1	–	−25.2
	$\delta^{15}N$ (‰)	2.0	1.8	–	1.0

The application of cluster analysis and PCA to the analytical data by Kropf *et al.* (2010) failed to yield useful results and so no interpretation was undertaken. LDA was performed for classification by geographic origin for each monofloral honey type, with significant discrimination observed when the first two discriminant functions for each were plotted (Fig. 10.6).

Several parameters were observed to be significant in the geographic discrimination of each monofloral honey type; proline, free acid content and

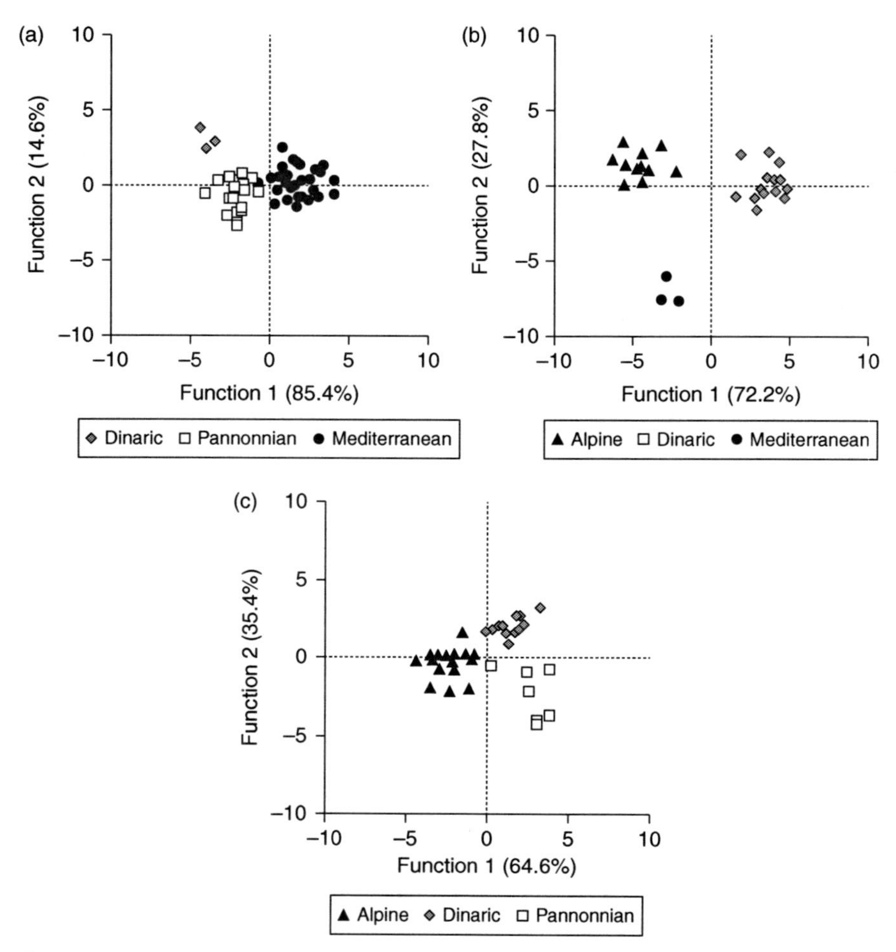

Fig. 10.6 The discriminant functions classifying geographic origin for each of the monofloral honey types, (a) black locust, (b) lime and (c) chestnut, investigated by Kropf *et al.* (2010).

$\delta^{13}C_{protein}$ values had major influence upon geographic discrimination for black locust honey. Geographic discrimination of lime honey was highly influenced by electrical conductivity and total ash content while sulphur, total ash and potassium content were observed to be the major influences for geographic discrimination of chestnut honey (Kropf *et al.*, 2010).

10.2.3 Botanical origin

The common approach of using stable isotope and/or trace element measurements to determine the geographic origin of honey work well when the botanical origin of the honey is known. It appears that the floral type has the greatest influence on these parameters and so a means of verifying botanical origin needs to be used in conjunction with these provenancing tools. The conventional means of determining the contributing plant is palynology (Louveaux *et al.*, 1978). However, this is a specialised technique with relatively few qualified practitioners. Recently attention has been focused on determining the amounts of metabolites and small molecules in honey. Much of this interest was motivated by the discovery that the concentration of methylglyoxal is responsible for the antimicrobial activity of manuka honey (Mavric *et al.*, 2008). Marceau and Yaylayan (2009) profiled the α-dicarbonyl contents of honey from different botanical origins and found considerable variation. The actual amounts of many α-dicarbonyls could be manipulated by heating/aging the honey so care is required in interpreting such data. Stephens *et al.* (2010) demonstrated a difference in phenolic profiles between manuka honey and the closely related kanuka. This is an important result as these two honey types cannot be differentiated using palynology and the manuka is thought to have much greater value as an antimicrobial agent than kanuka. The phenolic profiling was also effective in identifying heat treatment of the honey.

Donarski *et al.* (2010) used one-dimensional ^{1}H-NMR spectroscopy to measure and characterise specific biomarkers for sweet chestnut and strawberry tree honey. In a previous study (Donarski *et al.*, 2008) they had applied multivariate statistical methods to identify spectral regions that were characteristic of the geographic origin of the honey sample. Their approach to identifying floral type was to investigate the peaks in the NMR spectrum that were characteristic of Corsican honey and correlate these with specific honey types. Further work from the EU-TRACE programme developed specific real-time PCR systems for identifying plant species (Laube *et al.*, 2010). Honey is a challenging matrix for molecular techniques due to the very low levels of plant DNA it contains. Laube *et al.* (2010) developed specific primers and probes for the plant DNA of interest and were able to identify the contributing plants with a high degree of specificity. Several factors contribute to make the characterisation of honey composition in terms of the relative contributions of the plant species problematic using this approach. Therefore it is a qualitative means of identifying which plants were in the region where the nectar was collected. This information in itself can be useful in determining geographic origin but requires significant knowledge of the distribution of plant species.

10.3 Adulteration of honey

The major components of honey are the carbohydrates glucose and fructose, the addition of sugar syrups, such as high fructose corn syrup, is a common way of extending honey by adulteration. Several methods for the detection of honey adulteration with sugar syrups have been developed with the most utilised technique being high performance liquid chromatography; this method however does not detect adulteration to low levels. The use of stable carbon isotope ratio analysis for the detection of honey adulteration was pioneered by White and Winters (1989). The method is based on the different photosynthetic pathways utilised by monocotyledonous plants (corn, sugar cane) and dicotyledonous plants (flowering plants from which bees collect nectar). Dicotyledonous plants utilise the Calvin-Benson photosynthetic cycle and are denoted as C3 plants while monocotyledonous plants utilise the Hatch-Slack photosynthetic cycle and are denoted as C4 plants. Due to the different photosynthetic production pathways, carbon isotope ratios of C3 and C4 plant tissues and secretions differ with carbon isotope ratios commonly in the range of −21‰ to −32‰ for C3 plants and −12‰ to −19‰ for C4 plants. Honeys are mainly a product of C3 plant secretions; those that exhibited carbon isotope ratios greater than −23.5‰ were considered as suspect. Adulteration of honey became more sophisticated as a result of the use of stable carbon isotopes for detection. Companies involved in honey adulteration adapted by blending artificial sweeteners into honey that resulted in carbon isotope ratios less than −23.5‰. To combat this more sophisticated level of adulteration, comparison of carbon isotope ratios from the protein fraction of honey to the carbohydrate fraction was undertaken by White and Winters (1989). The carbon isotope values for honey proteins and carbohydrates should not differ by more than 1.0‰ if they come from the same source. The percentage of adulteration can also be estimated by the difference in carbon isotope ratios between protein and carbohydrate fractions (Padovan *et al.*, 2003). This test has become standard for the detection of honey adulteration with C4 sugar syrups (AOAC method 998.12), however in some situations, as have been reported for some European honeys and New Zealand manuka honey, this test can return false positive results. As such, development of this method is ongoing with some researchers having investigated possible refinements (Cabañero *et al.*, 2006; Rogers *et al.*, 2010), while others have investigated alternative methods for the detection of honey adulteration (Bertelli *et al.*, 2010; Kelly *et al.*, 2003).

Kelly *et al.* (2003) investigated the use of ATR-IR spectroscopy as a means of detecting the adulteration of honey by sugar syrups. Using three different adulterant solutions made of different proportions of fructose to glucose (0.7:1, 1.2:1 and 2.3:1), 25 honey samples were purposely adulterated to three different levels, 7%, 14% and 21%, resulting in a total of 225 adulterated honey solutions. A further 99 authentic honey samples were used as a calibration set for comparison. ATR-IR spectra in the range 6000–12 000 cm^{-1} were acquired for each sample and, in conjunction with the statistical methods of partial least squares regression (PLS) and k nearest neighbours (kNN), models were determined and evaluated for detecting adulteration. The changes in the concentrations of fructose and glucose upon adulteration with sugar syrup produced corresponding changes to the IR absorbances associated with those sugars, with major differences observed in the C–O, C–C and O–H stretching vibrational modes. PLS was undertaken on the spectral values in an effort to differentiate authentic samples from adulterated ones, the results of which were favourable with >90% of the adulterated honeys detected. The majority of misclassified samples were from honey that had been adulterated to the 7% level. The standard error of prediction for the model was calculated at 4.75%, two times which is equal to the 95% confidence interval limit for the model, 9.5%; Kelly *et al.* (2003) suggest that the model is not able to predict adulteration below this level. The statistical kNN method was applied in an effort to overcome the limitations encountered using PLS. Application of kNN resulted in an overall classification as adulterated or authentic honey of 92%, extension to the level of adulteration resulted in 100% of all honey adulterated at 21% being detected, 87–95% of honey adulterated to 14% detected and 75% of honey adulterated at 7% detected. Kelly *et al.* (2003) reported mid-infrared spectroscopic techniques as a potential tool for the rapid screening and detection of honey adulteration that was likely to be sensitive to a broad range of sugar adulterants.

An investigation into the utilisation of one- and two-dimensional high resolution nuclear magnetic resonance (NMR) spectroscopy as a means for the detection of honey adulteration by commercially available sugar syrups was undertaken by Bertelli *et al.* (2010). Sixty-three samples of authentic honey were obtained, a portion of which were adulterated with one of seven commercially obtained sugar syrups to a specific level of adulteration, either 10%, (20% or 40%, resulting in another 63 adulterated samples. ^{1}H-NMR and ^{1}H-^{13}C heteronuclear multiple bond correlation (HMBC) spectra were acquired for all 126 honey samples as well as samples of the seven sugar syrup adulterants. After characterisation of the NMR spectra, Bertelli *et al.* (2010) applied the multivariate statistical methods of factor analysis (FA), to reduce the number of spectroscopic variables analysed, and generalised discriminant analysis (GDA), to differentiate the different levels of adulterated samples from the authentic samples. Of the two NMR spectroscopic approaches investigated by Bertelli *et al.* (2010), the one-dimensional approach, ^{1}H-NMR, gave the better results in terms of discrimination: 95.2% correctly classified upon leave-one-out cross validation, compared to the two-dimensional approach of ^{1}H-^{13}C HMBC, 90.5% correctly classified upon leave-one-out cross validation. Overlap of the two lowest levels of adulteration was observed when the first two canonical discriminant functions were plotted for the two-dimensional approach which was not observed for the one-dimensional approach (Fig. 10.7). The authors determined that the canonical functions for the one-dimensional approach were highly correlated to the region of the NMR spectrum where the peaks arising from the anomeric hydroxyl protons of sugars are observed, suggesting that particular region of the NMR spectrum was the greatest discriminator in terms of adulteration. It was concluded that utilisation of high resolution ^{1}H-NMR, in conjunction with appropriate statistical methods, was a simple and rapid technique for the determination of sugar syrup adulteration of honey.

10.4 Conclusion and future trends

As discussed earlier, fraud in the honey industry is becoming more sophisticated. New techniques such as ultra-filtration to remove all pollen make

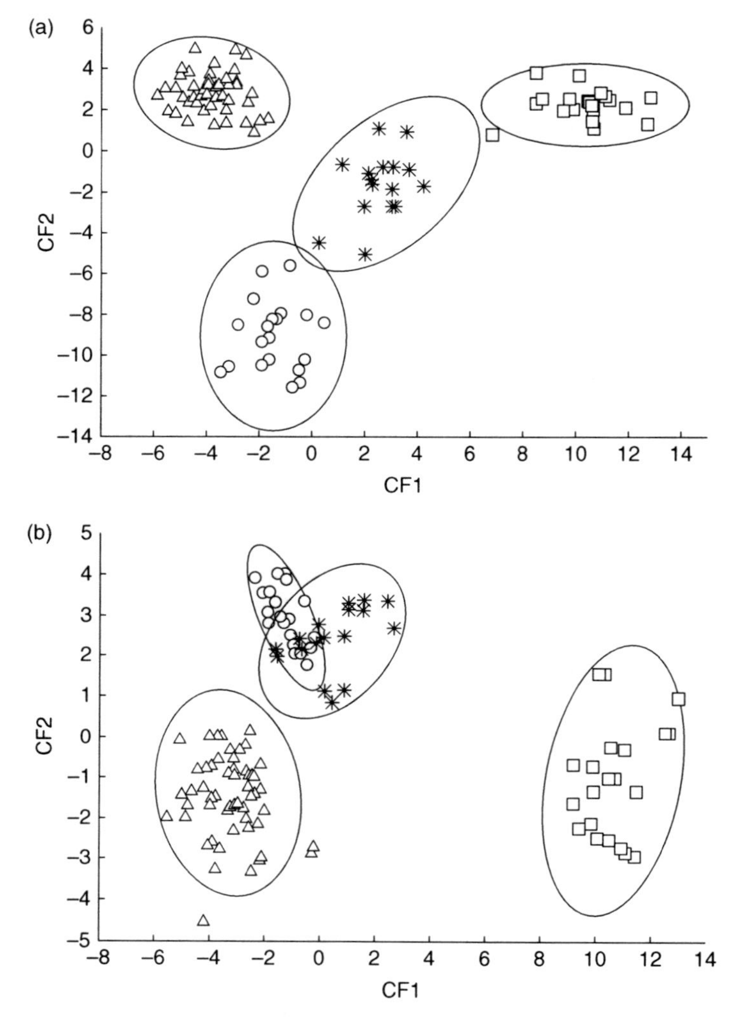

Fig. 10.7 Score plots of the first two canonical discriminant functions for 126 authentic and adulterated honey samples measured using (a) a one-dimensional NMR approach and (b) a two-dimensional NMR approach. Discrimination was based on the level of adulteration, authentic (Δ), 10% adulteration (○), (20% adulteration (*), 40% adulteration (□). Figure from Bertelli *et al.* (2010).

it difficult to determine the authenticity of some honey by traditional techniques such as palynology. Measures of properties inherent and unique to a particular honey type/origin are required. Chemical fingerprinting shows great promise for this purpose (Bogdanov *et al.*, 2007; Chudzinska and Baralkiewicz, 2010; Schellenberg *et al.*, 2010) but is currently hampered by the paucity of quality-assured data for reference. Multivariate statistical tools are being successfully applied in the interpretation of this type of data. However, most published studies have very small datasets (<100 samples) and so the multivariate approach must be treated with caution. Priority must be given to the populating of reference databases. The other aspect of the multivariate statistical treatment that is lacking is the inclusion of confidence intervals. Authors invariably present PCA or LDA plots with data clusters, or circle the data to imply a level of confidence in the discrimination.

It is crucial for the application of this technique to be able to apportion the floral and geographical components of the signal (Bogdanov *et al.*, 2007; Wang and Li, 2007). Other techniques may be applied to specific honey types, for example, phenolic profiling for manuka honey (Stephens *et al.*, 2010) or protein fingerprinting (Wang *et al.*, 2009). Again these techniques require a significant investment in the collection of authentic data to be used to define the specifications of the honey thus enabling the reliable detection of fraud.

The ultimate aim in this field is to move from chemical provenancing, which can only determine if a sample complies with the specification previously determined for a region, to geochemical provenancing. The principle of geochemical provenancing is that the chemical signal (trace elements and some stable isotopes) in the honey is derived from the geology of the production region. If the transfer functions (soil-plant-nectar) can be determined then maps of soil isotope/trace element composition could be used in a predictive manner to suggest regions of origin for an unknown sample. This requires large-scale databasing to produce maps along the lines of 'isoscapes' (e.g., Bowen, 2010) and the development of models to predict the production area soil composition given a particular honey composition.

10.5 References

ADAMS, C.J., C.H. BOULT, B.J. DEADMAN, J.M. FARR, M.N.C. GRAINGER, M. MANLEY-HARRIS and M.J. SNOW. (2008). Isolation by HPLC and characterisation of the bioactive fraction of New Zealand manuka honey. *Carbohydr. Res.*, **343**, 651–659.

ADAMS, C.J., M. MANLEY-HARRIS and P.C. MOLAN. (2009). The origin of methylglyoxal in New Zealand manuka honey. *Carbohydr. Res.*, **344**, 1050–1053.

ALLEN, K.L., P.C. MOLAN and G.M. REID. (1991). A survey of the antibacterial activity of some New Zealand honeys. *J. Pharm. Pharmacol.*, **43**, 817–822.

ALLISANDRAKIS, E., P.A. TARANTILIS, P.C. HARIZANIS and M. POLISSIOU. (2007). Comparison of the volatile composition in thyme honeys from several origins in Greece. *J. Agric. Food Chem.*, **55**, 8152–8157.

ANALYTICS, M. Available: http://ebookbrowse.com/sol-food-chloromphenicol-in-honey-0606-pdf-d119295023 (Accessed 23 March (2012).

ANKLAM, E. (1998). A review of the analytical methods to determine the geographical and botanical origin of honey. *Food Chem.*, **63**, 549–562.

BERTELLI, D., LOLLI, M., PAPOTTI, G., BORTOLOTTI, L., SERRA, G. and PLESSI, M. (2010). Detection of honey adulteration by sugar syrups using one-dimensional and two-dimensional high-resolution nuclear magnetic resonance. *J. Agric. Food Chem.*, **58**, 8495–8501.

BOGDANOV, S. (2004). Physico-chemical methods for the characterisation of unifloral honeys: a review. *Apidologie*, **35**, S4–S17.

BOGDANOV, S., M. HALDIMANN, W. LUGINBUHL and P. GALLMANN. (2007). Minerals in honey: environmental, geographical and botanical aspects. *J. Api. Res. Bee World*, **46**, 269–275.

BOWEN, G. (2010). Isoscapes: Spatial pattern in isotope biogeochemistry. *Ann. Rev. Earth. Planet. Sci.*, **38**, 161–187.

CABAÑERO, A.I., RECIO, J.L. and RUPÊREZ, M. (2006). Liquid chromatography coupled to isotope ratio mass spectrometry: a new perspective on honey adulteration detection. *J. Agric. Food Chem.*, **54**, 9719–9727.

CAMIN, F., K. WIETZERBIN, A.B. CORTES, G. HABERHAUER, M. LEES, and G. VERSINI. (2004). Application of multielement stable isotope ratio analysis to the characterisation of French, Italian and Spanish cheeses. *J. Agric. Food Chem.*, **52**, 6592–6601.

CAROLI, S., G. FORTE, A.L. IAMICELI and B. GALOPPI. (1999). Determination of essential and potentially toxic trace elements in honey by inductively coupled plasma-based techniques. *Talanta*, **50**, 327–336.

CHUDZINSKA, M. and D. BARALKIEWICZ. (2010). Estimation of honey authenticity by multielements characteristics using inductively coupled plasma-mass spectrometry (ICP-MS) combined with chemometrics. *Food Chem. Toxicol.*, **48**, 284–290.

CLARK, I. D. (1997). *Environmental Isotopes in Hydrogeology*, New York, CRC Press.

COMETTO, P. M. P.F. FAYE, R.D. NARANJO, M.A. RUBIO and M.A.J. ALDAO. (2003). Comparison of free amino acids profile in honey from three Argentinian regions. *J. Agric. Food Chem.*, **51**, 5079–5087.

COTTE, J.F., H. CASABIANCA, S. CHARDON, J. LHERITIER and M.F. GRENIER-LOUSTALOT. (2004). Characterization of honey amino acid profiles using high-pressure liquid chromatography to control authenticity. *Anal. Bioanal. Chem.*, **380**, 698–705.

CRISS, R. E. (1999). *Principles of Stable Isotope Distribution*, New York, Oxford University Press.

DEVILLERS, J.J.C. DORÉ, M. MARENCO, F. POIRIER-DUCHÊNE, N. GALAND, and C. VIEL. (2002). Chemometrical analysis of 18 metallic and nonmetallic elements found in honeys sold in France. *J. Agric. Food Chem.*, **50**, 5998–6007.

DHARMANANDA, S. (2003). Available: http://www.itmonline.org/arts/bees.htm (Accessed 23 March (2012).

DONARSKI, J.A., S.A. JONES and A.J. CHARLTON. (2008). Application of cryoprobe 1Hnuclear magnetic spectroscopy and multivariate analysis for the verification of Corsican honey. *J. Agri. Food Chem.*, **56**, 5451–5456.

DONARSKI, J.A., S.A. JONES, M. HARRISON, M. DRIFFIELD and A.J. CHARLTON. (2010). Identification of botanical biomarkers found in Corsican honey. *Food Chem.*, **118**, 987–994.

DONER, L.W. (1977). The sugars of honey – a review. *J. Sci. Food Agri.*, **28**, 443–456.

FERNANDEZ-TORRES, R., J.L. PÉREZ-BERNAL, M.Á. BELLO-LÓPEZ, M. CALLEJÓN-MOCHÓN, J.C. JIMÉNEZ-SÁNCHEZ and A. GUIRAÚM-PÉREZ. (2005). Mineral content and botanical origin of Spanish honeys. *Talanta*, **65**, 686–691.

FŐRSTEL, H. (2007). The natural fingerprint of stable isotopes – use of IRMS to test food authenticity. *Anal. Bioanal. Chem.*, **388**, 541–544.

GONZALEZ-MIRET, M.L., A. TERRAB, D. HERNANZ, M.Á. FERNÁNDEZ-RECAMALES, and F.J. HEREDIA. (2005). Multivariate correlation between colour and mineral composition of honeys and by their botanical origin. *J. Agri. Food Chem.*, **53**, 2574–2580.

GONZALEZ-PARAMAS, A.M., J.A. BÁREZ, R.J. GARCIA-VILLANOVA, T.R. PALÁ, R.A. ALBAJAR and J.S. SÁNCHEZ. (2000). Geographical discrimination of honeys by using mineral composition and common chemical quality parameters. *J. Sci. Food Agri.*, **80**, 157–165.

HADJMOHAMMADI, M.R., S. NAZARI and K. KAMEL. (2009). Determination of Flavanoid Markers in Honey with SPE and LC using experimental design. *Chromatographia*, **69**, 1291–1297.

Health Canada (2004). Chloramphenicol in Honey (Online). Available: http://www.hc-sc.gc.ca/dhp-mps/vet/faq/faq_chloramphenicol_honey-miel-eng.php#a1 (Accessed 23 March 2012).

Infoshare. (2011). Harmonised Trade – Exports: Honey, Natural (Online). Wellington: Statistics New Zealand. Available: http://www.stats.govt.nz/infoshare/ (Accessed 6 June 2011).

JERKOVIC, I., Z. MARIJANOVIĆ, I. LJUBIČIĆ and M. GUGIĆ. (2010). Contribution of the bees and comb to honey volatiles: blank trial probe for the chemical profiling of honey diversity. *Chem. Biodiv.*, **7**, 1217–1230.

JONASSON, S., H. MEDRANO and J. FLEXAS. (1997). Variation in leaf longevity of *Pistacia lentiscus* and its relationship to sex and drought stress inferred from leaf delta 13C. *Funct. Ecol.*, **11**, 282–289.

KASKONIENE, V. and P.R. VENSKUTONIS. (2010). Floral markers in honey of various botanical and geographic origins: a review. *Comp. Rev. Food Sci. Food Safety*, **9**, 620–634.

KELLY, J.F.D., G. DOWNEY, and V. FOURATIER. (2003). Initial study of honey adulteration by sugar solutions using midinfrared (MIR) spectroscopy and chemometrics. *J. Agri. Food Chem.*, **52**, 33–39.

KROPF, U., M. KOROŠEC, J. BERTONCELJ, N. OGRINC, M. NEČEMER, P. KUMP and T. GOLOB. (2010). Determination of the geographical origin of Slovenian black locust, lime and chestnut honey. *Food Chem.*, **121**, 839–846.

LATORRE, M.J., R. PEÑA, C. PITA, A. BOTANA, S. GARCÍA and C. HERRERO. (1999). Chemometric classification of honeys according to their type. II. Metal content data. *Food Chem.*, **66**, 263–268.

LAUBE, I., H. HIRD, P. BRODMANN, S. ULLMANN, M. SCHÖNE-MICHLING, J. CHISHOLM and H. BROLL. (2010). Development of primer and probe sets for the detection of plant species in honey. *Food Chem.*, **118**, 979–986.

LEEDER, J. (2011). Honey laundering: the sour side of nature's golden sweetener (Online). The Globe and Mail. Available: http://www.theglobeandmail.com/news/technology/science/honey-laundering-the-sour-side-of-natures-golden-sweetener/article1859410/singlepage/#articlecontent (Accessed 16 August 2011).

LOUVEAUX, J., A. MAURIZIO and G. VORWOHL. (1978). Methods of melissopalynology. *Bee World*, **59**, 139–157.

LUSBY, P.E., A.L. COOMBES and J.M. WILKINSON. (2005). Bactericidal activity of different honeys against pathogenic bacteria. *Arch. Med. Res.*, **36**, 464–467.

MADEJCZYK, M. and D. BARALKIEWICZ. (2008). Characterization of Polish rape and honeydew honey according to their mineral contents using ICP-MS and F-AAS/AES. *Analyt. Chim. Acta*, **617**, 11–17.

MARCEAU, E. and V.A. YAYLAYAN. (2009). Profiling of a-dicarbonyl content of commercial honeys from different botanical origins: identification of 3,4,-Dideoxyglucosone-3-ene (3,4-DGE) and related products. *J. Agr. Food Chem.*, **57**, 10837–10844.

MAVRIC, E., S. WITTMANN, G. BARTH and T. HENLE. (2008). Identification and quantification of methylglyoxal as the dominant antibacterial constituent of manuka (Leptospermum scoparium) honey from New Zealand. *Mol. Nutr. Food Res.*, **52**, 483–489.

Mieliditalia. (2011). http://www.mieliditalia.it/index.php/en-fr-es-de/english/80246-huge-sequestration-of-illegal-honey (accessed 13 March 2013).

MOLAN, P.C. and K.M. RUSSELL. (1988). Non-peroxide antibacterial activity in some New Zealand honeys. *J. Api. Res.*, **27**, 62–67.

NECEMER, M., I.J. KOŠIR, P. KUMP, U. KROPF, M. JAMNIK, J. BERTONCELJ, N. OGRINC and T. GOLOB. (2009). Application of total reflection X-ray spectrometry in combination with chemometric methods for determination of the botanical origin of Slovenian honey. *J. Agric. Food Chem.*, **57**, 4409–4414.

NOZAL, M.J., J.L. BERNAL, L. TORIBIO, M. ALAMO and J.C. DIEGO. (2005). The use of carbohydrate profiles and chemometrics in the characterization of natural honeys of identical geographical origin. *J. Agric. Food Chem.*, **53**, 3095–3100.

PADOVAN, G.J., D. DE JONG, L.P. RODRIGUES and J.S. MARCHINI. (2003). Detection of adulteration of commercial honey samples by the $^{13}C/^{12}C$ isotopic ratio. *Food Chem.*, **82**, 633–636.

PISANI, A., G. PROTANO and F. RICCOBONO. (2008). Minor and Trace elements in different honey types produced in Siena County (Italy). *Food Chem.*, **107**, 1553–1560.

RASHED, M.N. and M.E. SOLTAN. (2004). Major and trace elements in different types of Egyptian mono-floral and non-floral bee honeys. *J. Food Comp. Anal.*, **17**, 725–735.

REBANE, R. and K. HERODES. (2008). Evaluation of the botanical origin of Estonian unifloral and polyfloral honeys by amino acid content. *J. Agric. Food Chem.*, **56**, 10716–10720.

RODRIGUEZ-OTERO, J.L., P. PASEIRO, J. SIMAL and A. CEPEDA. (1994). Mineral content of the honeys produced in Galicia (north-west Spain). *Food Chem.*, **49**, 169–171.

ROGERS, K.M., K. SOMERTON, P. ROGERS and J. COX. (2010). Eliminating false positive C4 sugar tests on New Zealand Manuka honey. *Rapid Comm. Mass Spec.*, **24**, 2370–2374.

RUIZ-MATUTE, A.I., M. BROKL, A.C. SORIA, M.L. SANZ and I. MARTÍNEZ-CASTRO. (2010). Gas chromatographic-mass spectrometric characterisation of tri and tetrasaccharides in honey. *Food Chem.*, **120**, 637–642.

SAPSTED, D. (2005). Beekeeper uncovered 'scam' over local honey. http://www.telegraph.co.uk/news/uknews/1503781/Beekeeper-uncovered-scam-over-local-honey.html (Accessed 13 March 2013).

SCHELLENBERG, A., S. CHMIELUS, C. SCHLICHT, F. CAMIN, M. PERINI, L. BONTEMPO, K. HEINRICH, S.D. KELLY, A. ROSSMANN, F. THOMAS, E. JAMIN and M. HORACEK. (2010). Multielement stable isotope ratios (H, C, N, S) of honey from different European regions. *Food Chem.*, **121**, 770–777.

SMITH, B.N. and S. EPSTEIN. (1971). Two categories of 13C/12C ratios for higher plants. *Plant Physiol.*, **47**, 380–384.

STEPHENS, J.M., R.C. SCHLOTHAUER, B.D. MORRIS, D. YANG, L. FEARNLEY, D.R. GREENWOOD and K.M. LOOMES. (2010). Phenolic compounds and methylglyoxal in some New Zealand manuka and kanuka honeys. *Food Chem.*, **120**, 78–86.

TERRAB, A., D. HERNANZ and F.J. HEREDIA. (2004). Inductively coupled plasma optical emission spectrometric determination of minerals in thyme honeys and their contribution to geographical discrimination. *J. Agri. Food Chem.*, **52**, 3441–3445.

THODE, H.G. (1991). Sulphur isotopes in nature and environment: an overview. In: KROUSE, H.R., GRINNENKO, V.A. (ed.) SCOPE 43 stable isotopes: *Natural and Anthropogenic Sulphur in the Environment.* Chichester: John Wiley & Sons.

TUZEN, M.S. SILICI, D. MENDIL and M. SOYLAK. (2007). Trace element levels in honeys from different regions of Turkey. *Food Chem.*, **103**, 325–330.

USFDA (2010). *Import Alert 36–03: Detention Without Physical Examination of Honey Due to Chloramphenicol.* U.S. Food and Drug Administration, Rockville, MD.

WANG, J., M.M. KLIKS, W. QU, S. JUN, G. SHI and Q.X. LI. (2009). Rapid determination of the geographical origin of honey based on protein fingerprinting and barcoding using MALDI-TOF-MS. *J. Agri. Food Chem.*, **57**, 10081–10088.

WANG, J. and Q. X. LI. (2007). Chemical composition, characterization, and differentiation of honey botanical and geographical origins. *Advan. Food Nutrit. Res.*, **62**, 89–137.

WHITE, J.W. and WINTERS, K. (1989). Honey protein as internal standard for stable carbon isotope ratio detection of adulteration of honey. *J. AOAC Int.*, **72**, 907–911.

YAO, L., Y. JIANG, R. SINGANUSONG, N. DATTA and K. RAYMONT. (2004). Phenolic acids and abscisic acid in Australian Eucalyptus honeys and their potential for floral authentication. *Food Chem.*, **86**, 169–177.

YARSAN, E. (2007). Contents of some metals in Honeys from different regions in Turkey. *Bull. Environ. Contam. Toxicol.*, **79**, 255–258.

Index

CPSIA information can be obtained at www.ICGtesting.com
Printed in the USA
BVOW02*0354030214

343683BV00005B/124/P

9 780857 092748